T0395922

Advances in Remediation Techniques for Polluted Soils and Groundwater

Advances in Remediation Techniques for Polluted Soils and Groundwater

Edited by

Pankaj Kumar Gupta

Wetland Hydrology Research Laboratory, Faculty of Environment,
University of Waterloo, Waterloo, ON, Canada

Basant Yadav

Department of Water Resources Development and Management (WRD&M),
Indian Institute of Technology Roorkee, Uttarakhand, India

Sushil Kumar Himanshu

Department of Food, Agriculture and Bioresources,
School of Environment, Resources and Development,
Asian Institute of Technology, Pathum Thani, Thailand

ELSEVIER

Elsevier
Radarweg 29, PO Box 211, 1000 AE Amsterdam, Netherlands
The Boulevard, Langford Lane, Kidlington, Oxford OX5 1GB, United Kingdom
50 Hampshire Street, 5th Floor, Cambridge, MA 02139, United States

British Library Cataloguing-in-Publication Data
A catalogue record for this book is available from the British Library

Library of Congress Cataloging-in-Publication Data
A catalog record for this book is available from the Library of Congress

ISBN: 978-0-12-823830-1

For Information on all Elsevier publications
visit our website at https://www.elsevier.com/books-and-journals

Publisher: Susan Dennis
Acquisitions Editor: Anita A. Koch
Editorial Project Manager: Judith Clarisse Punzalan
Production Project Manager: Sujatha Thirugnana Sambandam
Cover Designer: Matthew Limbert

Typeset by MPS Limited, Chennai, India

Working together
to grow libraries in
developing countries

www.elsevier.com • www.bookaid.org

Contents

LIST OF CONTRIBUTORS xi

ABOUT THE EDITORS xv

CHAPTER 1 Flow and movement of gaseous pollutants in the subsurface:
CO$_2$ dynamics at a carbon capture and storage site 1
Shachi and Anuradha Garg
1.1 Introduction 1
1.2 Worldwide CO$_2$ storage projects 2
1.3 Gaseous CO$_2$ in the subsurface 3
1.4 Factors affecting CO$_2$ migration in the subsurface 6
1.5 CO$_2$–brine–rock interaction in the subsurface 8
1.6 Potential risk associated with CO$_2$ leakage 10
1.7 Numerical modeling for investigating CO$_2$ dynamics 11
1.8 Modeling of CO$_2$ in subsurface: a case study 12
1.9 Conclusions and future prospective 14
Acknowledgments 15
References 15

CHAPTER 2 Column adsorption studies for the removal of chemical
oxygen demand from fish pond wastewater using
waste alum sludge 21
Muibat D. Yahya, Ibrahim A. Imam and
Saka A. Abdulkareem
2.1 Introduction 21
2.2 Materials and methods 23
2.3 Results and discussion 29
2.4 Conclusion 43
References 44

CHAPTER 3 Farm management practices for water quality improvement:
economic risk analysis of winter wheat production in the
Southern High Plains 49
Yubing Fan and Sushil Kumar Himanshu
3.1 Introduction 49
3.2 Data 50

Contents

3.3   Methods..53
3.4   Results and discussion 56
3.5   Conclusion 61
Appendix A...62
References 64

CHAPTER 4   Bioremediation of contaminated soils by bacterial
biosurfactants...67
Sabah Fatima, Muzafar Zaman, Basharat Hamid,
Faheem Bashir, Zahoor Ahmad Baba and
Tahir Ahmad Sheikh
4.1   Introduction..67
4.2   Bacterial biosurfactants and their classification.....................68
4.3   Role of bacterial biosurfactants in the bioremediation
of contaminated soils..70
4.4   Conclusion and future prospectus..................................78
References 79

CHAPTER 5   Evaluation of machine learning-based modeling
approaches in groundwater quantity and quality prediction.......87
Madhumita Sahoo
5.1   Overview...87
5.2   Popular Ml techniques used in groundwater modeling............91
5.3   Efficacy of ML-based modeling......................................93
5.4   Conclusion 96
References 97

CHAPTER 6   Microbial consortium for bioremediation of polycyclic
aromatic hydrocarbons polluted sites....................................105
Pankaj Kumar Gupta
6.1   Introduction...105
6.2   PAHs pollutants: source, toxicity, and metabolic pathways........106
6.3   Bioremediation of PAHs 107
6.4   Plant—microbes interactions 111
6.5   Microbes and their consortium to degrade PAHs...................113
6.6   Microbial degradation kinetics models 116
6.7   Conclusion and future recommendations...........................117
Acknowledgments...118
References 118

CHAPTER 7   Fate, transport, and bioremediation of PAHs in
experimental domain: an overview of current status and
future prospects..125
Pankaj Kumar Gupta, Manik Goel, Sanjay K. Gupta and
Ram N. Bhargava
7.1   Introduction...125

7.2   PAHs fate and transport mechanisms 127
7.3   Studies investigated PAHs behaviors in laboratory domain 128
7.4   Polishing PAH-polluted site using subsurface-constructed
      wetlands 131
7.5   Conclusive remark and future prospects............................... 135
References 136

CHAPTER 8   Mathematical modeling of contaminant transport
in the subsurface environment 141
Abhay Guleria and Sumedha Chakma
8.1   Introduction.. 141
8.2   Contaminant transport models for saturated
      porous media.. 143
8.3   Categorization of mathematical modeling studies
      related to Indian groundwater and soil systems.................... 148
8.4   Contaminant transport modeling in the subsurface
      environment using mobile—immobile model 158
8.5   Conclusion and future directions... 163
Acknowledgment 164
References 164

CHAPTER 9   Impacts of climatic variability on subsurface water
resources 171
Pankaj Kumar Gupta, Brijesh Kumar Yadav and
Devesh Sharma
9.1   Introduction.. 171
9.2   Impacts on atmospheric boundary 173
9.3   Impacts on water storage and flow pattern............................ 175
9.4   Impacts on ground—surface water interactions...................... 177
9.5   Impacts on subsurface water quality 178
9.6   Methodological framework for evaluating climate
      change impacts on subsurface 182
9.7   Conclusion and recommendations 183
Acknowledgment 184
References 185

CHAPTER 10   Microplastic in the subsurface system: Extraction and
characterization from sediments of River Ganga
near Patna, Bihar... 191
Rashmi Singh, Rakesh Kumar and Prabhakar Sharma
10.1   Introduction 191
10.2   Materials and method 199
10.3   Results and discussion.. 203
10.4   Conclusion 212
References 213

CHAPTER 11 Assessment of long-term groundwater variation
                in India using GLDAS reanalysis.. 219
                Swatantra Kumar Dubey, Preet Lal, Pandurang Choudhari,
                Aditya Sharma and Aditya Kumar Dubey
                11.1 Introduction 219
                11.2 Data used and methodology.. 221
                11.3 Results and discussion... 223
                11.4 Conclusion 229
                References 230

CHAPTER 12 Emerging contaminants in subsurface: sources,
                remediation, and challenges.. 233
                Anuradha Garg and Shachi
                12.1 Introduction 233
                12.2 Sources of emerging contaminants in groundwater 234
                12.3 Detection and analysis... 237
                12.4 Types of emerging contaminants 238
                12.5 Fate of emerging contaminants in groundwater.............. 240
                12.6 Potential risks associated with emerging
                      contaminants 242
                12.7 Remediation of emerging contaminants.......................... 244
                12.8 Challenges and scope 246
                12.9 Discussion and conclusion 247
                References 247

CHAPTER 13 Selenium and naturally occurring radioactive
                contaminants in soil—water systems.................................... 259
                Pankaj Kumar Gupta, Gaurav Saxena and Basant Yadav
                13.1 Introduction 259
                13.2 Selenium: distribution in Indian soil—water systems......... 260
                13.3 Naturally occurring radioactive material:
                      distribution in Indian soil—water systems 261
                13.4 Remedial measures... 262
                13.5 Field scale implications and future research.................... 264
                References 264

CHAPTER 14 Understanding and modeling the process of seawater
                intrusion: a review 269
                Lingaraj Dhal and Sabyasachi Swain
                14.1 Background 269
                14.2 Seawater intrusion process 270
                14.3 Measurement and monitoring of seawater intrusion 276
                14.4 Seawater intrusion modeling and prediction 278
                14.5 Management of seawater intrusion.................................. 280
                14.6 Seawater intrusion, climate change, and sea level rise......... 282
                14.7 Conclusion 283
                References 283

CHAPTER 15 Prioritization of erosion prone areas based on
a sediment yield index for conservation treatments:
A case study of the upper Tapi River basin 291
Santosh S. Palmate, Kumar Amrit, Vikas G. Jadhao,
Deen Dayal and Sushil Kumar Himanshu
    15.1   Introduction 291
    15.2   Study area...293
    15.3   Data used...293
    15.4   Methodology 298
    15.5   Results and discussion.....................................301
    15.6   Summary and conclusions................................305
    References 306

CHAPTER 16 Advances in hydrocarbon bioremediation products:
natural solutions 309
Pankaj Kumar Gupta
    16.1   Introduction 309
    16.2   Engineered constructed wetlands.......................310
    16.3   Native and specialized microbial communities...................312
    16.4   Biodiesels as biostimulators.................................313
    16.5   Phycoremediation 315
    References 315

CHAPTER 17 Nitrate-N movement revealed by a controlled in situ
solute injection experiment in the middle Gangetic
plains of India.. 319
Pankaj Kumar Gupta, Basant Yadav,
Kristell Le Corre and Alison Parker
    17.1   Introduction 319
    17.2   Study site 321
    17.3   Methodology 324
    17.4   Results and discussion.....................................325
    17.5   Conclusion 332
    CRediT authorship contribution statement 332
    Conflicts of interest 332
    Acknowledgment 333
    Data availability..333
    References 333

CHAPTER 18 Integrated water resources management in
Sikta irrigation system, Nepal 337
Rituraj Shukla, Ishwari Tiwari, Deepak Khare and
Ramesh P. Rudra
    18.1   Introduction 337
    18.2   Study area...338

18.3 Methodology/philosophy 340
18.4 Groundwater modeling 348
18.5 Result and discussion 350
18.6 Conclusion and recommendations 356
Acknowledgments 357
References 357

CHAPTER 19 Hydrocarbon pollution assessment and analysis
using GC—MS 361
Pankaj Kumar Gupta
19.1 Introduction 361
19.2 Previous works 361
19.3 GC—MS system: specification 362
19.4 Method of toluene analysis 369
19.5 Calibration 372
19.6 Mass spectrum of toluene 374
References 374

INDEX 379

List of contributors

Saka A. Abdulkareem
Federal University of Technology, Minna, Nigeria

Kumar Amrit
CSIR- National Environmental Engineering Research Institute, Mumbai, India

Zahoor Ahmad Baba
Division of Basic Science and Humanities, FOA, Sher-e-Kashmir University of Agricultural Sciences and Technology, Wadura, India

Faheem Bashir
Centre of Research for Development/Department of Environmental Science, University of Kashmir, Srinagar, India

Ram N. Bhargava
Department of Microbiology (DM), Babasaheb Bhimrao Ambedkar University, Lucknow, India

Sumedha Chakma
Department of Civil Engineering, Indian Institute of Technology Delhi, Delhi, India

Pandurang Choudhari
Department of Geography, University of Mumbai, Mumbai, India

Deen Dayal
Department of Water Resource Development and Management, Indian Institute of Technology, Roorkee, India

Lingaraj Dhal
Department of Water Resources Development and Management, Indian Institute of Technology Roorkee, Roorkee, India

Aditya Kumar Dubey
Department of Earth and Environmental Sciences, Indian Institute of Science Education and Research Bhopal, Bhopal, India

Swatantra Kumar Dubey
Department of Geology, Sikkim University, Gangtok, India

Yubing Fan
Texas A&M AgriLife Research, Vernon, TX, United States

Sabah Fatima
Department of Environmental Science, University of Kashmir, Srinagar, India

Anuradha Garg
Research Scholar, Department of Hydrology, IIT Roorkee, Roorkee, India

Manik Goel
Department of Hydrology, Indian Institute of Technology Roorkee, Roorkee, India

Abhay Guleria
Department of Civil Engineering, Indian Institute of Technology Delhi, Delhi, India

Pankaj Kumar Gupta
Wetland Hydrology Research Laboratory, Faculty of Environment, University of Waterloo, Waterloo, ON, Canada

Sanjay K. Gupta
Environmental Engineering, Department of Civil Engineering, Indian Institute of Technology Delhi, New Delhi, India

Basharat Hamid
Department of Environmental Science, University of Kashmir, Srinagar, India

Sushil Kumar Himanshu
Department of Food, Agriculture and Bioresources, School of Environment, Resources and Development, Asian Institute of Technology, Pathum Thani, Thailand

Ibrahim A. Imam
Federal University of Technology, Minna, Nigeria

Vikas G. Jadhao
Krishi Vigyan Kendra (KVK), Buldana, India; Department of Water Resource Development and Management, Indian Institute of Technology, Roorkee, India

Deepak Khare
Irrigation Department, Nepal; Indian Institute of Technology, Roorkee (IITR), India

Rakesh Kumar
School of Ecology & Environment Studies, Nalanda University, Rajgir, India

Preet Lal
Department of Geoinformatics, Central University of Jharkhand, Ranchi, India

Kristell Le Corre
Cranfield Water Science Institute, Cranfield University, Vincent Building, Cranfield, United Kingdom

Santosh S. Palmate
Texas A&M AgriLife Research Center at El Paso, El Paso, TX, United States

Alison Parker
Cranfield Water Science Institute, Cranfield University, Vincent Building, Cranfield, United Kingdom

Ramesh P. Rudra
School of Engineering (Water Resources Engineering), University of Guelph, Guelph, ON, Canada

Madhumita Sahoo
University of Alaska Fairbanks, Fairbanks, AK, United States; Gandhi Institute for Technology, Biju Pattnaik University of Technology, India

Gaurav Saxena
Laboratory for Microbiology, Department of Microbiology, Baba Farid Institute of Technology, Dehradun, India

Shachi
Research Scholar, Department of Hydrology, IIT Roorkee, Roorkee, India

Aditya Sharma
Department of Atmospheric Science, School of Earth Sciences, Central University of Rajasthan, Ajmer, India

Devesh Sharma
Department of Atmospheric Sciences, Central University of Rajasthan, Ajmer, India

Prabhakar Sharma
School of Ecology & Environment Studies, Nalanda University, Rajgir, India

Tahir Ahmad Sheikh
Division of Agronomy, FOA, Sher-e-Kashmir University of Agricultural Sciences and Technology, Wadura, India

Rituraj Shukla
School of Engineering (Water Resources Engineering), University of Guelph, Guelph, ON, Canada

Rashmi Singh
School of Ecology & Environment Studies, Nalanda University, Rajgir, India

Sabyasachi Swain
Department of Water Resources Development and Management, Indian Institute of Technology Roorkee, Roorkee, India

Ishwari Tiwari
Irrigation Department, Nepal

Basant Yadav
Department of Water Resources Development and Management (WRD&M), Indian Institute of Technology Roorkee, Roorkee, India

Brijesh Kumar Yadav
Department of Hydrology, Indian Institute of Technology Roorkee, Roorkee, India

Muibat D. Yahya
Federal University of Technology, Minna, Nigeria

Muzafar Zaman
Department of Environmental Science, University of Kashmir, Srinagar, India

About the editors

Dr. Pankaj Kumar Gupta is a contaminant hydrogeologist, interested in interdisciplinary research and teaching to understand multiscale interactions between different components of the subsurface environment, especially the soil–water–pollutant–microbes' system. He is a postdoctoral fellow in the faculty of environment, University of Waterloo, since Jan 2019. His current research is focused on investigating the behavior of pollutants in peatlands (Canada) and mineral aquifers (India) under dynamically fluctuating groundwater table conditions. Majority of his works focus on two areas: (1) understanding the occurrence of biogeochemical interactions when pollutants migrate into groundwater systems; and (2) developing remediation strategies. He received PhD from Department of Hydrology, Indian Institute of Technology Roorkee, India. He has been qualified national eligibility test (NET-Dec 2013) with junior research fellowship conducted by University Grant Commission (UGC), Government of India. He has also qualified Agricultural Scientist Recruitments Boards NET in Environmental Sciences (2014). He is an awardee of AGU Student Travel Grant (2017), JPGU Travel Grant (2018), EXCEED-SWINDON & DAAD Germany Grant (2018). He serves as an editorial board member for *SN Applied Sciences* and *Frontiers of Water*. He has published ~20 research papers, ~30 book chapters, and 2 popular science articles. He has performed many site restoration and remediation consultancy projects at industrial polluted sites in India.

Dr. Basant Yadav, PhD (Civil Engineering), is an assistant professor in the Department of Water Resources Development and Management (WRD&M), Indian Institute of Technology Roorkee, India. Before joining IIT Roorkee, Dr. Yadav worked as a postdoctoral fellow in Water Science Institute of Cranfield University (UK), working on groundwater quality and quantity management. Specific focus of Dr. Yadav's research includes flow and transport processes in groundwater and in the vadose zone; hydroinformatics; and managed aquifer recharge. Dr. Yadav's research emphasizes the Integration of experimental, numerical, and data-based modeling approaches for the better and efficient management of the groundwater quantity and quality. His work uses a range of numerical, statistical, and stochastic modeling approaches and experimental work to analyze the fate and transport of contaminants in saturated and unsaturated zones. Dr. Yadav is an agricultural engineer and received his MTech in hydrology jointly from Indian Institute of Technology Roorkee (India) and Technical University Stuttgart (Germany). He did his PhD from Indian Institute of Technology Delhi (India) on "Application of soft computing techniques in water quality and quantity modeling." He was awarded prestigious National Postdoctoral fellowship (NPDF) in 2017 to work on his postdoctoral project in Indian Institute of Science Bangalore, India. He has received scholarships like DAAD (Germany) and MHRD (India) for his masters and PhD work. He has authored/coauthored 17 research publications in high impact journals such as *Journal of Hydrology, Hydrological Science, Journal of Environmental Engineering, Journal of Hydrologic Engineering,* and *Measurement Journal.* Dr. Yadav has presented his work in highly reputed meeting like AGU (December 2014, USA), AOGS (August 2017, Singapore), and JpGU (June 2018, Japan). He is a member of the American Geophysical Union (AGU), Asia Oceania Geosciences Society (AOGS), and Japan Geoscience Union (JpGU). He is the recipient of the Rien van Genuchten Early-Career Award of Porous Media for a Green World 2020. He is the national secretary of the Indian Water Resources Society (IWRS) and a member of the Institution of Engineers (India). He is the reviewer of high repute journals like *Science of Total Environment, Hydrological Processes, Journal of Hydrology, Journal of Hydrologic Engineering, Water Resources Management, ISH Journal of Hydraulic Engineering,* and *Journal of Hydroinformatics.*

Dr. Sushil Kumar Himanshu is an assistant professor in the Department of Food, Agriculture and Bioresources, School of Environment, Resources and Development, Asian Institute of Technology (AIT), Klong Luang, Pathum Thani (Thailand). The specific focus of Dr. Himanshu's research includes precision farming, climate-resilient agriculture systems, on-farm irrigation water management, remote sensing and GIS applications in agriculture, applications of unmanned aerial systems (UAS) and wireless sensors in agricultural applications, big data analysis and applications, machine learning applications in agriculture, and hydrologic/cropping system modeling. He obtained his MTech degree in Hydrology and PhD in Water Resources Development and Management from the Indian Institute of Technology (IIT) Roorkee, India. His PhD research was focused on evaluating satellite-based precipitation estimates for hydrological modeling. He has received the best Water Resources Student award-2018 collectively by the Indian Water Resources Society and IIT Roorkee for his PhD research work. Before joining AIT, he was working as a postdoctoral research associate at the Texas A&M AgriLife Research (Texas A&M University System), USA, where his research was focused on developing and evaluating strategies that conserve soil and water, promote water use efficiency, and protect soil and water quality in diverse agro-ecosystems. He also worked as a research scientist at the National Remote Sensing Center of Indian Space Research Organization (ISRO) at Hyderabad, India, where he worked on the operationalization of national-level hydrological modeling framework for in-season hydrological water balance components at daily/weekly/fortnightly time steps. He has authored/coauthored more than 20 research publications in high impact journals such as *Journal of Hydrology, Scientific Reports-Nature, Soil and Tillage Research, Science of the Total Environment, CATENA, Agricultural Water Management, Journal of Hydrologic Engineering, Environmental Earth Sciences, Ecohydrology, Frontiers in Sustainable Food Systems*, and *Transactions of the ASABE*. He presented his research to several international conferences/meetings in various countries including the United States, Norway, UAE, and India. He is a member of the American Society of Agricultural and Biological Engineers (ASABE), Indian Water Resources Society (IWRS), American Society of Civil Engineers (ASCE), The Institution of Engineers (India), and International Association of Hydrological Sciences (IAHS). He reviewed more than 50 research articles from the high repute journals. Recently, he received an Outstanding Reviewer Award by the *Transactions of the ASABE Journal* for the year 2020.

Flow and movement of gaseous pollutants in the subsurface: CO_2 dynamics at a carbon capture and storage site

Shachi and Anuradha Garg

Research Scholar, Department of Hydrology, IIT Roorkee, Roorkee, India

1.1 Introduction

CO_2 is a greenhouse gas (GHG) posing severe threats on the environment and raising public concerns in developing and developed countries as per the fifth assessment report from IPCC. It contributes 72% of the greenhouse effect and a significant contributor to global climate change (IPCC, 2005). The present concentration of CO_2 in the atmosphere has crossed a remarkable value of ~ 400 ppm compared to the preindustrial era having only ~ 280 ppm (Orr, 2009; Shachi et al., 2019). The anthropogenic emissions of CO_2 have impacted global average temperature, change in the ocean pH level, sea-level rise, changes in precipitation pattern, changes in snow cover, leaching of soil nutrients, etc. (Stocker et al., 2013). To keep a check on global average temperature rise by $2°C$ till 2050 for reducing global warming, carbon capture and storage (CCS) is one such bridge technology to attain the same (Johns, 2017).

The CCS in deep geological formations aims to reduce the atmospheric concentration of CO_2, consisting of processes: capture and separation, transportation, and geological sequestration, occurring sequentially (Benson & Cole, 2008; Shukla et al., 2011). Geo-sequestration is the process of injection and storage of CO_2 in underground subsurface formations. It involves the operation of capturing CO_2 from stationary point sources such as power plants and cement factories, transportation, and compression of the captured CO_2 before injecting into the deep geological formations for storage (Koperna et al., 2017; Sun et al., 2016). The predominant aim of the geo-sequestration process is to mitigate the adverse effect of CO_2 in the atmosphere. The CCS may also apply to ocean and mineral sequestration; however, it is most promising to use in geo-sequestration, particularly in saline formations because of large storage capacity, the longevity of storage, along with the better understanding of the different storage mechanisms (Johns, 2017).

1

Advances in Remediation Techniques for Polluted Soils and Groundwater. DOI: https://doi.org/10.1016/B978-0-12-823830-1.00008-0

There are various CCS projects such as Sleipner (Norway), In-Salah (Algeria), and Weyburn (Canada), and the overall experience of these indicates the feasibility of geo-sequestration as a prominent option to mitigate the concentration of CO_2 in the atmosphere (Piao et al., 2018). The deep geological formations account for retaining $\sim 99\%$ of injected CO_2 for 1000 years (IPCC, 2005). The receiving geological formations selected for CO_2 storage should be situated at a depth of 800 to 1000 m, and CO_2 is injected in the supercritical state for optimum utilization of pore spaces (Dooley et al., 2006; Solomon, 2006). The geological strata selected for storing CO_2 include unmineable coal formations, inoperative oil and gas formations, and deep saline aquifers having brine or brackish water (Bentham & Kirby, 2005; Vishal et al., 2015). The sequestration of CO_2 can play a crucial role in enhancing fuel recovery, particularly oil and gas, enhanced oil recovery (EOR), enhanced gas recovery, and enhanced coal bed methane (Jafari et al., 2017; Wei et al., 2015). Thus this chapter provides a detailed account of the CO_2 injection into the subsurface formation and the various factors affecting the CO_2 migration.

1.2   Worldwide CO_2 storage projects

The global CO_2 storage potential of sedimentary formations is 320 Gt (Koide et al., 1993), and most of the deep saline aquifer sites are located in sedimentary formations (Shukla et al., 2010). Deep saline aquifers offer good storage potential worldwide (Celia et al., 2015), having the storage capacity in the range of ~ 2000 to 20,000 Gt CO_2. Considering the overall annual GHG emissions in the order of ~ 49 Gt CO_2/year (Pachauri et al., 2014), saline aquifers seem the most prominent sequestration option to control anthropogenic CO_2 release in the atmosphere (Bachu & Adams, 2003). Also, saline aquifers are associated with a wide range of permeability values, and more than 1 mD is recommended for storing CO_2 (Ringrose et al., 2013; Verdon et al., 2013). A significant reduction of CO_2 in the atmosphere can be obtained by injecting CO_2 at 10 Mt CO_2/year in such types of permeable formations (IPCC, 2014).

There are various CO_2 geo-sequestration projects planned worldwide to reduce carbon concentration in the atmosphere (Leung et al., 2014). The COP 21 summit held in Paris witnessed 195 countries that pledge to reduce carbon emissions to keep the global average temperature below the 20°C target (Ourbak & Tubiana, 2017; United Nations, 2015). Some major CO_2 sequestration projects are as follows:

Sleipner project: The Sleipner gas field, situated in the Norwegian sector operated by Statoil, initiated the concept of injecting CO_2 into an underlying saline formation of the North Sea for permanence since 1996 (Dubos-Sallée & Rasolofosaon, 2011). It is marked as the first offshore saline aquifer project in Norway. The

Utsira Sandstone formation of thickness around 200–300 m at Sleipner in the North Sea, almost 1 Mt of CO_2 per year, has been injected (Bachu, 2000; Bentham & Kirby, 2005). Since its beginning in 1996 to 2008, around 11 Mt of CO_2 was injected at a depth of 1000 m (Dubos-Sallée & Rasolofosaon, 2011). The Utsira formation is overlain by a shale caprock of thickness 200–300 m that safely stores the injected CO_2 (Goodman et al., 2013).

Weyburn project: It started injecting CO_2 in 2000 in the Midale field of south-central Saskatchewan, Canada, for enhancing oil recovery. Midale field consisted of carbonate reservoirs of Vuggy (limestone) and Marly (dolostone) beds, overlain by anhydrite caprock (Burrowes, 2001).

In-Salah project: The In-Salah project commenced in 2004, and since then, it is used for injecting 4000 tonnes of CO_2/day. It is located at a depth of 1800 m and injects CO_2 in carboniferous sandstone formation of thickness 20 m (Rutqvist et al., 2009). The geomechanical, geochemical, and geophysical investigations were carried out for observing the plume movement and dynamics using imaging techniques (Ringrose et al., 2013).

Otway Basin Pilot Project: The Kyoto Protocol signed by the Australian government in 2009 lays the foundation for the Otway Basin Pilot Project. The CO_2 was injected at a depth of 2000 m from the gas well into the saline formation of the Naylor field, Otway Basin at a flow rate of 150 tonnes/day for 2 years, to achieve the target of 100,000 tonnes of CO_2 (Cook, 2009). The integrity of the caprock was evaluated for fault reactivation by monitoring plume migration and geomechanical investigations.

Apart from these, CO_2 storage has been performed in abandoned coal mines since 1961 (Piessens & Dusar, 1961). From these project activities, geological sequestration is the most promising method for CO_2 mitigation from the atmosphere but has many challenges that need to be addressed by proper investigations. Modeling and experimental studies were performed to understand postinjection processes in the geological formations. However, complete geochemical interactions of the injected CO_2 in deep formations are still not very well understood. Therefore the investigation of CO_2 sequestration and its reactive transport in the subsurface is highly recommended (IEA, 2013). To investigate the geochemical interactions of the injected CO_2, understanding CO_2 geo-sequestration mechanisms is a prerequisite as described next.

1.3 Gaseous CO_2 in the subsurface

The injection of CO_2 in storage formation leads to geological and geochemical processes responsible for storing/trapping CO_2 safely. Fig. 1.1 shows the different formations used for safely storing CO_2 in the geological

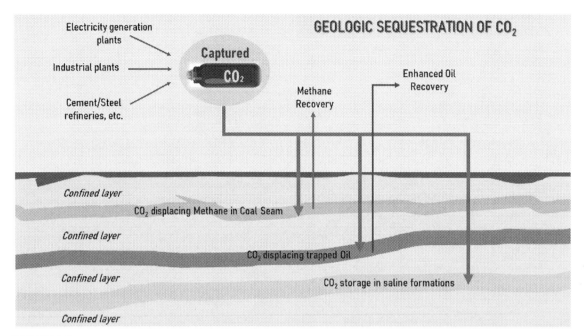

FIGURE 1.1
Different formation of geological sequestration process.

sequestration process. The injected CO_2 is stored using various trapping mechanisms for hundreds to thousands of years (Benson & Cole, 2008). There are four different trapping mechanisms (1) structural and stratigraphic trapping, the upward migration of CO_2 plume is interrupted by a sealing caprock unit (Espie, 2005; Ketzer et al., 2012); (2) capillary trapping, the injected CO_2 is rendered immobile as a disconnected or residual phase in pore spaces of the storage formation (Jalil et al., 2012; Qi et al., 2008); (3) solubility trapping when the injected CO_2 dissolves into the formation brine and forms a denser CO_2-saturated brine that moves slowly in the storage reservoir (Iglauer, 2011; Juanes et al., 2006); and (4) mineral trapping, when dissolved CO_2 reacts with formation rock to form stable carbonates and bicarbonates (Gallo et al., 2002; Li et al., 2006). In the deep geological formations, these trapping mechanisms interreact simultaneously with the injected CO_2, albeit they occur at different time intervals.

The captured CO_2 may also be compressed to the supercritical state before injecting in geological strata for optimum utilization of the available pore spaces (Gibson-Poole et al., 2008). The compressed supercritical CO_2 at a typical depth of 1 km behaves like a fluid, having density in the range of 290 to 841 $kg\,m^{-3}$ (Ivanovic et al., 2011). The density behavior depends widely on the pressure and temperature contrast, maintaining the CO_2/brine buoyant drive (Montesantos & Maschietti, 2020). The supercritical CO_2 is less dense than the resident brine that initially occupies the pore spaces (Orr, 2009). Lighter CO_2 thus rises and moves toward the sealing unit due to buoyancy force similar to the gaseous phase (Bachu, 2015). Nevertheless, supercritical CO_2 injection in the pore spaces of the geological formations is the most viably available sequestration option for secured storage, as CO_2 persists in this state for hundreds to thousands of years or even beyond that (Celia et al., 2015).

The CO_2 migration in the reservoir formation is a complex phenomenon that involves different parameters such as rock lithology, dynamics of the pore fluid, and brine−rock interaction (Saeedi, 2012). The moving front of injected CO_2 gets dissolved into the formation of brine and moves slowly and reaches beneath the sealing caprock. The CO_2 migration occurs under the influence of density contrast between the brine and the CO_2-saturated brine. The sealing caprock rendered the movement of rising CO_2 front and withstand stress variation and geochemical interactions caused by CO_2. This process occurs for a long time, from hundreds to thousands of years, until the injected CO_2 is finally immobilized and converted into mineral precipitates of carbonates and bicarbonates. Also, the reservoir rock is subjected to physical compaction and tension due to the dissolution/precipitation of minerals and reactivation or the initiation of faults/cracks due to pressure buildup caused by stress transition. This interferes with the caprock integrity and results in dynamic response and hence impacts efficiency.

The multiphase flow through a geological porous medium is complex and requires both qualitative and quantitative estimates. Darcy's law, derived by Henry Darcy in 1856, describes the simplest form of flow occurring through porous media. Previously, it was considered that the multiphase flow exclusively depends on fluid saturations. Still, later on, it was found to be dependent on wetting tendency, fluid and rock characteristics, interfacial tension, saturation history, overburden pressure, temperature, and flow direction (Saeedi, 2012). The multiphase flow behavior through the reservoir/rock system depends on fluid and rock characteristics. Thus, for the successful deployment of any CO_2 disposal project, a thorough investigation of different fluids involved in multiphase flow is required. The primary fluids are pure CO_2, a mixture of brine and aqueous CO_2 (binary mixture CO_2-H_2O), and a ternary mixture of $CO_2-H_2O-NaCl$.

Earlier the use of CO_2 was restricted to refrigeration industries and later used as a coolant in a nuclear power plant in a supercritical state and as a dry solvent for EOR. The early 1990s aims at reducing CO_2 concentration in the atmosphere by geological sequestration. The aqueous solution of CO_2 and

NaCl was widely considered for geological environments for understanding geochemical processes occurring during CO_2 sequestration. The dissolution of CO_2 in brine affects trapping mechanisms, thermophysical properties, and, hence, mutual solubilities of the CO_2−brine system.

1.4   Factors affecting CO_2 migration in the subsurface

There are various factors such as wetting tendency (wetting/nonwetting fluid), saturation history (drainage and imbibition), presence of the third phase, changes in stress and fluid properties with depth, flow parameters absolute and relative permeability, and dissolution/precipitation of minerals that control the multiphase flow characteristics of CO_2−brine−rock in the subsurface. These factors influence the efficiency of trapping mechanisms to sequester the injected CO_2 in the subsurface formation safely.

1.4.1   Wettability

It is the preferential fluid tendency in the presence of immiscible fluid to occupy interstitial voids of the porous medium. The successful implementation of geo-sequestration studies requires site investigation for containment and distribution as multiphase flow behavior is affected by wettability. For the CO_2 storage site, any leakage can occur if the net pressure exerted from beneath (by CO_2) exceeds the capillary entry pressure. The presence of CO_2 in the target brine formation can alter a porous medium's wetting behavior, affecting the fluid distribution of geological strata.

1.4.2   Saturation history

The wetting phase is displacing, called imbibition, and replaced in the porous medium called drainage. Saturation history or hysteresis effect is visible in the case of cyclic flooding. At a pore scale, the cause of hysteresis can be surface roughness and chemical heterogeneities. A change in contact angle is observed when the nonwetting phase displaced the wetting phase and vice versa due to surface roughness and chemical heterogeneities. Another cause of hysteresis can be entrapment of the nonwetting phase as disconnected blobs or ganglia in pore spaces. The subsequent drainage and imbibition process due to the presence of immobile phases results in different multiphase flow behavior.

1.4.3   Existence of third-phase saturation

The characteristic behavior of brine−rock system is greatly affected by the presence of mobile/immobile third phase. The third phase could be the trapped phase, which affects the permeability of both brine and CO_2. The inoperative oil/gas reservoirs used for storing CO_2 exist a trapped phase along with the formation of brine and injected CO_2. Hence, the multiphase behavior of a system containing the trapped phase is different from that of the pure phase, as the trapped hydrocarbon affects the injectivity of the CO_2 injection. The presence of a gas phase increases storage capacity because of its compressible nature compared to the liquid phase, which adversely affects the movement of injected CO_2. So, the presence of the trapped phase alters the performance of CO_2 storage in the subsurface.

1.4.4 Absolute and relative permeability

The permeability of a geological porous medium is a tensor quantity and is different in different directions and is expressed in Cartesian coordinate as:

$$k = \begin{bmatrix} k_{11} & k_{12} & k_{13} \\ k_{21} & k_{22} & k_{23} \\ k_{31} & k_{32} & k_{33} \end{bmatrix}$$

The permeability in the horizontal direction is generally higher than in the vertical direction; however, fissures and fractures also enhance the vertical value. It is also isotropic in the horizontal direction due to natural layering compared to a vertical direction where it is anisotropic. In particular, for a case of the Cartesian coordinate system, if two dimensions are represented in a horizontal plane and in the vertical direction, the above equation can be modified as:

$$k = \begin{bmatrix} k_h & 0 & 0 \\ 0 & k_h & 0 \\ 0 & 0 & k_v \end{bmatrix}$$

The directional dependence of permeability is not limited to absolute permeability but is valid for relative permeability also.

1.4.5 Dissolution and precipitation of minerals

The clogging of the pore spaces by displacement of fine particles, when subjected to CO_2 flooding, leads to formation change. The fine particles are mobilized due to the mineralogy of the formation, porosity/permeability of the network, pH, and viscosity thus transported to the porous system and significantly impairs the permeability. These fine particles could be clays such as kaolinite and illite and nonclays such as quartz, feldspars, carbonates, and salts, present in the sedimentary formations. The multiphase flow behavior is significantly impacted due to fine migration and thus results in dissolution/ precipitation of minerals, which either enhances or impairs the performance of the formation.

1.4.6 Change in stress pattern

With an increase in the depth of the subsurface formation for CO_2 storage, the petrophysical properties of the rock formation changes along with the stresses acting in the three dimensions. The porosity/permeability of the formation decreases with an increase in depth, and overburden pressure has an inverse relationship with depth. The effective net pressure of the reservoir formation changes with the injection of supercritical CO_2 due to the change in pore pressure, impacting the reservoir rock system's multiphase flow characteristics, especially the capillary trapping. Thus the shift in stress pattern not only alters the multiphase characteristics but also interferes with the storage capacity.

1.4.7 Change in fluid properties

The change in reservoir pore pressure due to fluid−rock interaction affects the thermophysical properties of the fluids in the pore space of the reservoir. The isothermal change that occurs during CO_2 injection alters the phase behavior along with the trapping mechanisms. The interfacial tension of the CO_2−brine system is impacted inversely due to increased pore pressure of the reservoir formation, thus influencing capillary trapping. With an increase in reservoir pressure, the solubility of CO_2 in brine increases, hence, solubility trapping. Further, mineral trapping is affected by the dissolution of CO_2 in brine.

1.5 CO_2−brine−rock interaction in the subsurface

CO_2 is injected into the saline formations with physiochemical equilibriums of fluid and rocks to reduce its concentration in the atmosphere (Zhu et al., 2019). The injected CO_2 interacts with brine and rocks and thus alters the equilibrium, leading to chemical reactions of dissolution and precipitation (Fischer et al., 2010; Wigand et al., 2008). The chemical interaction can impact the safety and stability of CO_2 storage and long-term storage capacity (Law & Bachu, 1996; Van der meer, 1993). These reactions are site-specific and depend on various factors such as temperature, pressure, brine salinity, and reservoir rock composition (Czernichowski-Lauriol et al., 2006; Rochelle et al., 2004). The laboratory experiments of brine−CO_2−rock interaction along with modeling offer a great understanding of the different mechanisms and processes related to geological storage of CO_2 for implementing the strategy effectively (Duan & Sun, 2003; Duan et al., 2006; Saeedi et al., 2016; Shachi et al., 2020). Hence, it is necessary to understand each formation's behavior before injecting CO_2 into the subsurface in the field. The summary of CO$_2$—brine—rock interaction studies performed on different reservoir rock samples is listed in Table 1.1.

The corrosive nature of CO$_2$ lays the foundation for accessing the interaction between CO$_2$—brine—rock (Hangx et al., 2013). Initial batch experiments focused on the long-term impact of CO$_2$ on host rock, which gradually moves to geochemical modeling involving complex 3D geometries. To have useful knowledge about these behaviors, in situ temperature and pressure experiments on CO$_2$/brine/rock systems can play a crucial role in CO$_2$ sequestration in the field. The injection of CO$_2$ into the formation changes the solid matrix's physics and chemistry and the resident fluid in the formation (Fischer et al., 2010). The study performed by Rochelle et al. (1999) stated the use of dry supercritical CO$_2$ as it causes little reaction than the dissolved form, which ultimately leads to dissolution and precipitation. It is because of the corrosive nature of CO$_2$ due to the formation of carbonic acid, the assessment of CO$_2$—rock interactions has been a research topic since CO$_2$ storage emerged (Gunter et al., 1997).

Table 1.1 Overview of CO$_2$—brine—rock interaction during geo-sequestration.

Type of rock formation and study	Observed changes in used samples	References
Sandstone/experiment	Dissolution of minerals SEM (Scanning electron microscope) micrographs show corrosion and precipitation	Fischer et al. (2010)
Sandstone, claystone, and mudstone/experiment	Pro tact but the sealing rock was found to have increased porosity, capillary and threshold diameter, and density. Also, larger pore size was obtained	Tarkowski et al. (2015)
Carbonates/experiment	Drop-in pH results in the mobilization of harmful trace elements, which is further adsorbed and precipitated	Kampman et al. (2014)
Sandstone/experiment	Reduction in permeability owing to salt precipitation and dissolution of fine particles clogged the pore spaces	Yu et al. (2017)
Carbonates/experiment	Dissolution and an increase in porosity and permeability of samples. XRD (X-ray diffraction) and SEM images suggest dominant solubility trapping	Azin et al. (2015)
Sandstone/experiment	Irreducible brine saturation and pore size distribution affect salt precipitation, which further affects the flow area by reducing it	Sokama-Neuyam and Ursin (2018)
Reservoir core plugs/ experiment and modeling	Porosity and permeability reduction were observed along with the disappearance of large pores	Xie et al. (2017)
Sandstone/experiment	Alteration in petrophysical properties of samples. Formation damage and fine migration is observed	Saeedi et al. (2016)
Sandstone and claystone/ experiment	The mineral composition was observed using XRD and SEM-EDS (Energy dispersive spectroscopy). Dissolution and precipitation of minerals were observed	Wdowin et al. (2014)
Carbonate/experiment	Porosity and permeability reduction along with the dissolution and precipitation of minerals were observed	Shachi et al. (2020)

1.6   Potential risk associated with CO$_2$ leakage

The risk of leakage associated with CO$_2$ due to its less dense nature than the resident brine in the targeted reservoir enables it to move toward the caprock (Weir, 1995; Xu et al., 2003; Xu et al., 2007). The leakage of CO$_2$ from the storage formation or the brine by the injected CO$_2$ is a challenging task associated with the geological storage of CO$_2$. The geological discontinuities such as faults and wells act as potential sources for leakage of CO$_2$ (Chang et al., 2008; Pan & Oldenburg, 2020). The leakage allows the movement of CO$_2$ in shallower geological units and further to the atmosphere, thus enhancing the pollution of underground water, soil, and air.

There are adequate studies that emphasize selecting suitable geological storage sites (Carpenter & Aarnes, 2013; Oldenburg & Unger, 2003; Oldenburg et al., 2009). However, the injected CO$_2$ increases the formation pressure that will change effective stresses, enhance permeability, and propagate movement along faulted zone with seismicity, resulting in leakage from the storage reservoir (Oldenburg et al., 2017; Pruess, 2006; Rutqvist & Tsang, 2002). Some studies performed for accessing risk and hydraulic fracturing associated with geological sequestration of CO$_2$ under simplified conditions (Durucan et al., 2011; Fu et al., 2017; Morris et al., 2011). The hydromechanical analysis performed by Figueiredo et al., (2017) also addressed the hydraulic fracturing and permeability alteration due to high-pressure fluid injection.

The targeted geological formations such as saline aquifers must have an overlying low-permeability caprock and offers excellent storage potential (Celia et al., 2015). The essential requirement for CO$_2$ storage is caprock integrity owing to wettability effect of reservoir/caprock system (Basirat et al., 2017; Yang et al., 2020), to prevent further CO$_2$ leakage from the storage zone. The injection of supercritical CO$_2$ in the reservoir results in increased pore pressure and temperature drop and affecting formation rock stresses (Doughty & Oldenburg, 2020). An increase in reservoir pressure, likely due to fluid injection, may initiate and further propagate hydraulic fractures (Zhou & Burbey, 2014). The reservoir pressure buildup should not exceed the estimated reservoir fracturing pressure to avoid fracturing of caprock. However, the high injection pressure is preferred for economic consideration, and thus, there is jeopardy between enhancing injection efficiency and reducing leakage risk for low-permeability reservoirs. Most of the earlier CO$_2$ projects are implemented at sites with high-permeability reservoirs and favorable injection and storage conditions, such as Sleipner (Norway, North sea) and Snohvit project (Eiken et al., 2011).

1.7 Numerical modeling for investigating CO_2 dynamics

Numerical modeling is required to predict the migration pattern of CO_2 through the geological reservoirs to estimate storage capacity, geochemical interactions, and pressure buildup due to CO_2 injection (Mykkeltvedt, 2014). Mathematically, CO_2 injection involves flow and transport processes that are highly complex and are described using a coupled system of partial differential equations that are nonlinear. The investigation related to CO_2 injection by considering all geological conditions is further complicated due to involved thermal, mechanical, chemical, and hydrodynamic processes at a large spatial and temporal scale. However, simplified numerical modeling dealing with CO_2 movement and associated risk during its geo-sequestration can help immensely finalize a particular site for its suitability for storage and controlling relevant operational parameters during the project implementation.

Geological carbon sequestration is associated with injecting CO_2 deep beneath the surface in large quantities to limit its consequences in the atmosphere (Bentham & Kirby, 2005). The stability of CO_2 mass sequestrated in the subsurface depends on interactions of injected CO_2 with formation fluid and geological formation under existing site conditions. The density of CO_2 at a subsurface depth of 800 to 1000 m ranges from 250 to 800 kg m^{-3}, which initiates density-dependent flow during and after its geo-sequestration (Bachu & Adams, 2003; Celia et al., 2015). The injection of supercritical CO_2 in saline aquifers also commences the movement of two immiscible fluids, formation brine as a wetting phase and CO_2 as a nonwetting phase in the receiving complex (Khudaida & Das, 2014), which requires both physical and chemical parameters for modeling. These parameters include relationship between permeability, saturation, and capillary pressure (Bachu et al., 1994; Birkholzer et al., 2009; Juanes et al., 2006); CO_2 trapping mechanisms (Xu et al., 2003; Xu et al., 2007); and reactions of dissolution and precipitation (André et al., 2007; Coronado & Díaz-Viera, 2017). These parameters play an important role at the pore-scale level in predicting the movement of injected CO_2 in the target formation. Injection pressure and flow rate of CO_2 play a crucial rate at the reservoir scale in deciding the overall movement of CO_2 and its associated leakage risk.

The storage sites selected for injecting CO_2 are finalized based on their sufficient storage capacity, ease of injectivity, and presence of overlying impermeable caprock without any geological discontinuities for safely storing the injected volume (Grude et al., 2014). However, fracturing due to high reservoir pressure buildup during the CO_2 injection can initiate leakage of CO_2 and/or brine from the reservoir (Chang et al., 2008). Such types of leakages allow the movement of CO_2 in overlying shallower geological units and subsequently to the atmosphere and contaminate the overlying groundwater, soil, and air. Therefore avoiding any leakage of either the CO_2 from the storage formation or the brine by the injected CO_2 is a crucial task associated with the geological storage of CO_2.

The flow of CO_2 through saline aquifers is described using the following mass balance equation (Celia et al., 2015)

$$\sum_{\gamma} \left[\frac{\partial}{\partial t} \left(S_{\gamma} \phi \ \rho_{\gamma} \right) + \nabla . \left(\rho_{\gamma} \ \nu_{\gamma} \right) - \rho_{\gamma} \ q_{\gamma} \right] = 0 \tag{1.1}$$

where S_{γ} is the saturation level of phase γ (brine or CO_2), φ is porosity, ρ_{γ} is density, t is elapsed time, and ν_{γ} is the average pore velocity of the formation. The concentration of CO_2 in the domain is expressed by advection–diffusion equation as:

$$\frac{\partial C}{\partial t} + \nu . \nabla C = \nabla . (D \nabla C) \tag{1.2}$$

where C is the concentration of CO_2 at time t, D is the diffusion–dispersion coefficient, and ν is the velocity.

To solve these flow and transport equations, certain constitutive relations are required. The average pore velocity ν_{γ} which can be described by Darcy's law for multiphase flow is represented as: $\nu_{\gamma} = -\frac{k_{r\gamma}}{\mu_{\gamma}} K . (\nabla p_{\gamma} - \rho_{\gamma} g)$ and the relative permeability is given by $k_{r\gamma} = \frac{k_{\gamma}}{K}$, substituting in Eq. 1.1 we get:

$$\nu_{\gamma} = \frac{\partial \left(S_{\gamma} \phi \rho_{\gamma} \right)}{\partial t} - \nabla . \left(\rho_{\gamma} \frac{k_{r\gamma}}{\mu_{\gamma}} K \left(\nabla p_{\gamma} - \rho_{\gamma} g \right) \right) = 0$$

The CO_2–brine system is a case of two-fluid flow process in which a decrease in saturation level of brine results in an increase of saturation level of CO_2 (S) and thus enhancing capillary pressure (P_c). Capillary pressure is given by the difference of nonwetting phase pressure (P_n) and wetting phase pressure (P_w) as $p_n - p_w = p_c$

The flow and transport Eqs. 1.1 and 1.2 are coupled and solved to predict the migration behavior of injected CO_2 in the deep subsurface formation, as shown in Fig. 1.2.

1.8   Modeling of CO_2 in subsurface: a case study

A 2D physical domain representing a characteristic deep subsurface formation at a typical depth of deep saline formation was simulated to investigate the migration behavior of CO_2 under site-specific reservoir conditions. The supercritical CO_2 is injected at the bottom from a line source in a two-

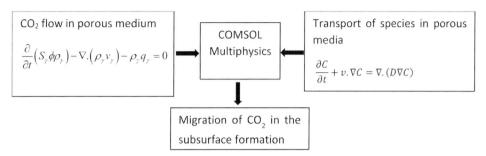

FIGURE 1.2
Schematic diagram showing the coupling of flow and transport equation for obtaining CO_2 migration.

Table 1.2 Summary of parameters used in the simulation.

Parameters	Value
Porosity	0.2
Permeability	10 mD
Density of brine	1200 kg m^{-3}
Density of CO_2	700 kg m^{-3}
Initial CO_2 concentration, C_0	0 mol m^{-3}
Normalized CO_2 concentration, C_s	1 mol m^{-3}

dimensional homogeneous porous medium. The summary of the parameters used in the simulation is listed in Table 1.2.

The model domain is located at a depth of 950 m from the ground surface between two impermeable rock layers with a permeability of 10 mD and 20% porosity, as shown in Fig. 1.3A. It extends 100 m in the horizontal direction and 50 m vertically. The domain is considered fully saturated with brine before the CO_2 injection. The right boundary is considered axial symmetric, top and bottom, expected injection points; boundaries are considered no flux type, as shown in Fig. 1.3B. The pressure of 10 MPa is specified at the left boundary. The nonwetting fluid CO_2 is injected at the bottom of the study domain in a line source of 20 m. The simulation is performed for 20 years for the injection of supercritical CO_2 in the geological formation.

The objective of the simulation is to predict the movement of CO_2 from the injection point to its surrounding areas in deep saline aquifers using the COMSOL simulator. In this simulation, brine is considered the wetting phase, and CO_2 is considered the nonwetting phase, injected in the saline formation fully saturated with brine. The CO_2 injection leads to the drainage of brine from the saturated pores followed by CO_2 invasion.

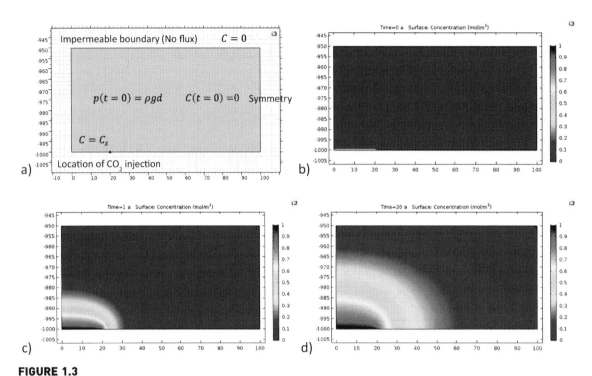

FIGURE 1.3

(A) Model domain for performing simulation showing the boundary conditions, (B) initial condition when the CO_2 is injected, and migration of CO_2 after (C) 1 and (D) 20 years.

At the start of the simulation, CO_2 is absent in the domain. After 1 year, the maximum concentration of CO_2 can be observed at the injection wells, and CO_2 has traveled around 15 m in the vertical direction and 25 m in the horizontal direction. Further, after 20 years, the CO_2 has migrated 25 m in the vertical direction and around 40 m in the horizontal direction.

1.9   Conclusions and future prospective

CCS is an emerging technique to reduce global warming and its adverse effect on the atmosphere. The carbon sequestration projects worldwide indicate the feasibility of geo-sequestration as a prominent option to mitigate the concentration of CO_2 in the atmosphere. In recent years, the potential of geo-sequestration has been widely explored in sedimentary formations along with inoperative/depleted oil and gas formations (offshore/onshore). The behavior of gaseous CO_2 in the subsurface formation depends on the trapping mechanisms that play an essential role in safe storage for a long time. The migration of CO_2 in subsurface formation is greatly affected by wetting tendency, saturation history, presence of the third phase, change in stress pattern, dissolution/precipitation of minerals, fluid properties, and porosity/permeability of the formation. Numerical modeling for CO_2 migration may help finalize a particular site for its suitability for storage and control relevant operational parameters during the project implementation. However, a thorough assessment of CO_2 leakage risk analysis is thus needed for complex network reservoirs having a broad spectrum of porosity and permeability values in real field applications. Moreover, a better understanding of the impact of CO_2 flow rate that controls the formation pressure buildup may provide insights for selecting the suitable receiving formation that can safely store CO_2 for a longer time duration without causing any leakage and fracturing.

Acknowledgments

Shachi is grateful to the Indian Ministry of Human Resource Development (MHRD) for financial support, and Anuradha is thankful to University Grant Commission (UGC) for financial assistance, as JRF/SRF.

References

André, L., Audigane, P., Azaroual, M., & Menjoz, A. (2007). Numerical modeling of fluid-rock chemical interactions at the supercritical CO2-liquid interface during CO2 injection into a carbonate reservoir, the Dogger aquifer (Paris Basin, France). *Energy Conversion and Management*, 48(6), 1782−1797.

Azin, R., Mehrabi, N., Osfouri, S., & Asgari, M. (2015). Experimental study of CO_2—Saline aquifer-carbonate rock interaction during CO_2 sequestration. *Procedia Earth and Planetary Science*, 15, 413−420.

Bachu, S. (2000). Sequestration of CO2 in geological media: Criteria and approach for site selection in response to climate change. *Energy Conversion and Management*, 41(9), 953−970.

Bachu, S. (2015). Review of CO_2 storage efficiency in deep saline aquifers. *International Journal of Greenhouse Gas Control*, 1−15, Elsevier Ltd.

Bachu, S., & Adams, J. J. (2003). Sequestration of CO_2 in geological media in response to climate change: Capacity of deep saline aquifers to sequester CO_2 in solution. *Energy Conversion and Management*, 44(20), 3151−3175.

Bachu, S., Gunter, W. D., & Perkins, E. H. (1994). Aquifer disposal of CO2: hydrodynamic mineral trapping. *Energy Conversion and Management*, 35(4), 269−279.

Basirat, F., Yang, Z., & Niemi, A. (2017). Pore-scale modeling of wettability effects on CO_2−brine displacement during geological storage. *Advances in Water Resources*. Elsevier Ltd.

Benson, S. M., & Cole, D. R. (2008). *CO_2 sequestration in deep sedimentary formations.*

Bentham, M., & Kirby, G. (2005). CO_2 storage in saline aquifers. *Oil & Gas science and Technology*, 60(3), 559−567.

Birkholzer, J. T., Zhou, Q., & Tsang, C. (2009). *Large-scale impact of CO_2 storage in deep saline aquifers: A sensitivity study on pressure response in stratified systems.* 3, 181−194.

Burrowes, G. (2001). Investigating sequestration potential of carbonate rocks during tertiary recovery from a billion barrel oil field, Weyburn, Saskatchewan: The geoscience framework

(IEA Weyburn CO2 monitoring and Storage Project). *Energy and Mines, Miscellaneous Report, 1*(Figure 2), 64—71.

Carpenter, M., & Aarnes, J. (2013). New certification framework for carbon dioxide storage sites. *Energy Procedia, 37,* 4879—4885, Elsevier B.V.

Celia, M. A., Bachu, S., Nordbotten, J. M., & Bandilla, K. W. (2015). Status of CO2 storage in deep saline aquifers with emphasis on modeling approaches and practical simulations. *Water Resources Research, 51*(9), 6846—6892.

Chang, K. W., Minkoff, S. E., & Bryant, S. L. (2008). *Modeling leakage through faults of CO_2 stored in an aquifer.*

Cook, P. J. (2009). Demonstration and deployment of carbon dioxide capture and storage in Australia. *Energy Procedia, 1*(1), 3859—3866, Elsevier.

Coronado, M., & Díaz-Viera, M. A. (2017). Modeling fines migration and permeability loss caused by low salinity in porous media. *Journal of Petroleum Science and Engineering, 150* (November 2016), 355—365, Elsevier.

Czernichowski-Lauriol, I., Rochelle, C., Gaus, I., Azaroual, M., Pearce, J., & Durst, P. (2006). Geochemical interactions between CO2, pore-waters and reservoir rocks—Lessons learned from laboratory experiments, field studies and computer simulations. *Advances in the geological storage of carbon dioxide NATO science series: IV: Earth and Environmental Sciences* (pp. 157—174), 65(III).

Dooley, J. J., Dahowski, R. T., Davidson, C. L., Wise, M., Gupta, N., Kim, S., & Malone, E. (2006). *Carbon dioxide capture and geologic storage.*

Doughty, C., & Oldenburg, C. M. (2020). CO_2 plume evolution in a depleted natural gas reservoir: Modeling of conformance uncertainty reduction over time. *International Journal of Greenhouse Gas Control, 97*(February), 103026, Elsevier.

Duan, Z., & Sun, R. (2003). An improved model calculating CO2 solubility in pure water and aqueous NaCl solutions from 273 to 533K and from 0 to 2000 bar. *Chemical Geology, 193* (3—4), 257—271.

Duan, Z., Sun, R., Zhu, C., & Chou, I. M. (2006). An improved model for the calculation of CO_2 solubility in aqueous solutions containing Na +, K +, Ca2 +, Mg2 +, Cl-, and SO42-. *Marine Chemistry, 98*(2—4), 131—139.

Dubos-Sallée, N., & Rasolofosaon, P. N. J. (2011). Estimation of permeability anisotropy using seismic inversion for the CO2 geological storage site of Sleipner (North Sea). *Geophysics, 76*(3).

Durucan, S., Shi, J. Q., Sinayuc, C., & Korre, A. (2011). In Salah CO2 storage JIP: Carbon dioxide plume extension around KB-502 well—New insights into reservoir behaviour at the In Salah storage site. *Energy Procedia, 4,* 3379—3385, Elsevier.

Eiken, O., Ringrose, P., Hermanrud, C., Nazarian, B., Torp, T. A., & Høier, L. (2011). Lessons Learned from 14 years of CCS Operations: Sleipner, In Salah and Snøhvit. *Energy Procedia, 4,* 5541—5548.

Espie, A. A. (2005). CO2 capture and storage: Contributing to sustainable world growth. *International petroleum technology conference.*

Figueiredo, B., Tsang, C. F., Rutqvist, J., & Niemi, A. (2017). The effects of nearby fractures on hydraulically induced fracture propagation and permeability changes. *Engineering Geology, 228*(August), 197—213, Elsevier.

Fischer, S., Liebscher, A., & Wandrey, M. (2010). CO_2-brine-rock interaction—First results of long-term exposure experiments at in situ P-T conditions of the Ketzin CO_2 reservoir. *Chemie der Erde - Geochemistry, 70*(S3), 155—164.

Fu, P., Settgast, R. R., Hao, Y., Morris, J. P., & Ryerson, F. J. (2017). The influence of hydraulic fracturing on carbon storage performance. *Journal of Geophysical Research: Solid Earth, 122* (12), 9931—9949.

Gallo, Y. L., Couillens, P., & Manai, T. (2002). CO_2 sequestration in depleted oil or gas reservoirs. *International conference on health, safety and environment in oil and gas exploration and production*, 1390–1392.

Gibson-Poole, C.M., Edwards, S., Langford, R.P., & Vakarelov, B. (2008). Review of geological storage opportunities for carbon capture and storage (CCS) in Victoria. *PESA Eastern Australasian Basins Symposium III*, 1–20.

Goodman, A., Bromhal, G., Strazisar, B., Rodosta, T., Guthrie, W. F., Allen, D., & Guthrie, G. (2013). Comparison of methods for geologic storage of carbon dioxide in saline formations. *International Journal of Greenhouse Gas Control, 18*, 329–342.

Grude, S., Landrø, M., & Dvorkin, J. (2014). Pressure effects caused by CO_2 injection in the Tubåen Fm., the Snøhvit field *International Journal of Greenhouse Gas Control* (27, pp. 178–187). Elsevier Ltd.

Gunter, W. D., Wiwehar, B., & Perkins, E. H. (1997). Aquifer disposal of CO2-rich greenhouse gases: Extension of the time scale of experiment for CO_2-sequestering reactions by geochemical modelling. *Mineralogy and Petrology, 59*(1), 121–140.

Hangx, S., Linden, A. V. D., Marcelis, F., & Bauer, A. (2013). International journal of greenhouse gas control the effect of CO_2 on the mechanical properties of the Captain Sandstone: Geological storage of CO 2 at the Goldeneye field (UK). *International Journal of Greenhouse Gas Control, 19*, 609–619, Elsevier Ltd.

IEA. (2013). *Technology roadmap—Carbon capture and storage.* 59.

Iglauer, S. (2011). *Dissolution trapping of carbon dioxide in reservoir formation brine—A carbon storage mechanism.*

IPCC. (2005). *IPCC, 2005: Special report on carbon capture and storage.*

IPCC. (2014). *Climate change 2014: Mitigation of climate change. Cambridge University Press.*

Ivanovic, J., Ristic, M., & Skala, D. (2011). Supercritical CO_2 extraction of *Helichrysum italicum*: Influence of CO_2 density and moisture content of plant material. *Journal of Supercritical Fluids, 57*(2), 129–136, Elsevier B.V.

Jafari, M., Cao, S. C., & Jung, J. (2017). Geological CO2 sequestration in saline aquifers: Implication on potential solutions of China's power sector. *Resources, Conservation and Recycling, 121*, 137–155, Elsevier B.V.

Jalil, M. A. A., Masoudi, R., Darman, N. H., & Othman, M. (2012). Study the CO_2 injection and sequestration in depleted M4 carbonate gas condensate reservoir, Malaysia. *Society of petroleum engineers—carbon management technology conference 2012, 1*, 68–81.

Johns, C. (2017). *Carbon sequestration—Why and how? Independent strategic analysis of Australia's global interests.*

Juanes, R., Spiteri, E. J., Orr, F. M., & Blunt, M. J. (2006). Impact of relative permeability hysteresis on geological CO_2 storage. *Water Resources Research, 42*(12), 1–13.

Kampman, N., Bickle, M., Wigley, M., & Dubacq, B. (2014). Fluid flow and CO_2-fluid-mineral interactions during CO_2-storage in sedimentary basins. *Chemical Geology, 369*, 22–50.

Ketzer, J. M., Iglesias, R. S., & Einloft, S. (2012). Reducing greenhouse gas emissions with CO_2 capture and geological storage. In W.-Y. Chen, J. Seiner, T. Suzuki, & M. Lackner (Eds.), *Handbook of climate change mitigation. Springer.*

Khudaida, K. J., & Das, D. B. (2014). A numerical study of capillary pressure—Saturation relationship for supercritical carbon dioxide (CO_2) injection in deep saline aquifer. *Chemical Engineering Research and Design, 92*(12), 3017–3030.

Koide, H. G., Tazaki, Y., Noguchi, Y., Iijima, M., Ito, K., & Shindo, Y. (1993). Carbon dioxide injection into useless aquifers and recovery of natural gas dissolved in fossil water. *Energy Conversion and Management, 34*(9–11), 921–924.

Koperna, G. J., Gupta, N., Godec, M., Tucker, O., Riestenberg, D., & Cumming, L. (2017). The grand challenge of carbon capture and sequestration. *Journal of Petroleum Technology, 69*(01), 39−41.

Law, D. H., & Bachu, S. (1996). Hydrogeological and numerical analysis of CO_2 disposal in deep aquifers in the Alberta sedimentary basin. *Energy Conversion and Management, 37*(95), 1167−1174.

Leung, D. Y. C., Caramanna, G., & Maroto-Valer, M. M. (2014). An overview of current status of carbon dioxide capture and storage technologies. *Renewable and Sustainable Energy Reviews, 39,* 426−443, Elsevier.

Li, Z., Dong, M., Li, S., & Huang, S. (2006). CO_2 sequestration in depleted oil and gas reservoirs-caprock characterization and storage capacity. *Energy Conversion and Management, 47*(11−12), 1372−1382.

Montesantos, N., & Maschietti, M. (2020). Supercritical carbon dioxide extraction of lignocellulosic bio-oils: The potential of fuel upgrading and chemical recovery. *Energies, 13*(7).

Morris, J. P., Hao, Y., Foxall, W., & McNab, W. (2011). A study of injection-induced mechanical deformation at the In Salah CO_2 storage project. *International Journal of Greenhouse Gas Control, 5*(2), 270−280, Elsevier Ltd.

Mykkeltvedt, T. S. (2014). *Numerical solutions of two-phase flow with applications to CO_2 sequestration and polymer flooding.* University of Bergen.

Oldenburg, C. M., & Unger, A. J. A. (2003). On leakage and seepage from geologic carbon sequestration sites. *Vadose Zone Journal, 2*(3), 287.

Oldenburg, C. M., Dobson, P. F., Wu, Y., Cook, P. J., Kneafsey, T. J., Nakagawa, S., Ulrich, C., Siler, D. L., Guglielmi, Y., Ajo-Franklin, J., Rutqvist, J., Daley, T. M., Birkholzer, J. T., Wang, H. F., Lord, N. E., Haimson, B. C., Sone, H., Vigiliante, P., Roggenthen, W. M., . . . & Heise, J. (2017). *Hydraulic fracturing experiments at 1500 m depth in a deep mine: Highlights from the kISMET project.* 42nd workshop on geothermal reservoir engineering. 1−9.

Oldenburg, C. M., Nicot, J., & Bryant, S. L. (2009). Energy Procedia case studies of the application of the certification framework to two geologic carbon sequestration sites. *Energy Procedia, 1*(1), 63−70, Elsevier.

Orr, F. M. (2009). Onshore geologic storage of CO_2. *Science (New York, N.Y.), 325*(5948), 1656−1658.

Ourbak, T., & Tubiana, L. (2017). Changing the game: The Paris Agreement and the role of scientific communities. *Climate Policy, 17*(7), 819−824, Taylor & Francis.

Pachauri, R. K., Allen, M. R., Barros, V., Broome, J., Cramer, W., Christ, R., Church, J., Clarke, L., Dahe, Q., & Dasgupta, P. (2014). *Summary for policymakers. Climate change 2014: Synthesis report.* Contribution of Working Groups I, II and III to the Fifth Assessment Report of the Intergovernmental Panel on Climate Change.

Pan, L., & Oldenburg, C. M. (2020). Mechanistic modeling of CO_2 well leakage in a generic abandoned well through a bridge plug cement-casing gap. *International Journal of Greenhouse Gas Control, 97*(February), 103025, Elsevier.

Piao, J., Han, W.S., Choung, S.,& Kim, K. (2018). *Dynamic behavior of CO_2 in a wellbore and storage formation: Wellbore-coupled and salt-precipitation processes during geologic CO_2 sequestration.* 2018.

Piessens, K., & Dusar, M. (1961). *CO_2-sequestration in abandoned coal mines.* Belgium.

Pruess, K. (2006). *On CO_2 behavior in the subsurface, following leakage from a geologic storage reservoir.* (1), 1−12.

Qi, R., LaForce, T. C., & Blunt, M. J. (2008). Design of carbon dioxide storage in oilfields. *Proceedings—SPE annual technical conference and exhibition.* 3(September), 1730−1741.

Ringrose, P. S., Mathieson, A. S., Wright, I. W., Selama, F., Hansen, O., Bissell, R., Saoula, N., & Midgley, J. (2013). The In Salah CO_2 storage project: Lessons learned and knowledge transfer. *Energy Procedia, 37*, 6226−6236, Elsevier B.V.

Rochelle, C. A., Czernichowski-Lauriol, I., & Milodowski, A. E. (2004). The impact of chemical reactions on CO_2 storage in geological formations: A brief review. *Geological Society, London, Special Publications, 233*(1), 87−106.

Rochelle, C. A., Pearce, J., & Holloway, S. (1999). The underground sequestration of carbon dioxide: containment by chemical reactions in the deep geosphere. *Geological Society Special Publication, 157*(1990), 117−129.

Rutqvist, J., & Tsang, C. F. (2002). A study of caprock hydromechanical changes associated with CO_2-injection into a brine formation. *Environmental Geology, 42*(2−3), 296−305.

Rutqvist, J., Vasco, D. W., & Myer, L. (2009). Coupled reservoir-geomechanical analysis of CO_2 injection at In Salah, Algeria. *Energy Procedia, 1*(1), 1847−1854, Elsevier.

Saeedi, A. (2012). *Multiphase flow during CO_2 Geo-sequestratio*n.

Saeedi, A., Delle, C., Esteban, L., & Xie, Q. (2016). Flood characteristic and fluid rock interactions of a supercritical CO_2, brine, rock system: South West Hub, Western Australia. *International Journal of Greenhouse Gas Control, 54*, 309−321, Elsevier Ltd.

Shachi., Gupta, P. K., & Yadav, B. K. (2019). Aspects of CO_2 injection in geological formations and its risk assessment. In R. N. Bhargava (Ed.), *Ecological implications and management* (pp. 83−100). Springer.

Shachi., Yadav, B. K., Rahman, M. A., & Pal, M. (2020). Migration of CO_2 through carbonate cores: Effect of salinity, pressure, and cyclic brine-CO_2 injection. *Journal of Environmental Engineering, 146*(2).

Shukla, R., Ranjith, P., Haque, A., & Choi, X. (2010). A review of studies on CO_2 sequestration and caprock integrity. *Fuel, 89*(10), 2651−2664, Elsevier Ltd.

Shukla, R., Ranjith, P. G., Choi, S. K., & Haque, A. (2011). Study of caprock integrity in geosequestration of carbon dioxide. *International Journal of Geomechanics, 11*(4), 294−301.

Sokama-Neuyam, Y. A., & Ursin, J. R. (2018). The coupled effect of salt precipitation and fines mobilization on CO2 injectivity in sandstone. *Greenhouse Gases: Science and Technology, 8*(6), 1066−1078.

Solomon, S. (2006). *Criteria for intermediate storage of carbon dioxide in geological formations.* (1).

Stocker, T. F., Dahe, Q., Plattner, G. -K., Alexander, L. V., Allen, S. K., Bindoff, N. L., Bréon, F. -M., Church, J. A., Cubash, U., Emori, S., Forster, P., Friedlingstein, P., Talley, L. D., Vaughan, D. G., & Xie, S. -P. (2013). *Technical summary. Climate change 2013: The physical science basis.* Contribution of Working Group I to the Fifth Assessment Report of the Intergovernmental Panel on Climate Change.

Sun, Y., Li, Q., Yang, D., & Liu, X. (2016). Laboratory core flooding experimental systems for CO2 geosequestration: An updated review over the past decade. *Journal of Rock Mechanics and Geotechnical Engineering, 8*(1), 113−126.

Tarkowski, R., Wdowin, M., & Manecki, M. (2015). Petrophysical examination of CO2-brine-rock interactions—Results of the first stage of long-term experiments in the potential Zaosie Anticline reservoir (central Poland) for CO_2 storage. *Environmental Monitoring and Assessment, 187*(1).

United Nations. (2015). *Adoption of the Paris Agreement.*

Van der meer, L. G. H. (1993). The conditions limiting CO_2 storage in aquifers. *Energy Conversion and Management, 34*(9), 959−966.

Verdon, J. P., Kendall, J. M., Stork, A. L., Chadwick, R. A., White, D. J., & Bissell, R. C. (2013). Comparison of geomechanical deformation induced by megatonne-scale CO2 storage at

Sleipner, Weyburn, and In Salah. *Proceedings of the National Academy of Sciences of the United States of America, 110*(30).

Vishal, V., Ranjith, P. G., & Singh, T. N. (2015). An experimental investigation on behaviour of coal under fluid saturation, using acoustic emission. *Journal of Natural Gas Science and Engineering, 22*, 428−436, Elsevier B.V.

Wdowin, M., Tarkowski, R., & Franus, W. (2014). Determination of changes in the reservoir and cap rocks of the Chabowo Anticline caused by CO_2-brine-rock interactions. *International Journal of Coal Geology, 130*, 79−88, Elsevier B.V.

Wei, N., Li, X., Fang, Z., Bai, B., Li, Q., Liu, S., & Jia, Y. (2015). Regional resource distribution of onshore carbon geological utilization in China. *Journal of CO_2 Utilization, 11*, 20−30.

Weir, G. J. (1995). Reservoir storage and containment of greenhouse gases. *Transport in Porous Media, 36*(6), 531−534.

Wigand, M., Carey, J. W., Schütt, H., Spangenberg, E., & Erzinger, J. (2008). Geochemical effects of CO_2 sequestration in sandstones under simulated in situ conditions of deep saline aquifers. *Applied Geochemistry, 23*(9), 2735−2745.

Xie, Q., Saeedi, A., Delle Piane, C., Esteban, L., & Brady, P. V. (2017). Fines migration during CO_2injection: Experimental results interpreted using surface forces. *International Journal of Greenhouse Gas Control, 65*(February), 32−39, Elsevier.

Xu, T., Apps, J. A., & Pruess, K. (2003). Reactive geochemical transport simulation to study mineral trapping for CO_2 disposal in deep arenaceous formations. *Journal of Geophysical Research: Solid Earth, 108*(B2).

Xu, T., Apps, J. A., Pruess, K., & Yamamoto, H. (2007). Numerical modeling of injection and mineral trapping of CO_2 with H_2S and SO_2 in a sandstone formation. *Chemical Geology, 242* (3−4), 319−346.

Yang, Z., Chen, Y., & Niemi, A. (2020). Gas migration and residual trapping in bimodal heterogeneous media during geological storage of CO_2. *Advances in Water Resources, 142*, Elsevier Ltd, (October 2019).

Yu, Z., Liu, K., Liu, L., Yang, S., & Yang, Y. (2017). An experimental study of CO_2-oil-brine-rock interaction under in situ reservoir conditions. *Geochemistry Geophysics Geosystems*, 1−26.

Zhou, X., & Burbey, T. J. (2014). Fluid effect on hydraulic fracture propagation behavior: A comparison between water and supercritical CO_2-like fluid. *Geofluids, 14*(2), 174−188.

Zhu, T., Ajo-franklin, J., Daley, T. M., & Marone, C. (2019). Dynamics of geologic CO2 storage and plume motion revealed by seismic coda waves. *Proceedings of the National Academy of Sciences of the United States of America, 116*(7), 2464−2469.

Column adsorption studies for the removal of chemical oxygen demand from fish pond wastewater using waste alum sludge

Muibat D. Yahya, Ibrahim A. Imam and Saka A. Abdulkareem

Federal University of Technology, Minna, Nigeria

2.1 Introduction

2.1.1 Fish pond wastewater

One of the most prevailing activities in the aquaculture industry of today is fish production. With increase in population, agricultural activities, in general, and fish production, in particular, have gained active involvement of individuals setting up small and medium fish-farming enterprise in Nigeria. This coupled with government intervention and support for both small-scale and large-scale fish farming makes Nigeria the largest aquaculture producer in Sub-Saharan Africa [Akankali et al., 2011; Food & Agriculture Organization (FAO), 2017]. Fish production supplies more than 30% of the total animal protein consumed in Nigeria and has no doubt engineered the crescive rate of employment in the country (Emmanuel et al., 2014). Despite the numerous credits associated with the aquaculture industry, its adversity in terms of environmental pollution resulting from the wastewater it generates is seldom evaluated. The rising quest for the research into wastewater treatment technologies has been triggered by the creation of stringent regulations standard and increasing environmental awareness (Anijiofor et al., 2018). Teodorowicz et al. (2006) reported that fish farm pollutants include metabolic by-products of the fish such as feces and fish-feeding methods. Wastewater from fish pond contains pollutants which include suspended solid; total phosphorus; volatile solids; soluble reactive nutrients such as phosphorus, nitrogen, ammonia, nitrate, and nitrites (Ansah et al., 2013; Da et al., 2015; Schwartz & Boyd, 1994); and biological pollutants such as biochemical oxygen demand and chemical oxygen demand (COD) (Anijiofor et al., 2018), which leads to the eutrophication and squalor of the receiving water body. Pollutants are directly or indirectly introduced into the fish pond through activities such as fish feeding, excretion, and application

Advances in Remediation Techniques for Polluted Soils and Groundwater. DOI: https://doi.org/10.1016/B978-0-12-823830-1.00006-7

of organic and inorganic fertilizers (Ansah et al., 2013), which increases the amount of micro-organisms, soluble and insoluble substances in the wastewater that will in turn increase biological oxygen demand (BOD) and COD pollution load in the receiving water body.

Wastewater from fish pond, depending on the duration of the water in the pond, season, and type of feed and fertilizer used, have varying pollutant concentration. However, an aggregate of random pollutants that are chemically oxidized in water is known to be COD (Geerdink et al., 2017). The high concentration of pollutants (COD in particular) present in the effluent of fish ponds is largely due to the manufactured feed and fertilizers necessary for boosting fish production in aquaculture (Chatvijitkul et al., 2017). This can cause organic or nutrient pollution to the water bodies receiving the effluent and making water obtained from such source costly to treat for end use (Tucker & Hargreaves, 2008). The reduction of COD from fish pond wastewater before discharging is important for controlling pollution load of surface water within the locality of fish pond. The threshold value for COD in fish pond wastewater as specified by National Environmental Standards and Regulation Enforcement Agency is 30 mg L^{-1} before discharging into the environment [National Environmental Standard Regulation & Enforcement Agency (NESREA) Federal Republic of Nigeria official gazette, 2011].

2.1.2 Methods of removing chemical oxygen demand

Various treatment methods have been used to remove COD from wastewaters among which coagulation/precipitation (Chaudhari et al., 2010); membrane separation (Dong et al., 2013); sorption (Dridi-Dhaouadi & Mhenni, 2014); thermolysis (Sahu & Chaudhari, 2015); filtration (Mohamed, 2015); and combined coagulation, ozonation, and sequencing batch reactor (SBR) (Zou, 2017) have been reported to reduce the concentration of COD from wastewater significantly. All methods reported have intrinsic merits and demerits, depending on the concentration of pollutant in the wastewater, treatment objective, process cost, and simplicity (Xu et al., 2015). However, adsorption technique due to its ease of operation, operational cost, flexibility, and efficiency has emerged as one of the methods extensively applied for the removal of mineral and organic micropollutants from wastewater with activated carbon being the major adsorbent employed (Sugashini & Begum, 2013; Kalavathy et al., 2010). Energy consumption and regeneration of activated carbon are high (Lin & Juang, 2009). Thus attention is focused on cheap materials, mostly assumed to be waste materials, which are relatively abundant and can selectively take up molecules from aqueous solution. These materials are open to modification procedure that could enhance their sorption performance (Ali et al., 2012).

2.1.3 Waste alum sludge

Waste alum sludge are the by-products of water treatment plant when salts of aluminum are primarily employed for coagulation and flocculation (Yang et al., 2006; Zhao et al., 2011). Alum sludge is extensively generated as the by-products of treatment plants due to the increasing use of aluminum salts since it is effective, cheap, and readily available (Dassanayake et al., 2015). Conventionally, sludge irrespective of where it is generated from is seen as waste material that is mostly managed by land filling, farmland application, and incineration. The increased production of sludge and its unfeasible management as waste have paved way for its application as an adsorbent for the removal of various types of contaminants (Devi & Saroha, 2017; Kong et al., 2014; Xu et al., 2015). The use of sludge to produce adsorbents is reportedly being studied for the removal of various pollutants such as organics from wastewater (Pan et al., 2011), phosphate from wastewater (Chen et al., 2013), COD from leachate (He et al., 2014), ammonia–nitrogen from wastewater (Yunnen et al., 2016), and ammonia from wastewater (Yang et al., 2016) and for the treatment of diary wastewater (Suman et al., 2017). According to Zhao et al. (2009), the average removal efficiency for COD and BOD using dewatered alum sludge is 73.3% ± 15.9% and 82.9% ± 12.3%, respectively. Yang et al. (2015) prepared an adsorbent using sludge obtained from drinking waterworks which shows ammonium removal of 90%. Removal of turbidity, COD, and BOD from the synthetic dairy wastewater was around 93%, 65%, and 67%, respectively, when water treatment sludge was used as coagulant (Suman et al., 2017). These studies demonstrate the effectiveness of alum sludge for the removal of COD from wastewater. In light of the above studies, this study investigates the continuous adsorptive removal of COD from fish pond wastewater using waste alum sludge obtained from water treatment plant in a fixed-bed column. The effects of process parameters such as flow rate and bed height were studied and the results obtained confirm that waste alum sludge can be a potential adsorbent for removal of COD from fish pond wastewater.

2.2 Materials and methods

2.2.1 Preparation of the adsorbent

2.2.1.1 Procedure for the preparation of adsorbent from sludge

The sludge exploited for this study was collected from the sedimentation basin of Niger State Water Board, Chanchaga Treatment Plant (Niger State, Nigeria) which treats water sourced from dam constructed at the same location. The sludge was transformed into adsorbent materials by thermal pyrolysis. The sludge obtained was allowed to settle for 12 h and then decanted to separate the slurry from the clear water zone. The slurry was air-dried for 7 days under ambient condition and the mass of the dried sludge was weighed before it was further dried in an oven at 60°C for 30 min after which it was again weighed. Oven-drying and subsequent reweighing was continued until two successive measurements showed no difference in the weight measured. The sludge was then pulverized and sieved so as to achieve homogeneity in particle size distribution. The homogenize sample is then fired in a furnace for 45 min at a constant heating rate of $30°C \, min^{-1}$.

Fish pond effluent was obtained from Musgola Farm, Lapai Gwari, Minna Local Government Area of Niger State, the major activity of which is centered on the production of catfish. The fish pond, which is earthen, has the dimension of 60 by 30 ft. The effluent sample was collected at 5:45 p.m. on October 27, 2017, which is toward the end of raining season for the year. The atmospheric conditions (temperature and humidity) were mild. The temperature of the sample collected was 29°C, same as the ambient temperature and the temperature of the fish pond's wastewater from which the sample was collected. Wastewater from Musgola Farm is not subjected to any form of treatment before it is being discarded. The collected effluent sample was rid of suspended particulate matter by filtration.

2.2.1.2   Characterization of sludge-derived adsorbent
Fourier-transform infrared analysis

Fourier-transform infrared (FT-IR) analysis was carried out using FT-IR spectrophotometer (FTIR-8400, Shimadzu). Results of the alum sludge revealed the various functional groups present on its surface. Fig. 2.1 shows the infrared radiation spectra of the alum sludge before and after adsorption. The spectra show the complex chemical bond structure on the surface of the alum sludge constitutes mixture of inorganic and organic matter. The peaks observed at 3842 and $3742 \, cm^{-1}$ indicate -OH bonding to inorganic matter, contrary to -OH and -NH stretch correlation for organic matter which has the maximum peak value of $3276 \, cm^{-1}$ (Hossain et al., 2011).

Braunuer−Emmett−Telller surface area analysis

Braunuer−Emmett−Telller (BET) surface area analysis is used to determine specific surface area and pore size distribution of an adsorbent by calculating the amount probing gas (nitrogen gas) physically adsorbed on the surface of the solid at liquid nitrogen temperature. Information regarding how the physical structure of an adsorbent will influence the manner with which adsorbent interacts with its environment can be deduced from BET surface analysis. Properties such as adsorption rate, moisture retention, and shelf life of the adsorbent can be correlated to adsorbent's specific surface area (Hanaor et al., 2014). Surface area of the prepared waste alum sludge was

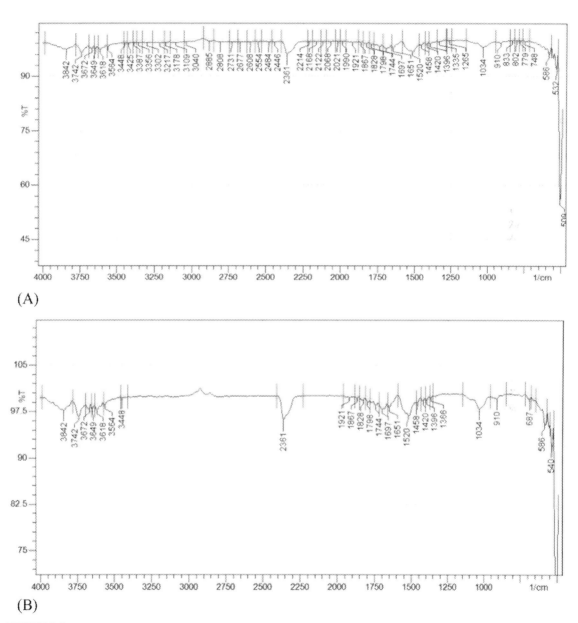

FIGURE 2.1

FT-IR spectra of drinking water treatment sludge (A) before adsorption and (B) after adsorption. *FT-IR*, Fourier-transform infrared.

studied using nitrogen gas adsorption at the temperature 77.35K and the results are given in Table 2.1.

The surface area of the sludge was obtained as $139 \ m^2 \ g^{-1}$, pore volume as $0.126 \ cm^3 \ g^{-1}$, and pore radius as 5.228 nm. Since the pore radius is within the range of 2 to 50 nm, the sample is said to be mesoporous.

2.2.2 Fixed-bed column studies

The continuous adsorption studies were carried out in glass column of 12 mm internal diameter and height of 200 mm incorporated with medium porous disk at its bottom to serve as adsorbent support. Flow through the column was achieved by using peristaltic pump, which has the capacity to pump a low-viscous incompressible fluid in the range of $0.0002-20 \ mL \ min^{-1}$. The column was set up as shown in Fig. 2.2, the downstream of the peristaltic pump was connected to the column in such a way that the column is operated in a downward delivery pattern and the upstream (suction) of the pump hose is dipped into the effluent. A constant influent concentration of $72 \ mg \ L^{-1}$ was fed into the column at a desired flow rate of $5 \ mL \ min^{-1}$ for a varied bed depth of 50, 75, and 100 mm. Samples were collected from the column's exit and analyzed (using HACH digestion solution, HACH COD reactor model

Table 2.1 Multipoint Braunuer–Emmett–Telller adsorption parameters for waste alum sludge.

Surface area ($m^2 \ g^{-1}$)	C constant	Pore volume ($cm^3 \ g^{-1}$)	Pore radius (Å)
139	7.47	0.126	52.28

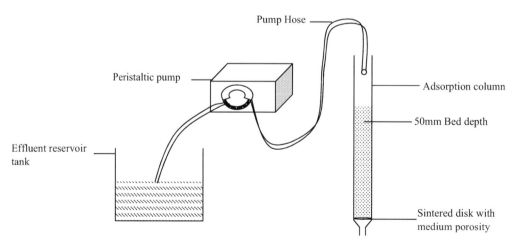

FIGURE 2.2
Schematic setup for fixed-bed adsorption column.

45600−00 and HACH calorimeter model Dr/890). For the bed depth was fixed at 50 mm, the adsorption characteristics for the varied influent flow rate of 3, 5, and 7 mL min^{-1} were studied.

2.2.2.1 Column data analysis

The occupation or binding of atoms, ions, or molecules of a gas or liquid stream (adsorbate) onto the surface of a solid material (adsorbent) is termed adsorption. As adsorption process proceeds, adsorbates are being accumulated on the surface of the adsorbent until a film of the adsorbate is formed on the adsorbent in an amount sufficient enough to make the adsorbent lose its capacity to retain anymore adsorbate on its surface, the adsorbent is said to be exhausted or saturated at this point (Chen et al., 2013). The maximum quantity of adsorbate molecules an adsorbent could accumulate on its surface at equilibrium conditions is the sorption capacity of the adsorbent. Sorption capacity of an adsorption is determined by the nature of its precursor, surface properties, type of adsorbate molecule, pore structure, and so on (Javaid et al., 2015).

The performance of fixed-bed adsorption is best described by the fractional concentration time plot (breakthrough curve). The breakthrough curve is a representation that shows the points where pollutant molecule starts to appear at the outlet end in trace amount and the point where the outlet concentration of the pollutant almost equals to the inlet concentration in a fixed-bed column (Sugashini & Begum, 2013). The breakthrough point is the point on the fractional concentration time plot where 5% of the inlet concentration of the pollutant molecule is observed at the outlet end of the adsorption column. The pollutant concentration at the outlet gradually increases to the point where it is 95% of the inlet concentration. At this point, the bed is said to be saturated/exhausted (Patel & Vashi, 2015). For the successful design and operation of fixed-bed continuous adsorption process, it is paramount that the characteristic breakthrough curve at specified operating conditions should be predictable. Breakthrough curve can be used to determine the operating life span of the bed and regeneration times for specific process condition (Ahmad & Hameed, 2010).

The breakthrough curve is the graph of C_t/C_i against time (t), where C_t and C_i in mgL^{-1} is the amount of COD in the outlet stream at time t and its concentration in the inlet stream, respectively (Sugashini & Begum, 2013).

The mass transfer zone Δt is mathematically expressed in Eq. (2.1) as:

$$\Delta t = t_s - t_b \tag{2.1}$$

where t_b represents the breakthrough time [i.e., the time at which the adsorbate concentration in the column outlet stream rapidly rises to a significant value, and t_s represents bed saturation time (time at which the adsorbate concentration in the column outlet stream exceeds 99% of the columns inlet stream concentration)] (Kalavathy et al., 2010).

The length of mass transfer zone of the column bed, Z_{mt}, can be obtained from the breakthrough plot and also calculated from Eq. (2.2):

$$Z_{mt} = Z\left(1 + \frac{t_b}{t_e}\right) \tag{2.2}$$

where Z stands for the column bed depth in mm (Futalana et al., 2011).

The total adsorption column capacity q_{total} is calculated from Eq. (2.3) by the following relationship:

$$q_{total} = \frac{QA}{1000} = \frac{Q}{1000}\int_{t=0}^{t=t_{total}} C_{ads}dt \tag{2.3}$$

where A represents area under the breakthrough curve for the graph of $C_{ads} = (C_0 - C_t)$ against time (mg.min mL^{-1}), Q stands for the volume inlet flow rate in mL min^{-1}, and t_{total} is the time of flow in minute (Futalana et al., 2011).

The total amount of adsorbate sent into the column, m_{total}, is calculated from Eq. (2.4):

$$m_{total} = \frac{C_oQt_{total}}{1000} \tag{2.4}$$

The total percent adsorbate removal (column performance) with respect to flow volume can be calculated from Eq. (2.5):

$$\text{Total removal}(\%) = \frac{q_{tot}}{m_{total}} \tag{2.5}$$

Adsorption is an equilibrium process and equilibrium studies of adsorption gives information about the amount of adsorbent required for a unit mass of pollutant under the system conditions (Aksu & Gonen, 2004). The equilibrium adsorption capacity ($q_{eq(exp)}$) is calculated from Eq. (2.6):

$$q_{eq(exp)} = \frac{q_{tot}}{m} \tag{2.6}$$

where m is the mass of the adsorbent in gram (Sugashini & Begum, 2013).

The unadsorbed adsorbate concentration at equilibrium in the column (C_{eq}) in mg L^{-1} can be calculated from Eq. (2.7):

$$C_{eq} = \frac{m_{total} - q_{tot}}{V_{eff}} \times 1000 \tag{2.7}$$

where V_{eff} is the effluent volume which is the product of volumetric flow rate and total flow time.

2.2.2.2 Error analysis

Error analysis is the most suitable optimization method to evaluate the best fitted model for experimental data. In linear regression analysis, the different forms of equation resulting from assumptions used for their formulation have an effect on the coefficient of regression (R^2); the nonlinear regressive analysis provides a better preference in avoiding such discrepancy (Han et al., 2009). The mathematical equation of least square of error (or SS error) is expressed in Eq. (2.8) as:

$$SS = \frac{\sum \left[\left(\frac{C_t}{C_o}\right)_c - \left(\frac{C_t}{C_o}\right)_e \right]^2}{N}$$

(2.8)

where $\left(\frac{C_t}{C_o}\right)_c$ is ratio of effluent and influent concentrations obtained from the models and $\left(\frac{C_t}{C_o}\right)_e$ is ratio of effluent and influent concentration obtained from experiment and N is the number of the experimental points.

2.2.2.3 Chemical oxygen demand analysis

COD was analyzed using HACH reactor (model 45600−00) and HACH programmed calorimeter (model Dr/890) in parts per million (ppm). The reactor was used to digest the mixture of HACH dichromate digestion solution and wastewater sample. The value of COD concentration is obtained using HACH calorimeter.

Each COD vial to be used was labeled for each sample to be analyzed and a blank vile was made using deionized water. 2 mL of the sample was measure and transferred into each vial containing dichromate digestion solution prepared by HACH Company with catalog number (cat. 212259). The digestion solution is effective for COD concentration range of 20 to 1500 ppm. The mixture was then placed in the fume cabinet of COD reactor and was digested for 2 h after which the sample was withdrawn and allowed to cool to ambient temperature. The blank vile was placed into the vile compartment at the surface top of the HACH programmed calorimeter for calibration. The cooled sample was then placed in the vile compartment to determine COD value of the sample.

2.3 Results and discussion

The raw sample collected from Musgola Farms was first tested for COD contamination within 24 h of collection. The COD value for the raw sample was 72 ppm (mg L^{-1}). The wastewater sample was then treated with the waste alum sludge for 1 h to assess the sludge's affinity for COD contaminants. The threshold limit for COD in wastewater before discharging into receiving water body in Nigeria is 30 mg L^{-1} as specified by National Environmental

Standards and Regulation Enforcement Agency [National Environmental Standard Regulation & Enforcement Agency (NESREA) Federal Republic of Nigeria official gazette, 2011]. The COD concentration for the treated wastewater sample was then analyzed and the value of 2 ppm was obtained. With over 95% removal of COD by waste alum sludge on first trial, the experiment was advanced to study the effect of flow rate (3, 5, and 7 mL min^{-1}) and bed height (50, 75, and 100 mm) on the removal COD in a fixed-bed column. Data obtained were fitted to several continuous system adsorption models and the alignment pattern of each model to the experiment data was analyzed using linear regression and SS error analysis. Results obtained showed that efficiency of the waste alum sludge for the removal COD contaminants depend on both the flow rate of the wastewater and bed height of the adsorbent. At higher flow rate, the breakthrough time and bed exhaustion time were shorter as compared to what were obtained at lower flow rate. The efficiency of the column also depends on the bed height; the percentage of COD removal, breakthrough time, and bed exhaustion time increased as the height of the bed was increased from 50 to 75 and 100 mm.

2.3.1 Dynamic adsorption studies

Analysis of column data obtained for the fixed-bed adsorption of COD on waste alum sludge is summarized in Table 2.1. Changing flow rate from 3 to 5 mL min^{-1}, the magnitude of equilibrium adsorption capacity (q_e) reduces with negligible difference (i.e., from 40.35 to 39.68 mg g^{-1}, respectively), as the flow rate increases to 7 mL min^{-1}, equilibrium adsorption capacity increased to 53.98 mg g^{-1}. Contrary to the previous trend, for the bed depth of 50, 75, and 100 mm, the equilibrium adsorption capacity of the bed was 39.68, 35.35, and 28.97 mg g^{-1}, respectively, which implies that the equilibrium adsorption capacity decreases with increasing bed height which can be attributed to broadened mass transfer zone consequent to varied bed depth with subsequent reduction in adsorbate molecule in the solution since the influent concentration was kept constant. The trend of results obtained in this experiment is in conformity with findings presented by Patel and Vashi (2015), in which adsorbents prepared from powdered neem leaf were used to adsorb COD from dyeing mill wastewater. The column efficiency expressed in terms of percentage removal of COD attained the optimum value of 70.28% at the process condition of 3 mL min^{-1} flow rate and 50 mm bed depth with influent concentration of 72 mg L^{-1}. The percentage removal shows declining trend at higher flow rates and increases with increasing bed depth. This trend conforms to the result presented by Yahya et al. (2020a) and Sugashini and Begum (2013) in which adsorption of adsorbate molecules decreases with increasing flow rate and/or with decreasing bed depth in which adsorption of adsorbate molecules decreases with increasing flow rate and/or with decreasing bed depth.

2.3.2   Influence of operational variables

2.3.2.1   Effect of influent flow rate

In a continuous system, influent flow rate into an adsorption column deter-mines the average time spent by an adsorbate molecule in the column. Higher flow rates permit smaller time of interaction between the adsorbent and adsorbate molecule, while lower flow rates tend to allow more time in that case. The effect of flow rate on fixed-bed adsorption of COD was studied by varying the influent flow rate from 3, 5, to 7 mL min^{-1} at influent concen-tration of 72 mg L^{-1} with a fixed bed depth of 50 mm. Fig. 2.3 shows the breakthrough curve for the varying flow rate where it can be observed that the breakthrough time increases with increase in flow rates. Results as presented indicate that the time for bed saturation reduces as flow rate is increased since lower time is available for the premium interaction between the adsorbate molecule and the adsorbent that would allow complete loading of adsorbate molecules in the adsorbent pores at higher flow rates. For 3 mL min^{-1}, the breakthrough was observed at about 540 min and 99% of the influent con-centration was detected at the time of 1170 min. For 5 and 7 mL min^{-1} flow rates, the breakthrough time and bed saturation time were obtained as 270 and 810 min and 120 and 810 min, respectively. The shape of the break-through curve is influenced by the flow rate; at low flow rate the break-through curve tends to be less steep (characterized by smooth S-shaped curve). This is because at low flow rate, more time of contact is provided so that the adsorbate molecules could properly diffuse into the pores of the adsorbent, thereby increasing the adsorption efficiency of the bed (Ahmad &

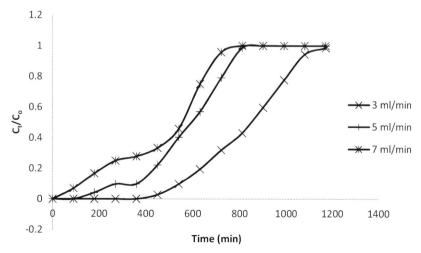

FIGURE 2.3

Breakthrough curve for adsorption of COD at varying inlet flow rates. *COD*, chemical oxygen demand.

Hameed, 2010; Tan et al., 2008) Thus breakthrough time, length of mass transfer zone, bed exhaustion time, percentage removal, and adsorption capacity were higher for the flow rate of 3 mL min^{-1} when compared to those obtained at the varied flow rate of 5 and 7 mL min^{-1}. Similar report was given by Ahmad et al. (2012) and Yahya et al. (2020b) that the column performed better at a lower flow rate which resulted into a longer exhaustion and breakthrough time. Sur and Mukhopadhyay (2018) confirmed that decrease in the velocity of the industrial effluent sufficiently increases the adsorption of the COD from the gas effluent passed through the fluidized bed of the adsorbent in their study on the removal of COD from electroplating industry effluent using *Pseudomonas aeruginosa* and *Pseudomonas putida*. Patel and Vashi (2015), Sugashini and Begum (2013), Han et al. (2009), and Nwabanne and Igbokwe (2012) also reported similar trend in their study.

2.3.2.2 *Effect of bed depth*

The breakthrough curve for the varied bed height of 50, 75, and 100 mm for 5 mL min^{-1} flow rate and influent concentration of 72 mg L^{-1} is shown in Fig. 2.4. It can be seen that the breakthrough time and time of bed exhaustion increase with increasing bed depth. For the bed height of 100 mm, the breakthrough was observed at 530 min and 97% of the inlet concentration was obtained at 1170 min. For the bed height of 75 mm, breakthrough was observed at about 360 min and the bed was completely exhausted at 1080 min. When the bed height was at 50 mm, the breakthrough was observed at a time less than 360 min and the bed was exhausted at 810 min. Higher bed depth translates to increase in the amount of adsorbent, which

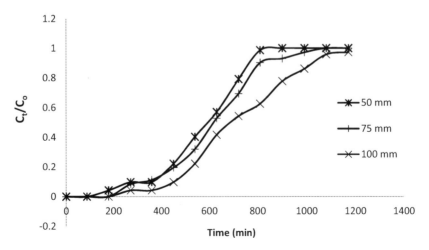

FIGURE 2.4
Breakthrough curve for adsorption of COD at different bed depth. *COD*, chemical oxygen demand.

Table 2.2 Column data analysis for various parameters.

Z (mm)	Q (mL min^{-1})	C_o(mg L^{-1})	q_{total} (mg)	W_{total} (mg)	q_e (mg g^{-1})	% Removal	Z_m (mm)
50	3	72	177.53	252.72	40.35	70.25	30.77
50	5	72	174.60	324.00	39.68	53.89	38.61
50	7	72	237.51	408.24	53.98	58.18	44.44
75	5	72	219.15	388.80	35.35	56.37	53.68
100	5	72	257.85	421.20	28.97	61.22	61.54

broadens the length of the mass transfer zone in the column. The increase in adsorption with increase in bed height was a result of the increase in the doses of adsorbent in the larger bed which provides greater adsorption sites (Vijayaraghavan et al., 2004; Zulfadhly et al., 2001). The effect of change in bed depth as shown in Table 2.2 confirms that increase in bed depth results to decline in adsorption capacity, increase in percentage removal of COD, and extension in length of mass transfer zone (Z_m). The gradual drop in the equilibrium adsorption capacity implies that insufficiency in the concentration COD requires to interact with the adsorbent since the concentration of influent stream is kept constant for the varying bed depth. The broadened mass transfer zone leads to the increase in the percentage removal of COD from the influent stream due to increase in active site as bed depth increases. These findings concur to the outcome of Patel and Vashi (2015), Sugashini and Begum (2013).

2.3.3 Kinetics studies on the adsorption of chemical oxygen demand on the sludge adsorbent

2.3.3.1 Thomas kinetics model

Thomas model, developed by H.C. Thomas (1944), assumes plug flow behavior in the bed and uses Langmuir kinetics of adsorption−desorption process for equilibrium (Balci et al., 2011; Han et al., 2009). This model can be used to calculate the maximum adsorption capacity of a bed at certain process conditions. Thomas model can be expressed in its linear form as follows in Eq. (2.9):

$$\ln\left(\frac{C_i}{C_t} - 1\right) = \frac{k_{TH}q_o m}{Q} - k_{TH}C_i t \tag{2.9}$$

where k_{TH} is the Thomas rate constant in mL min^{-1}.mg, q_o is the equilibrium uptake per gram of the adsorbent in mg L^{-1}, C_i in mg L^{-1} is the inlet concentration, C_t in mg L^{-1} is the outlet concentration at time t, m is the mass of adsorbent in gram, Q is the flow rate in mL min^{-1}, and t is the time of flow in minute (Ahmad & Hameed, 2010).

Table 2.3 Constants of Thomas adsorption model for various operating condition.

Z (mm)	Q (mL min^{-1})	C$_o$ (mg L^{-1})	k$_{TH}$ (L min^{-1}.g)	q$_o$ (mg g^{-1})	R^2	SS
50	3	72	0.125	43.02	0.970	0.0033
50	5	72	0.139	45.18	0.902	0.0021
50	7	72	0.097	55.73	0.872	0.0048
75	5	72	0.117	25.89	0.897	0.0310
100	5	72	0.137	32.57	0.987	0.0260

The Thomas model is suitable for adsorption process when external and internal diffusion were not the limiting step of the process (Banerjee et al., 2012; Chen et al., 2012).

The values of Thomas constants obtained from various operating condition are shown in Table 2.3. The value k_{TH} is obtained from the slope of $\ln\left(\frac{C_o}{C_t} - 1\right)$ time profile and q_o is obtained from the intercept of the plot. The regression analysis coefficient ranging from 0.870 to 0.987 ($0.987 < R^2 > 0.870$) shows that Thomas model fits to the experimental data obtained from continuous adsorption of COD using drinking water sludge as adsorbent. There is no significant difference in the Thomas model kinetic constant (k_{TH}) at the various operating conditions. The adsorption capacity, q_o, increases as the influent flow rate increases. The adsorption capacity is also affected by the bed depth, at 50 mm bed depth the value of q_o is 45.18 mg g^{-1} and at the bed depth of 75 and 100 mm, the values of q_o were obtained 25.89 and 32.57 mg g^{-1}, respectively. The k_{TH} value increased with increase in flow rate and bed height. The maximum adsorption capacity q_o was observed to increase with increase in flow rate but to decrease with increase in bed height. This trend was similarly reported by Radhika and Palanivelu (2006) and Lavinia et al. (2015). The variation in adsorption capacity of COD on the waste alum sludge with respect to flow rate and bed depth implies that the driving force for the adsorption of COD on waste alum sludge is the difference in concentration of COD at the surface of the adsorbent and in the solution (Ahmad et al., 2012). Application of least square error analysis on experimental data revealed that at higher bed depth, error value tends to rise above 1%, data obtained from Thomas model significantly drift away from experimental data. The R^2 values ranged from 0.987 to 0.872 indicates good linearity. This further indicates that the Thomas model equation of linear regression analysis describes the breakthrough data under the condition studied (Ahmad et al., 2012; Sekhula et al., 2012). Fig. 2.5 shows comparative breakthrough curve for experimental data and theoretical data obtained using Thomas model. It was revealed from the plot that experimental data aligned more at lower flow rate and higher bed depth confirming that Thomas model assumes plug flow behavior bed.

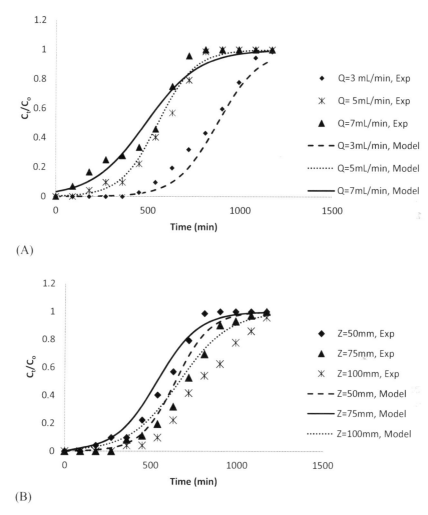

FIGURE 2.5

Comparison of breakthrough curves for experimental and theoretical data using Thomas model influent concentration of $C_o = 72$ mg L^{-1} for (A) varying flow rate and (B) varying bed depth.

2.3.3.2 Yoon–Nelson kinetics model

Scientists, namely Y.H. Yoon and J.H. Nelson in 1984, developed a simple theoretical model which does not concentrate on the type of adsorbent, properties of adsorbate, or the physical features of the fixed bed (Patel, 2019). Yoon–Nelson model that is based on the decline in the probability of uptake of an adsorbate molecule with time is proportional to the probability of the adsorbate uptake and the adsorbate penetration through the adsorbent. This model is less complicated than other models and there are no detailed data regarding the characteristic of the adsorbate, choice of adsorbent, and the physical properties of the bed (Aksu & Gonen, 2004; Han et al., 2009). Yoon–Nelson model in a linearized form for any single-component system is expressed in Eq. (2.10) as:

$$\ln\left(\frac{C_t}{C_i - C_t}\right) = k_{YN}t - \tau k_{YN} \tag{2.10}$$

where k_{YN}, the rate velocity constant, is in min^{-1}, τ represents the time required to obtain half of the inlet concentration of the adsorbate in the column outlet stream (50% breakthrough) in minute. The constants of Yoon–Nelson model obtained at various operating parameter are shown in Table 2.4.

The graph of $\ln\left(\frac{C_t}{C_o - C_t}\right)$ was plotted against time and the Yoon–Nelson constant k_{YN} and the time for 50% breakthrough τ were calculated for the different operating conditions. This study observed that the Yoon–Nelson constant decreases as the flow rate and bed height are increased. This trend conforms to the results reported by Patel and Vashi (2012). This observation was further confirmed by Ahmad et al. (2012) in the removal of COD from cotton textile wastewater using activated water. The regression coefficient reveals that Yoon–Nelson constants fit with experimental data ($R^2 > 0.87$), similarly reported by Radhika and Palanivelu (2006) and Olivares et al. (2013). τ obtained from the model is similar to τ_{exp} and thus further confirms that Yoon–Nelson model could be used to predict 50% breakthrough time for the continuous sorption of COD on sludge derived from drinking water plant (Bankole et al., 2017; Ligaray et al., 2018). Fig. 2.6 shows a comparative breakthrough curve for experimental data and theoretical data using Yoon–Nelson model. It can be seen that the experimental data fits more at 5 mL min^{-1} when varying flow rate and 75 mm when the bed depth was varied.

2.3.3.3 Adams–Bohart kinetics model

In 1920 G. Adams and E.Q. Bohart in an effort to study charcoal–chlorine gas adsorption proposed a continuous kinetic model which is useful to other continuous adsorption system (Han et al., 2009; Patel, 2019). Adams–Bohart

Table 2.4 Constants of Yoon–Nelson model at various operating conditions.

Z (mm)	Q (mL min⁻¹)	C_o (mg L⁻¹)	k_{YN} (L min⁻¹)	τ (min)	R^2	τ_{exp} (min)	SS
50	3	72	0.009	876.333	0.970	910	0.0033
50	5	72	0.010	552.600	0.902	680	0.0021
50	7	72	0.007	485.429	0.872	620	0.0048
75	5	72	0.008	644.250	0.985	720	0.0018
100	5	72	0.007	791.000	0.987	850	0.0065

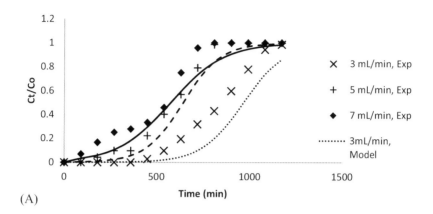

(A)

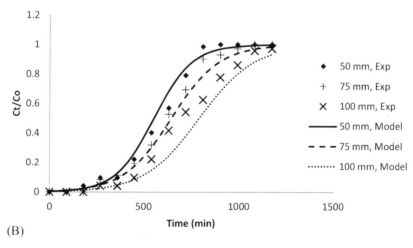

(B)

FIGURE 2.6

Comparison of breakthrough curves for experimental and theoretical data using Yoon–Nelson Model for influent concentration of $C_o = 72$ mg L^{-1} for (A) varying flow rate and (B) varying bed depth.

model assumes that adsorption rate is proportional to residual strength of the adsorbent and the concentration of the adsorbate molecule (Aksu & Gonen, 2004). The model is used for describing initial part of breakthrough curve. It is expressed in Eq. (2.11):

$$\ln\left(\frac{C_t}{C_o}\right) = k_{AB}C_i t - k_{AB}N_o\frac{Z}{F} \tag{2.11}$$

where k_{AB} (L mg^{-1}.min) is the kinetic constant, F (cm min^{-1}) is the linear velocity, Z (cm) is the depth of the column, and N_o (mg L^{-1}) is the saturation concentration. For description of the initial part of the adsorption

Table 2.5 Constants of Adams–Bohart model at various operating conditions.

Z (mm)	Q (mL min⁻¹)	C_o (mg L⁻¹)	k_AB (L mg⁻¹.min)10⁻³	N_o (mg L⁻¹)10³	R²	SS
50	3	72	0.111	35.303		0.0031
					0.881	
50	5	72	0.069	52.017		0.00096
					0.968	
50	7	72	0.056	60.549		0.0023
					0.946	
75	5	72	0.069	33.584		0.00064
					0.93	
100	5	72	0.083	27.636		0.0048
					0.871	

process, Adams–Bohart model was investigated using data obtained from the adsorption of COD on drinking water treatment sludge. N_o and k_{AB} were calculated for 20% breakthrough time for different adsorption conditions and the values obtained were summarized in Table 2.5.

The values of N_o are within the range of 40.97 to 77.88 with the values decreasing with increasing bed height but the reverse at higher flow rate. Decrease in adsorption capacity N_o with increasing bed height was further reported by Ahmad et al. (2012). The regression coefficient ($R^2 > 0.871$) implies that the experimental data obtained fits to Adams–Bohart model (Ahmad et al., 2012; Nayl et al., 2017; Olivares et al., 2013). Fig. 2.7 shows the curve for 20% breakthrough for experimental data and theoretical data obtained using Adams–Bohart model with close alignment of data observed at 3 mL min⁻¹ flow rate (plot A) and 100 mm bed depth (plot B). It can be seen that simulation of the whole breakthrough curve is effective with the Yoon–Nelson model at lower flow rate and at higher bed height (Ahmad et al., 2012).

2.3.3.4 Bed-depth service time model

Scientist, Hutchins (1973) proposed an approached based on Adams and Bohart quasichemical rate law (Patel, 2019). The justification of this model is that equilibrium is not achieved immediately in bed; hence, the rate of sorption of adsorbate molecule is proportional to both the residual capacity of the adsorbent and the influent concentrations (Sugashini & Begum, 2013). This is the most widely used continuous adsorption model. In Hutchins' modification of Adams–Bohart model, only three fixed-bed tests are required to collect the necessary data for predictions. Also, Hutchins' approach allows for the value of other flow rates to be computed from the first flow rate by multiplying the slope by the ratio of the first flow rate and

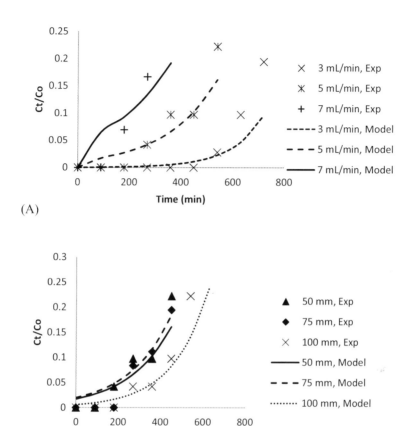

FIGURE 2.7

Comparison of 20% breakthrough curves for experimental and theoretical data using Adams–Bohart model for influent concentration of $C_0 = 72$ mg L^{-1} for (A) varying flow rate and (B) varying bed depth.

new flow rate (Akpen et al., 2015). This model is effective in predicting the relationship between bed depth, Z, and service time, t, in terms of influent concentration and adsorption parameters (Han et al., 2009). The linearized form of bed-depth service time (BDST) model is expressed in Eq. (2.12):

$$tz_s = \frac{n_o Z}{C_o F} - \frac{1}{k_a C_o} \ln\left(\frac{C_o}{C_t} - 1\right) \tag{2.12}$$

where t_s is the service time in *minutes*, n_o (mg g^{-1}) is the adsorption capacity, and k_a is the BDST rate constant in L mg^{-1}.min. Experimental results obtained from the adsorption of COD from fish pond effluent in a fixed bed of drinking water treatment sludge were fitted into BDST to determine the adsorption capacity and kinetic constant.

It can be seen from Table 2.6 that the models kinetic constant (K_a) gives no significant difference as the influent flow rate and bed depth of the adsorbent were varied. The adsorption capacity (N_o) increased with increase in flow rate of the effluent at constant initial concentration of 72 mg L^{-1} and bed depth of 50 mm. As the bed depth was gradually increased, the adsorption capacity was observed to be decreasing at a steady rate. This relationship may be due to the increase in the adsorbent active site. This was similarly reported by López-Cervantes et al. (2018). Adsorption capacity decreases due to the increase in adsorbent active site. Efficient adsorbents are characterized by high N_o values. The regression coefficient ($R^2 > 0.872$) and low least square of error analysis ($SS < 0.5\%$) show that the experimental data best fit to the BDST model. Fig. 2.8 shows the BDST plot for the adsorption of COD on drinking water treatment sludge at different breakthrough concentration. For early breakthrough (i.e., 10% breakthrough), the service time for the varied bed depth gives a higher linear plot having the regression coefficient of

Table 2.6 Bed-depth service time parameters at various conditions.

Z (mm)	Q (mL min^{-1})	C$_o$ (mg L^{-1})	K$_a$ (L mg^{-1}.min) $\times 10^{-3}$	N$_o$ (mg g^{-1})	R^2	SS
50	3	72	0.14	30.728	0.970	0.0039
50	5	72	0.159	33.851	0.902	0.0018
50	7	72	0.120	39.833	0.872	0.0048
75	5	72	0.122	25.309	0.985	0.0018
100	5	72	0.109	22.786	0.987	0.0042

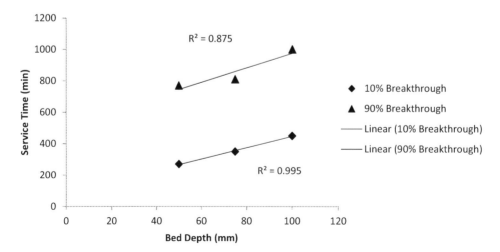

FIGURE 2.8
BDST plot for COD adsorption. *BDST*, bed-depth service time; *COD*, chemical oxygen demand.

0.995. For 90% breakthrough, the regression coefficient was obtained as 0.875. High-value regression coefficient obtained indicates good validity of BDST model as applied to the adsorption of COD using drinking water treatment sludge. Results obtained by fitting experimental data to BDST model is in agreement with those obtained by Bharati and Ramesh (2013) and Kumar and Bandyopadhyay (2006).

2.3.3.5 Clark kinetics model

The Clark model was developed by scientist, R.M. Clark in 1987 for breakthrough curves, which assumes that (1) the rate of adsorption of pollutant molecules onto an adsorbent depends on action of outer mass transfer and complies with Freundlich adsorption isotherm and (2) behavior of flow in column is piston type (Patel, 2019). Clark solved all the mass transfer equations in the system. Clark's solution of the system is expressed in Eq. (2.13) as:

$$\frac{C_t}{C_o} = \left[\frac{1}{1+Ae^{-rt}}\right]^{\frac{1}{n-1}} \tag{2.13}$$

where A and r are constants of the kinetic equation determined by either linear or nonlinear regression analysis and n is obtained from Freundlich equation as given by Eq. (2.14):

$$q_e = K_F C_e^{1/n} \tag{2.14}$$

where K_F is a constant representing relative adsorption capacity of the adsorbent $(mg\,g^{-1})\,(mg\,L^{-1})n$, and the constant $1/n$ indicates the intensity of the adsorption (Medvidovic et al., 2008; Han et al., 2009). Prior to continuous adsorption studies, batch adsorption was carried out and it was discovered that Freundlich model is valid for the adsorption of COD on drinking water treatment sludge, the degree of adsorption obtained as a Freundlich constant is 1.2, that is, $n = 1.2$, and was used to calculate the Clark model constants for different operating conditions.

It has been summarized in Table 2.7 that as the flow rate increases from 3, 5 to 7 mL min^{-1}, the value of r increases from 0.007 to 0.008 min^{-1} with

Table 2.7 Parameters of the Clark model at different conditions ($n = 1.2$).

Z (mm)	Q (mL min^{-1})	C$_o$ (mg L^{-1})	r (min^{-1})	A	R^2	SS
50	3	72	0.007	46.53	0.919	0.00204
50	5	72	0.007	7.09	0.810	0.00225
50	7	72	0.008	1.66	0.795	0.00726
75	5	72	0.006	7.05	0.958	0.00353
100	5	72	0.005	7.13	0.967	0.00614

constant A decreasing from 46.53, 7.09 to 1.66, respectively. Also for the bed depth of 50, 75, and 100 mm, the value of r obtained was 0.007, 0.006, and 0.005 min^{-1} and the constant A was obtained as 7.09, 7.05, and 7.13, respectively, which implies that r and A decrease with increase in the bed depth of the system, leading to elongation of bed exhaustion time. Since the regression coefficients of the experimental data obtained range between

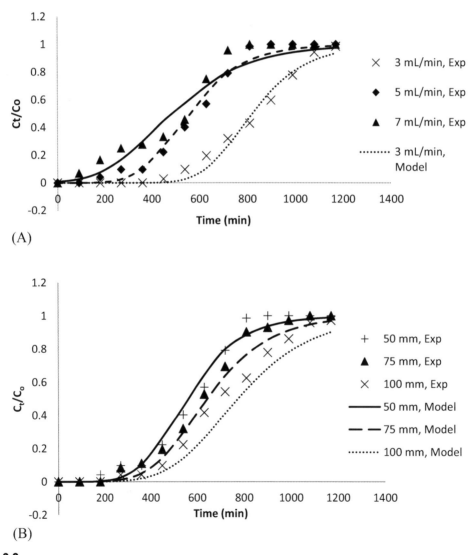

FIGURE 2.9

Comparison of breakthrough curves for experimental and theoretical data using Clarks model for influent concentration of $C_0 = 72$ mg L^{-1} for (A) varying flow rate and (B) varying bed depth.

0.795 and 0.967, the adsorption of COD on drinking water treatment sludge can be described using drinking water treatment sludge as adsorbent. Also, on the account of least square of error analysis, the error quotient between the experimental data and theoretical data obtained via Clark model is small since the SS error obtained is less than 1% for all operating conditions. The trend observed when fitting experimental data to Clarks model is similar to the report presented by Dorado et al. (2014), Han et al. (2009), and Aksu and Gonen (2004). Fig. 2.9 shows the breakthrough curve for both experimental data and theoretical data obtained using Clarks model. Data fits more at flow rate of 3 and 5 mL min^{-1} in plot A. Also data tends to fit more at 75 and 100 mm in plot B.

2.4 Conclusion

The following conclusion can be drawn on the basis of the experimental results obtained from this study: Waste alum sludge can be used for the treatment of fish pond effluent, especially, for the removal of COD pollutants. Since untreated wastewater from fish ponds, as uncovered from this study, contains COD contaminant higher than its acceptable limit, adsorption treatment using waste alum sludge, as confirmed by result obtained, will be efficient to reduce the COD level in fish pond wastewater in a remarkable cost-efficient operation due to abundant availability of the adsorbent precursor. Also, the entire operation seeks to sanitize the environment through the utilization of industrial waste to treat agricultural waste. Regeneration and postutilization of spent adsorbent were not in scope of this research. Further research into the aspect of the spent adsorbent regeneration and postutilization may be required to fully understand the feasibility of operation in using waste alum sludge adsorbent for the treatment of fish pond wastewater. Characterization of the sludge revealed a large surface area of 139 cm^2 g^{-1}. The FT-IR analysis indicates -OH bonding to inorganic matter and, for the loaded sludge, the appearance of vibration of sulfur–hydrogen interaction (S–H) in the sludge matrix. The breakthrough time and column exhaustion time are dependent on the bed depth and influent flow rate. These parameters are important in scaling up industrial or pilot plant operation, which this study had achieved. The kinetics isotherms for continuous adsorption models were tested using the Thomas, Yoon–Nelson, Adams–Bohart, BDST, and Clark models. The observed linearity value (R^2) suggested that data are well fitted into these models. The Clark model best described the adsorption of COD on waste alum sludge in column adsorption on the account of having the least of square error analysis applied to all the tested models.

References

Ahmad, A. A., & Hameed, B. H. (2010). Fixed-bed adsorption of reactive azo-dye onto granular activated carbon prepared from waste. *Journal of Hazardous Materials*, *175*(1−3), 298−303. Available from https://doi.org/10.1016/j.jhazmat.2009.10.003.

Ahmad, A. A., Idris, A., & Hameed, B. H. (2012). Color and COD reduction from cotton textile processing wastewater by activated carbon derived from solid waste in column mode. *Desalination and Water Treatment*, *41*(1−3), 224−231. Available from https://doi.org/10.1080/19443994.2012.664717.

Akankali, J. A., Abowei, J. F. N., & Eli, A. (2011). Pond fish culture practices in Nigeria. *Advance Journal of Food Science and Technology*, *3*(3), 181−195.

Akpen, G. D, Nwaogazie, I. L., & Leton, T. G. (2015). Column studies on the removal of chromium from waste water by mango seed shell activated carbon. *Journal of Science and Technology (Ghana)*, *35*(2), 1−12.

Aksu, C., & Gonen, F. (2004). Biosorption of phenol by immobilized activated sludge in a continuous packed bed: Prediction of breakthrough curves. *Process Biochemistry*, *39*(5), 599−613. Available from https://doi.org/10.1016/S0032-9592(03)00132-8.

Ali, I., Asim, M., & Khan, T. A. (2012). Low cost adsorbents for the removal of organic pollutants from wastewater. *Journal of Environmental Management*, *113*(30), 170−183. Available from https://doi.org/10.1016/j.jenvman.2012.08.028.

Anijiofor, S. C., Daud, N. N., Idrus, S., & Man, H. C. (2018). Recycling of fishpond wastewater by adsorption of pollutants using aged refuse as an alternative low-cost adsorbent. *Sustainable Environment Research*, 1−7. Available from https://doi.org/10.1016/j.serj.2018.05.005.

Ansah, Y. B., Frimpong, E. A., & Amisah, S. (2013). Characterization of potential aquaculture pond effluents, and physico-chemical and microbial assessment of effluent-receiving waters in central Ghana. *African Journal of Aquatic Science*, *38*(2), 185−192. Available from https://doi.org/10.2989/16085914.2013.767182.

Balci, B., Keskinkan, O., & Avci, M. (2011). Use of BDST and an ANN model for prediction of dye adsorption efficiency of *Eucalyptus camaldulensis* barks in fixed-system. *Expert Systems with Applications*, *38*, 949−956. Available from https://doi.org/10.1016/j.eswa.2010.07.084.

Banerjee, K., Ramesh, S. T., Gandhimathi, R., Nidheesh, P. V., & Bharathi, K. S. (2012). A Novel Agricultural Waste Adsorbent, Watermelon Shell for the Removal of Copper from Aqueous Solutions. *Journal of Environmental Health Science Engineering*, *3*(2), 143−156. Available from https://doi.org/10.5829/idosi.ijee.2012.03.02.0396.

Bankole, M. T., Abdulkareem, A. S., Tijani, J. O., Ochigbo, S. S., Afolabi, A. S., & Roos, W. D. (2017). Chemical oxygen demand removal from electroplating wastewater by purified and polymer functionalized carbon nanotubes adsorbents. *Water Resources and Industries*, *18*, 33−50. Available from https://doi.org/10.1016/j.wri.2017.07.001.

Bharati, K. S., & Ramesh, S. P. T. (2013). Fixed bed adsorption studies on biosorption of crystal violet from aqueous solution by *Citrullus lanatus rind* and *Cyperus rotundus*. *Applied Water Science*, *3*(4), 673−687. Available from https://doi.org/10.1007/s13201-013-0103-4.

Chatvijitkul, S., Boyd, C. E., Davis, D. A., & McNevin, A. A. (2017). Pollution potential indicators for feed-based fish and shrimp culture. *Aquaculture (Amsterdam, Netherlands)*, *477*. Available from https://doi.org/10.1016/j.aquaculture.2017.04.034.

Chaudhari, P. K., Majumdar, B., Choudhary, R., Yadav, D. K., & Chand, S. (2010). Treatment of paper and pulp mill effluent by coagulation. *Environmental Technology*, *31*(4), 357−363. Available from https://doi.org/10.1080/09593330903486665.

Chen, S., Yue, Q., Gao, B., Li, Q., Xu, X., & Fu, K. (2012). Adsorption of hexavalent chromium from aqueous solution by modified corn stalk: A fixed-bed column study. *Bioresource Technology*, *113*, 114−120. Available from https://doi.org/10.1016/j.biortech.2011.11.110.

Chen, T. C., Shih, Y. J., Chang, C. C., & Huang, Y. H. (2013). Novel adsorbent of removal phosphate from TFT LCD wastewater. *Journal of Taiwan Institute Chemical Engineering*, *44*(1), 61−66. Available from https://doi.org/10.1016/j.jtice.2012.09.008.

Da, C. T., Phuoc, L. H., Duc, H. N., Troell, M., & Berg, H. (2015). Use of wastewater from striped catfish (*Pangasianodon hypophthalmus*) pond culture for integrated rice−fish−vegetable farming systems in the Mekong Delta, Vietnam. *Agroecology and Sustainable Food Systems*, *39*(5), 580−597. Available from https://doi.org/10.1080/21683565.2015.1013241.

Dassanayake, K. B., Jayasinghe, G. Y., Surapaneni, A., & Hetherington, C. (2015). A review on alum sludge reuse with special reference to agricultural applications and future challenges. *Waste Management*, *38*, 325−335. Available from https://doi.org/10.1016/j.wasman.2014.11.025.

Devi, P., & Saroha, A. K. (2017). Utilization of sludge based adsorbents for the removal of various pollutants: A review. *Science of the Total Environment*, *578*, 16−33. Available from https://doi.org/10.1016/j.scitotenv.2016.10.220.

Dong, B., Chen, H., Yang, Y., He, Q., & Dai, X. (2013). Treatment of printing and dyeing wastewater using MBBR followed by membrane separation process. *Desalination and Water Treatment*, *52*(22−24), 4562−4567. Available from https://doi.org/10.1080/19443994.2013.803780.

Dorado, A. D., Gamisans, X., Valderrama, C., Sole, M., & Lao, C. (2014). Cr(III) removal from aqueous solutions: a straightforward model approaching of the adsorption in a fixed-bed column. *Journal of Environmental Science and Health Part a-Toxic/Hazardous Substances & Environmental Engineering*, *49*(2), 179−186. Available from https://doi.org/10.1080/10934529.2013.838855.

Dridi-Dhaouadi, S., & Mhenni, M. F. (2014). Effect of dye auxiliaries on chemical oxygen demand and colour competitive removal from textile effluents using *Posidonia oceanica*. *Chemistry and Ecology*, *30*(6), 579−588. Available from https://doi.org/10.1080/02757540.2013.878336.

Emmanuel, O., Chinenye, A., Oluwatobi, A., & Peter, K. (2014). Review of Aquaculture production and management in Nigeria. *American Journal of Experimental Agriculture*, *4*(10), 1137−1151. Available from https://doi.org/10.9734/AJEA/2014/8082.

Food and Agriculture Organization (FAO). (2017). *Fishery and aquaculture country profile—The Federal Republic of Nigeria*. Online source <http://www.fao.org/fishery/facp/NGA/en>.

Futalana, C. M., Kanb, C. C., Dalidac, M. L., Pascuad, C., & Wan, M. W. (2011). Fixed-bed column studies on the removal of copper using chitosan immobilized on bentonite. *Carbohydrate Polymers*, *83*(2), 697−704. Available from https://doi.org/10.1016/j.carbpol.2010.08.043.

Geerdink, R. B., Sebastiaan., van den Hurk, R., & Epemaa, O. J. (2017). Chemical oxygen demand: Historical perspectives and future challenges. *Analytica Chimica Acta*, *9611*, 1−11. Available from https://doi.org/10.1016/j.aca.2017.01.009.

Han, R., Wang, Y., Zhao, X., Wang, Y., Xie, F., Cheng, J., & Tang, M. (2009). Adsorption of methylene blue by phoenix tree leaf powder in a fixed-bed column: experiments and prediction of breakthrough curves. *Desalination*, *245*(1−3), 284−297. Available from https://doi.org/10.1016/j.desal.2008.07.013.

Hanaor, D. A. H., Ghadiri, M., Chrzanowski, W., & Gan, Y. (2014). Scalable surface area characterization by electrokinetic analysis of complex anion adsorption. *Langmuir: The ACS Journal of Surfaces and Colloids*, *30*(50), 15143−15152. Available from https://doi.org/10.1021/la503581e.

He, Y., Liao, X., Liao, L., & Shu, W. (2014). Low-cost adsorbent prepared from sewage sludge and corn stalk for the removal of COD in leachate. *Environment Science Pollution Resources*, *21*, 8157−8166. Available from https://doi.org/10.1007/s11356-014-2755-5.

Hossain, M. K., Strezov, V., Chan, K. Y., Ziolkowski, A., & Nelson, P. F. (2011). Influence of pyrolysis temperature on production and nutrient properties of wastewater sludge biochar. *Journal of Environmental Management*, *92*(1), 223−228. Available from https://doi.org/10.1016/j.jenvman.2010.09.008.

Hutchins, R. (1973). New method simplifies design of activated-carbon systems. *Chemical Engineering Journal*, *80*(19), 133−138.

Javaid, A., Nor Aishah, S. A., & Khurram, S. (2015). A review on removal of pharmaceuticals from water by adsorption. *Desalination and Water Treatment*, *57*(27), 12842−12860. Available from https://doi.org/10.1080/19443994.2015.1051121.

Kalavathy, H., Karthik, B., & Miranda, L. R. (2010). Removal and recovery of Ni and Zn from aqueous solution using activated carbon from *Hevea brasiliensis*: Batch and column studies. *Colloids Surface*, *78*(2), 291−302. Available from https://doi.org/10.1016/j.colsurfb.2010.03.014.

Kong, L., Xiong, Y., Sun, L., Tian, S., Xu, X., Zhao, C., Luo, R., Yang, X., Shih, K., & Liu, H. (2014). Sorption performance and mechanism of a sludge-derived char as porous carbon-based hybrid adsorbent for benzene derivatives in aqueous solution. *Journal of Hazardous Material*, *274*, 205−211. Available from https://doi.org/10.1016/j.jhazmat.2014.04.014.

Kumar, U., & Bandyopadhyay, M. (2006). Fixed bed column study for Cd (II) removal from wastewater using treated rice husk. *Journal of Hazardous Materials*, *129*(1−3), 253−259. Available from https://doi.org/10.1016/j.jhazmat.2005.08.038.

Lavinia, T., Carmen, P., Carmen, T., & Ovidiu, T. (2015). Fixed Bed Column Study on the removal of chromium (III) ions from aqueous solutions by using hemp fibers with improved sorption performance. *Cellulose Chemistry and Technology*, *49*(2), 219−229.

Ligaray, M., Futalan, C. M., de Luna, M. D., & Wan, M.-W. (2018). Removal of chemical oxygen demand from thin-film transistor liquid-crystal display wastewater using chitosan-coated bentonite: isotherm, kinetics and optimization studies. *Journal of Cleaner Production*, *175*, 145−154. Available from https://doi.org/10.1016/j.jclepro.2017.12.052.

Lin, S. H., & Juang, R. S. (2009). Adsorption of phenol and its derivatives from water using synthetic resins and low-cost natural adsorbents: A review. *Journal of Environmental Management*, *90*(3), 1336−1349. Available from https://doi.org/10.1016/j.jenvman.2008.09.003.

López-Cervantes, J., Sánchez-Machado, D. I., Sánchez-Duarte, R. G., & Correa-Murrieta, M. A. (2018). Study of a fixed-bed column in the adsorption of an azo dye from an aqueous medium using a chitosan−glutaraldehyde biosorbent. *Adsorption science and Technology*, *36*(1−2), 215−232. Available from https://doi.org/10.1177/0263617416688021.

Medvidovic, N. V., Peric, J., & Trgo, M. (2008). Testing of breakthrough curves for removal of lead ions from aqueous solutions by natural zeolite-clinoptilolite according to the Clark kinetic equation. *Separation Science and Technology*, *43*(4), 944−959. Available from https://doi.org/10.1080/01496390701870622.

Mohamed, R. (2015). Efficiency of new Miswak, titanium dioxide and sand filters in reducing pollutants from wastewater. *Beni-Suef University Journal of Basic and Appliers Sciences*, *4*(1), 47−51. Available from https://doi.org/10.1016/j.bjbas.2015.02.007.

National Environmental Standard Regulation & Enforcement Agency (NESREA) Federal Republic of Nigeria official gazette (2011).

Nayl, A. A., Elkhashab, R. A., Malah, T. E., Yakout, S. M., El-Khateeb, M. A., Ali, M. M. S., & Ali, H. M. (2017). Adsorption studies on the removal of COD and BOD from treated sewage using activated carbon prepared from date palm waste. *Environmental Science Pollution Research*, *24*(28), 22284−22293. Available from https://doi.org/10.1007/s11356-017-9878-4.

Nwabanne, J. T., & Igbokwe, P. K. (2012). Adsorption performance of packed bed column for the removal of lead (II) using oil palm fibre. *International Journal of Applied Science and Technology*, *2*(5), 106−114.

Olivares, J. C., Alonso, C. P., Díaz, C. B., Nuñez, F. U., Chaparro-Mercado, M. C., & Bilyeu, B. (2013). Modeling of lead (II) biosorption by residue of allspice in a fixed-bed column. *Chemical Engineering Journal, 228*, 21−27. Available from https://doi.org/10.1016/j.cej.2013.04.101.

Pan, Z., Tian, J., Xu, G., Li, J., & Li, G. (2011). Characteristics of adsorbents made from biological, chemical and hybrid sludges and their effect on organics removal in wastewater treatment. *Water Resources, 45*(2), 819−827. Available from https://doi.org/10.1016/j.watres.2010.09.008.

Patel, H. (2019). Fixed-bed column adsorption study: A comprehensive review. *Applied Water Science, 9*(45), 1−17. Available from https://doi.org/10.1007/s13201-019-0927-7.

Patel, H., & Vashi, R. T. (2012). Fixed bed column adsorption of ACID Yellow 17 dye onto Tamarind Seed Powder. *Canadian Journal of Chemical Engineering, 90*, 180−185. Available from https://doi.org/10.1002/cjce.20518.

Patel, H., & Vashi, R. T. (2015). Characterization and column adsorptive treatment for COD and colour removal using activated neem leaf powder from textile wastewater. *Journal of Urban and Environmental Engineering, 9*(1), 45−53. Available from https://doi.org/10.4090/juee.2015.v9n1.045053.

Radhika, M., & Palanivelu, K. (2006). Adsorptive removal of chlorophenols from aqueous solution by low cost adsorbent-kinetics and isotherm analysis. *Journal Hazardous Material, 38*(1), 116−124. Available from https://doi.org/10.1016/j.jhazmat.2006.05.045.

Sahu, O. P., & Chaudhari, P. K. (2015). Removal of color and chemical oxygen demand from sugar industry wastewater using thermolysis processes. *Desalination and Water Treatment, 56*(7), 1758−1767. Available from https://doi.org/10.1080/19443994.2014.956797.

Schwartz, M. F., & Boyd, C. E. (1994). Channel catfish pond effluents. *The Progressive Fish-Culturist, 56*(4), 273−281. Available from https://doi.org/10.1577/1548-8640(1994)056 < 0273:CCPE > 2.3.CO;2.

Sekhula, M. M., Okonkwo, J. O., Zvinowanda, C. M., Agyei, N. N., & Abdul, J. C. (2012). Fixed bed column adsorption of Cu (II) onto maize tassel-PVA beads. *Journal of Chemical Engineering & Process Technology, 3*, 2. Available from https://doi.org/10.4172/2157-7048.1000131.

Sugashini, S., & Begum, K. M. M. S. (2013). Performance of ozone treated rice husk carbon (OTRHC) for continuous adsorption of Cr (VI) ions from synthetic effluent. *Journal of Environmental Chemical Engineering, 1*(1−2), 79−85. Available from https://doi.org/10.1016/j.jece.2013.04.003.

Suman, A., Ahmad, T., & Ahmad, K. (2017). Dairy wastewater treatment using water treatment sludge as coagulant: a novel treatment approach. *Environment, Development and Sustainability, 20*(4), 1615−1625. Available from https://doi.org/10.1007/s10668-017-9956-2.

Sur, D. H., & Mukhopadhyay, M. (2018). Process parametric study for COD removal of electroplating industry effluent. *Biotech, 8*, 84. Available from https://doi.org/10.1007/s13205-017-1059-0.

Tan, I. A. W., Ahmad, A. L., & Hameed, B. H. (2008). Adsorption of basic dye using activated carbon prepared from oil palm shell: Batch and fixed bed studies. *Desalination, 225*(1−3), 13−28. Available from https://doi.org/10.1016/j.desal.2007.07.005.

Teodorowicz, M., Gawronska, H., Lossow, K., & Lopata, M. (2006). Impact of trout farms on water quality in the Marozka river (Mazurian Lakeland, Poland). *Archives of Polish Fisheries, 14*(2), 243−255.

Thomas, H. (1944). Heterogeneous ion exchange in a flowing system. *Journal of the American Chemical Society, 66*, 1664−1666.

Tucker, C. S., & Hargreaves, J. A. (2008). *Environmental best management practices for aquaculture.* Ames, Iowa: Wiley-Blackwell.

Vijayaraghavan, K., Jegan, J., Palanivelu, K., & Velan, M. (2004). Removal of nickel(II) ions from aqueous solution using crab shell particles in a packed bed up flow column. *Journal of*

Hazardous Materials, 113(1−3), 223−230. Available from https://doi.org/10.1016/j.jhazmat.2004.06.014.

Xu, G., Yang, X., & Spinosa, L. (2015). Development of sludge-based adsorbents: Preparation, characterization, utilization and its feasibility assessment. *Journal of Environmental Management, 151*, 221−232. Available from https://doi.org/10.1016/j.jenvman.2014.08.001.

Yahya, M. D., Abubakar, H., Obayomi, K. S., Iyaka, Y. A., & Suleiman, B. (2020a). Simultaneous and continuous biosorption of Cr and Cu (II) ions from industrial tannery effluent using almond shell in a fixed bed column. *Results in Engineering, 6*(1). Available from https://doi.org/10.1016/j.rineng.2020.100113.

Yahya, M. D., Yohanna, I., Auta, M., & Obayomi, K. S. (2020b). Remediation of Pb (II) ions from Kagara gold mining effluent using cotton hull adsorbent. *Scientific African, 8*, e00399. Available from https://doi.org/10.1016/j.sciaf.2020.e00399.

Yang, L., Wei, J., Liu, Z., Wang, J., & Wang, D. (2015). Material prepared from drinking water-works sludge as adsorbent for ammonium removal from wastewater. *Applied Surface Science, 330*, 228−236. Available from https://doi.org/10.1016/j.apsusc.2015.01.017.

Yang, X., Xu, G., Yu, H., & Zhang, Z. (2016). Preparation of ferric-activated sludge-based adsorbent from biological sludge for tetracycline removal. *Bioresource Technology, 211*, 566−573. Available from https://doi.org/10.1016/j.biortech.2016.03.140.

Yang, Y., Tomlinson, D., Kennedy, S., & Zhao, Y. Q. (2006). Dewatered alum sludge: A potential adsorbent for phosphorus removal. *Water Science Technology, 54*(5), 207−213. Available from https://doi.org/10.2166/wst.2006.564.

Yunnen, C., Changshi, X., & Jinxia, N. (2016). Removal of ammonia nitrogen from wastewater using modified activated sludge. *Poland Journal of Environmental Study, 25*(1), 419−425. Available from https://doi.org/10.15244/pjoes/60859.

Zhao, Y. Q., Babatunde, A. O., Hu, Y. S., Kumar, J. L. G., & Zhao, X. H. (2011). Pilot field-scale demonstration of a novel alum sludge-based constructed wetland system for enhanced wastewater treatment. *Process Biochemical, 46*(1), 278−283. Available from https://doi.org/10.1016/j.procbio.2010.08.023.

Zhao, Y. Q., Zhao, X. H., & Babatunde, A. O. (2009). Use of dewatered alum sludge as main substrate in treatment reed bed receiving agricultural wastewater: Long-term trial. *Bioresource Technology, 100*(2), 644−648. Available from https://doi.org/10.1016/j.biortech.2008.07.040.

Zou, X. (2017). Advanced treatment of sodium dithionite wastewater using the combination of coagulation, catalytic ozonation and SBR. *Environmental Technology, 38*(9), 2489−2507. Available from https://doi.org/10.1080/09593330.2017.1349188.

Zulfadhly, Z., Mashitah, M. D., & Bhatia, S. (2001). Heavy metals removal in fixed-bed column by the macro fungus *Pycnoporus sanguineus*. *Environmental Pollution, 112*(3), 463−470. Available from https://doi.org/10.1016/S0269-7491(00)00136-6.

Farm management practices for water quality improvement: economic risk analysis of winter wheat production in the Southern High Plains

Yubing Fan[1] and Sushil Kumar Himanshu[2]

[1]Texas A&M AgriLife Research, Vernon, TX, United States, [2]Department of Food,
Agriculture and Bioresources, School of Environment, Resources and Development, Asian
Institute of Technology, Pathum Thani, Thailand

3.1 Introduction

Conservation tillage retains crop residues on the soil surface, which helps mitigate soil erosion, increase soil organic matter and nutrients, and enhance water storage capacity of the soil (Claassen et al., 2018; DeLaune et al., 2015; Rusu, 2014; Triplett & Dick, 2008). Appropriate tillage management can also show substantial environmental benefits, for example, through affecting the transport of nitrogen (N) and phosphorus (P) in runoff and groundwater (Bosch et al., 2005; DeLaune & Sij, 2012; Djodjic et al., 2002; King et al., 2015; Schelde et al., 2006). Compared to conventional tillage, conservation tillage practices help reduce the transport of particulate N (PN) and P (PP) in sediment and organic material of surface runoff (Endale et al., 2010; Gaynor & Findlay, 1995; Tiessen et al., 2010; Zhao et al., 2001), while conservation tillage may be associated with greater dissolved N [nitrate-N (NO_3-N)] and P (DP) levels (Daryanto et al., 2017; Kleinman et al., 2009; Smith & Cassel, 1991). In addition, tillage operations may affect the transport of bioavailable P (BAP) in surface runoff (Mueller et al., 1984). As a result, conservation tillage systems can have substantial implications for water quality and environmental sustainability (Locke et al., 2015; Lozier et al., 2017).

Wheat production can leave straws on the soil surface after harvesting, and conservation tillage may be combined to minimize soil erosion, especially given the rainy and windy weather in later spring and early summer in the Southern High Plains (SHP). In this regard, tillage effects on nutrient

Advances in Remediation Techniques for Polluted Soils and Groundwater. DOI: https://doi.org/10.1016/B978-0-12-823830-1.00004-3

discharges are of great significance in maintaining soil fertility for wheat cultivation and controlling nonpoint source pollution (Franzluebbers et al., 1994). Studies have reported the influence of conservation tillage on wheat production (Decker et al., 2009), soil biophysical properties (DeLaune et al., 2015), and soil water dynamics (Patrignani et al., 2012) in the SHP. An economic evaluation of environmental benefits from using conservation tillage is still limited (Fan et al., 2019); in particular, producers' farming decisions may be influenced by their risk attitude (Fan et al., 2020; Williams et al., 2012). Risk-averse producers may prefer conservation tillage practices because they can reduce operating costs, increase crop yields, and mediate adverse environmental impacts of agricultural production (Fathelrahman et al., 2011; Nail et al., 2007). Therefore it is necessary to comparatively evaluate the risk-adjusted profits of practicing alternative tillage systems (Bosch et al., 2005; Claassen et al., 2018), and more importantly, to incorporate their associated environmental benefits to better promote soil health in the SHP.

This study uses Monte Carlo simulation to draw stochastic data points on wheat yield based on a 12-year field experiment. The unique field data facilitate us to account for the combined environmental effects of alternative tillage practices on both surface runoff and groundwater quality. An enterprise budget is developed to estimate the production cost and revenue by incorporating the economic benefits associated with nutrient discharges in surface runoff and groundwater. Economic risk analysis is carried out using the stochastic efficiency with respect to a function (SERF) approach, and the net returns under alternative tillage methods are compared across different risk aversion levels of wheat producers by combining the economic benefits of nutrient discharges from winter wheat production.

3.2 Data

This study uses simulated wheat yield data based on three tillage practices in a field experiment from 1979 to 1990. The field experiment was conducted at El Reno, Oklahoma, located in the Reddish Prairies of the SHP (Fig. 3.1). The major soil type was Bethany silt loam (fine, mixed, thermic), and the average annual rainfall was 740 mm at El Reno. This study focuses on three watersheds E6, E7, and E8 (Table 3.1), all of which had been in native grassland and converted to conventionally tilled wheat in 1978 (Sharpley & Smith, 1994). E7 was managed with no-till wheat since 1984. Conventionally tilled wheat was managed with moldboard plowing at E6 and sweeps at E8. Fertilizers N and P were applied when planting winter wheat, and additional N was surface broadcast in spring according to soil

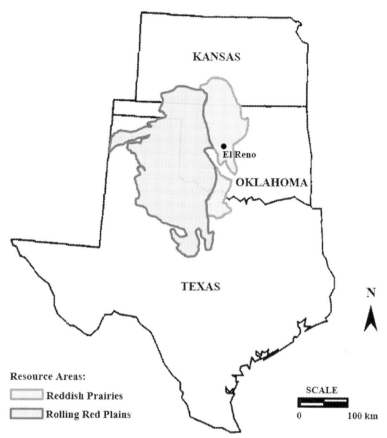

FIGURE 3.1
A schematic map showing the resource areas.

Table 3.1 Tillage treatments and fertilizer application (data from (Sharpley and Smith, 1994).

Watersheds	Time period	Tillage methods	Fertilizer application (kg ha^{-1})	
			N	P
E6	1979–90	Moldboard, disk	76	12
E8	1984–90	Sweeps, disk	56	13
E7	1979–90	No-till	56	13

test. No-till wheat was also planted with fertilizer broadcast on the soil surface. The average annual fertilizer application was 76 and 12 kg ha^{-1} for N and P, respectively, under the moldboard tillage at E6. N and P fertilizers

Table 3.2 Mean annual runoff, sediment, and nutrient discharge in the three tillage treatments (data from Sharply and Smith, 1994).

Tillage methods	Runoff (cm)	Sediment discharge (kg ha^{-1} year^{-1})	Dissolved P	Particulate P	Bioavailable P	Nitrate N	Total N
Moldboard	14.01	9540	0.23	4.33	1.1	2.04	28.93
Sweeps	11.2	4486	0.25	2.61	0.87	3.83	19.44
No-till	15.82	336	0.98	0.57	1.3	2.01	9.24

were 56 and 16 kg ha^{-1}, respectively, under both sweeps tillage at E8 and no-till at E7. Herbicides were used to control weed in the no-till wheat.

Further, the evaluation of water quality changes incorporates nutrient concentrations in both runoff and groundwater. Groundwater sampling wells were installed parallel to the direction of flow down at the center of each watershed. The water table depth of the wells was 10−25 m. Watershed runoff was measured using precalibrated flumes equipped with water-level recorders that automatically collected 5−15 samples during each runoff event. Appropriate procedures were followed to determine NO_3-N, ammonium-N (NH_4-N), DP, total Kjeldahl N (TKN), and total P (TP) in each sample. Total N (TN) was calculated as the sum of NO_4-N and TKN, and PN as the difference between TKN and NH_4-N. PP was calculated as the difference between TP and DP. Further, bioavailable PP (BPP) was obtained by subtraction of DP from BAP, and suspended sediment concentration of runoff was determined. More details about nutrient determination can be found in Sharpley and Smith (1994).

Table 3.2 shows a summary of the average annual runoff and nutrient discharge in the three tillage treatments. The surface runoff was 14.01, 11.20, 15.82 cm for moldboard, sweeps, and no-till, respectively. Sediment discharges were much higher under moldboard and sweeps tillage, 9540 and 4486 kg ha^{-1} year^{-1}, respectively, compared to 336 kg ha^{-1} year^{-1} under no-till. Overall, DP was much higher in no-till, 0.98 kg ha^{-1} year^{-1}, compared to 0.23 and 0.25 kg ha^{-1} year^{-1} for moldboard and sweeps tillage, respectively. However, PP was 0.57 kg ha^{-1} year^{-1} in no-till, which was much lower than 4.33 and 2.61 kg ha^{-1} year^{-1} in moldboard and sweeps tillage, respectively. The three tillage treatments showed similar BAP and nitrate N levels. No-till had a lower TN amount, 9.24 kg ha^{-1} year^{-1}, compared to the 28.93 and 19.44 kg ha^{-1} year^{-1} for moldboard and sweeps tillage, respectively. Nutrient prices are used to quantify the economic benefits of different tillage practices (Cartwright & Kirwan, 2015; Nail et al., 2007).

3.3 Methods

3.3.1 Monte Carlo simulation

Wheat yields and prices were simulated using multivariate empirical (MVE) distribution, which follows Monte Carlo protocols to account for correlation among stochastic variables (Richardson et al., 2000). MVE simulation can also provide consistent estimation for variables with high variation (Richardson et al., 2008); for example, high-yield variation can be caused by high weather variability in the SHP. This study carried out MVE simulations for wheat yields under three tillage practices based on field-level data (Sharpley & Smith, 1994). In addition, the simulation for wheat prices was conducted based on the annual prices in Oklahoma during 1984–2019 (NASS, 2021). Each simulation was carried out to generate 500 iterations of data points (Richardson et al., 2008). To conduct the stochastic simulations, the following step-wise procedure was followed:

1. Estimate yield range and yield distribution. The yield range was determined based on field data, and the Gray–Richardson–Klose–Schumann (GRKS) model (Richardson et al., 2008) was used to estimate yield distributions. The GRKS model can be specified as:

$$\Phi = GRKS(Min, Middle, Max) \tag{3.1}$$

   where Φ refers to the simulated distribution which follows a GRKS function with specified minimum, middle, and maximum values.

2. Estimate correlation matrix. A linear correlation matrix can be specified for the wheat yields under each tillage practice. For example, the wheat yields under the three tillage practices j, (j: 1 = moldboard, 2 = sweeps, and 3 = no-till) have a correlation matrix R with correlation r between each pair of the yields:

$$R_j = \begin{bmatrix} 1 & r_{12} & r_{13} \\ r_{21} & 1 & r_{23} \\ r_{31} & r_{32} & 1 \end{bmatrix} \tag{3.2}$$

3. Estimate correlated uniform standard deviates (CUSDs). To estimate CUSDs, we first took the correlation matrix's square root and multiplied a vector of independent standard normal deviates. CUSDs were then obtained by converting the standard normal deviates with an inverse transformation of a standard normal distribution (Richardson et al., 2008). Incorporating the CUSDs in the stochastic simulation avoids under- or overestimating the mean and variance of yields and prices if they are correlated.

3. Generate random variables. An empirical distribution function was employed using CUSDs in an inverse transformation of the empirical distribution:

$$\tilde{Y}_j = f\left(\bar{Y}_j, \sigma_j, CUSD_j\right) \text{ for } j = \text{three tillage practices} \tag{3.3}$$

where tilde ($\sim$) denotes a stochastic variable; $f(\cdot)$ represents a multivariate empirical function that follows a normal distribution; dash ($-$) denotes the variable mean of the estimated yield data or historical price data. The stochastic simulation follows a uniform distribution with N (500) intervals, and at least one value is randomly selected within each interval. This ensures that the simulation considers all corresponding areas of the probability distributions.

5. Evaluate model. To evaluate the validity of the simulated data, two-sample t-test can be used to determine whether the correlation coefficients of the historical and simulated matrices are statistically different at the 95% confidence level. The two-sample Hotelling T^2 test determines whether the mean vectors of the historical and simulated data are equal. The complete homogeneity test determines whether the mean vectors and covariance matrices are equivalent simultaneously.

3.3.2 Cost and profit estimation

The profitability of three tillage practices was evaluated by an enterprise budget developed by Oklahoma Cooperative Extension Service (OSU, 2021). Production cost was estimated using data on field operations and input use (Sharpley & Smith, 1994). Appropriate modifications were made on other input use and their prices to reflect the actual production payment in 2019. The operating cost for tillage practice j was calculated using:

$$OC_j = f(\text{Planting, tillage, input, labor, harvesting, etc.}) \tag{3.4}$$

where OC refers to the operating cost, including the expenses on seed, fertilizer, pesticide, herbicide, energy, labor, machinery repairs, maintenance, interest, and harvest costs.

$$\tilde{\pi}_j = \tilde{Y}_j \times \tilde{P} - OC_j \tag{3.5}$$

where $\tilde{\pi}_j$ is the estimated net return under each tillage practice; $\tilde{Y}_j$ is the simulated seed yield; and $\tilde{P}$ is the simulated wheat price.

In addition, the farm profit incorporating water quality benefits can be expressed as:

$$\tilde{\pi}_j^{wq} = \tilde{\pi}_j + b^{wq} \tag{3.6}$$

where $\tilde{\pi}_j^{wq}$ is the net benefit that combines farm income with the estimated benefit associated with water quality improvement, b^{wq}. The water quality benefit was estimated with nutrient changes in runoff and groundwater as well as their prices.

3.3.3 Stochastic efficiency approach

Uncertainties associated with weather, input use, yield, and price can affect farm income. Farmers with varying attitudes toward risks may have different preferences for alternative tillage practices; especially, the tillage practices may affect water quality in different ways. Risk-averse producers are more likely to choose a tillage practice that results in a smaller variation in farm profit. To examine the risk-adjusted profit, we follow Anderson and Dillon (1992) and use the absolute risk aversion coefficient (ARAC, r_a) to measure producers' risk attitude. The ARAC can be expressed by:

$$r_a(w) = \frac{r_r(w)}{w} \tag{3.7}$$

where $r_r(w)$ is the relative risk aversion coefficient (RRAC) for a certain amount of farm income w (Hardaker et al., 2004). According to Anderson and Hardaker (2003), five RRAC levels can be specified with 0, 1, 2, 3, and 4 for risk-neutral, somewhat risk-averse, rather risk-averse, very risk-averse, and extremely risk-averse, respectively. The average farm profit is equal to $\$115 \text{ ha}^{-1}$, which determines the upper bound of ARAC. Therefore the five corresponding absolute risk aversion levels are 0, 0.0087, 0.0174, 0.0261, and 0.0348, respectively.

We used the SERF approach to rank the profitability of three tillage practices across various risk aversion levels. Certainty equivalent (CE) is used to measure the guaranteed amount of money at which a producer would be willing to accept instead of taking a risky alternative action:

$$CE(w, r(w)) = U^{-1}(w, r(w)) \tag{3.8}$$

where $U(\cdot)$ represents the utility function. A producer is assumed to prefer a risky outcome with a higher CE value at a certain risk aversion level, r (Lien et al., 2007). This analysis adopts a negative exponential utility function which facilitates more efficient CE estimation with constant absolute risk aversion (Hardaker et al., 2015; Schumann et al., 2004).

A utility weighted risk premium (RP) is calculated at a certain risk aversion level of wheat farmers. An RP value is a difference in the CEs of adopting a specific tillage practice relative to a baseline practice.

$$RP_{BA} = CE_B - CE_A \tag{3.9}$$

The value of RP reflects the minimum amount of money that a farmer will have to receive before switching from practice A to practice B, for instance, from moldboard plowing to no-till production, at a certain risk aversion level. The value of RP also represents the risk-adjusted profit gain from adopting an alternative tillage practice. We use the Simulation and Econometrics to Analyze Risk (Simetar) software (Richardson et al., 2008) to conduct the stochastic simulations and economic risk analysis.

3.4   Results and discussion

3.4.1   Validation results

Simulated wheat yields were evaluated by comparing the observed and simulated data series. Fig. 3.2 shows the boxplot of wheat yields under the three tillage practices. Table 3.A1 in the appendix shows the validation test results. Sweeps and moldboard plowing have a higher wheat yield, 2690 and 2385 kg ha^{-1}, respectively, compared to 1895 kg ha^{-1} for no-till, while no-till shows a more substantial variation than other tillage practices. For the joint distribution of wheat yields under the three tillage practices, the two-sample Hotelling T^2 test shows an insignificant result (i.e., $P>.05$), which indicates that the mean vectors of the simulated and experimental yields are equal. The complete homogeneity test also shows that the variances of the simulated and experimental yields are not significantly different. In addition, the mean wheat price is $0.2139 kg^{-1}, and the two-sample t-test shows that it is not significantly different from the adjusted historical prices from 1984 to 2019 (BLS, 2021; NASS, 2021).

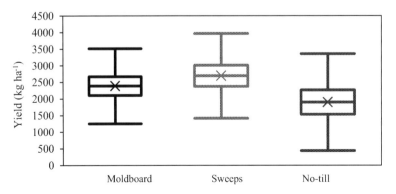

FIGURE 3.2
Boxplot of simulated wheat yields under three tillage methods.

3.4.2 Profit distributions

Net return of wheat production under each tillage practice was estimated by deducting the operating cost from total revenue. The operational costs are about \$413, \$397, and \$337 ha^{-1} for moldboard, sweeps, and no-till, respectively (Table 3.A2). In particular, moldboard plowing and sweeps have a higher cost for custom hire on labor and field operations, \$145.47 ha^{-1}, compared to \$70.50 ha^{-1} for no-till practice. In addition, moldboard plowing shows a higher expense on fertilizer, \$71.76 ha^{-1}, compared to sweeps and no-till at \$55.62 ha^{-1}. Overall, no-till reduces the operating costs by about \$76 ha^{-1} compared with moldboard plowing.

Fig. 3.3 shows the cumulative distribution functions of net returns under the three tillage practices (also see the boxplots of the net returns in Fig. 3.A1). The net return distribution of sweeps lies to the right of the other two distributions, followed by moldboard plowing. This suggests that sweeps tillage could have a higher probability of getting a high income. The long right tale of the sweeps distribution indicates that sweeps tillage has a larger chance of getting a very high-income level.

Further, stoplight charts can help better understand the net return distributions of the alternative tillage practices. Stoplight charts illustrate the probabilities of farm profits being less than a lower target value and greater than an upper target value for all risky alternatives (Richardson, 2010). This study uses \$0 and \$220 ha^{-1} as the lower and upper cutoff values, respectively (Fig. 3.4). These cutoff values roughly correspond to the 25th and 75th percentiles of the simulated iterations' net returns. Fig. 3.4 shows that sweeps tillage has the highest probability of obtaining a net return greater than \$220 ha^{-1}, that is, 0.39. The probability of obtaining a profit greater than \$220 ha^{-1} is 0.22 and 0.14 for moldboard plowing and no-till, respectively.

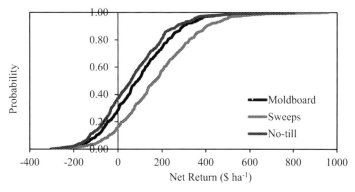

FIGURE 3.3

Cumulative distribution functions of net profits under three tillage methods.

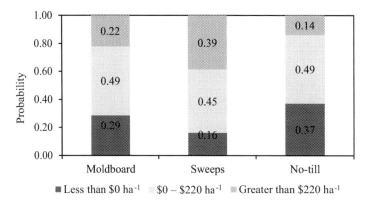

FIGURE 3.4
Stoplight chart of net returns under three tillage methods.

For the mid-income level $0 and $220 ha^{-1}, both moldboard plowing and no-till have the same higher probability, 0.49, compared to 0.45 for sweeps tillage. Overall, for a net return greater than $0 ha^{-1}, sweeps tillage has the highest probability, and thus wheat farmers would more likely practice sweeps tillage to achieve a high profit.

3.4.3 Comparison of alternative tillage practices

Fig. 3.5 shows the CEs and RPs of winter wheat production under three tillage practices. The tillage practices are ranked as the risk aversion level goes from risk-neutral (ARAC = 0) to extremely risk-averse (ARAC = 0.0348). The highest CE or RP value at each risk aversion level presents the most risk-efficient tillage practice for winter wheat producers.

Fig. 3.5A shows that sweeps tillage performs the best among the three tillage practices across all risk aversion levels. This suggests that sweeps tillage is consistently the most preferred practice regardless of producers' risk attitude. Fig. 3.5B shows that the RP gets lower as a producer becomes less risk-averse. For instance, RP is about $80 ha^{-1} at risk-neutral, and it goes down to $40 ha^{-1} at extremely risk-averse. No-till is even less preferred than moldboard plowing if a producer is risk-neutral, somewhat risk-averse, and rather risk-averse, but it becomes more preferred over moldboard if the producer becomes very and extremely risk-averse. This indicates that wheat producers may not adopt no-till within the initial couple of decades after converting from native grassland to wheat production if those producers are highly risk-averse.

In addition, the RP difference between sweeps and no-till becomes smaller as a producer becomes more risk-averse; for instance, the RP value of sweeps is

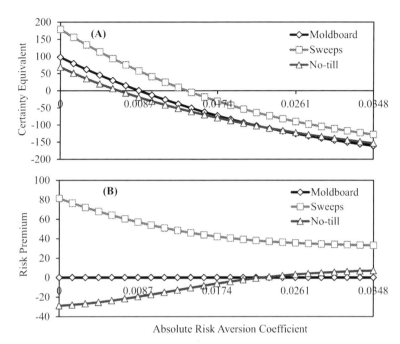

FIGURE 3.5

Results from SERF analysis under negative exponential utility function: (A) certainty equivalent ($ ha^{-1}) and (B) risk premium ($ ha^{-1}) relative to moldboard plowing tillage. *SERF*, stochastic efficiency with respect to a function.

about $110 ha^{-1} greater than that of no-till at ARAC = 0 (risk-neutral), and the difference becomes $15 ha^{-1} at ARAC = 0.0348 (extremely risk-averse). This suggests that producers would be willing to receive less payment as compensation before shifting from sweeps tillage to no-till if they become more risk-averse. In a risk analysis of annual rotations of winter wheat with soybeans, grain sorghum, and corn, Williams et al. (2012) found no-till wheat–soybean rotation was the most preferred system regardless of producers' risk aversion level. Fan et al. (2020) also showed that no-till with a wheat cover crop was the most preferred by risk-neutral and slightly risk-averse cotton producers and that conventional tillage was consistently the least preferred system.

3.4.4 Evaluation of water quality improvement

The risk-adjusted farm profits are also evaluated by combining the estimated benefits from water quality improvement with farm production income. Price assumptions are used to estimate the multiple benefits of alternative tillage practices (Table 3.A3). Fig. 3.6 shows the RP results based on a baseline price level, a lower price level, and two higher price levels in the subplots (A)−(D), respectively. Fig. 3.A2 in the appendix presents the corresponding

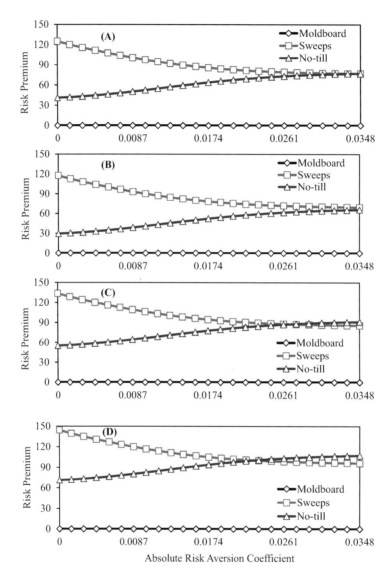

FIGURE 3.6

Risk premiums ($ ha^{-1}) relative to moldboard plowing, after combining with the estimated economic benefits of water quality improvement in four price scenarios: (A) baseline price, (B) low price, (C) higher price, and (D) very high price.

CE values. Fig. 3.6A shows that substantial water quality benefits associated with no-till dramatically improve its RP. No-till becomes the second most preferred practice following sweeps tillage for risk-neutral, somewhat risk-averse, and rather risk-averse producers. At the same time, no-till is equally preferred to sweeps tillage for very risk-averse and extremely risk-averse producers. Further, an overall comparison across the subplots (A) to (D) shows that lower nutrient prices decrease the preference ranking of no-till, even for highly risk-averse producers (Fig. 3.6B), while higher prices make no-till more likely to be preferred over sweeps tillage for moderate or even less risk-averse producers (Fig. 3.6C and D). This suggests that producers' preference of no-till is influenced by estimated benefits associated with water quality improvement (Owens et al., 2002; Tiessen et al., 2010). Endale et al. (2010) also confirmed that the runoff was reduced by no-till, and Benham et al. (2007) concluded that no-till resulted in less edge-of-field pollutant loss than strip-till. Nail et al. (2007) showed that conservation tillage systems were more likely to retain profitability with a higher nutrient price than intensive tillage.

3.5 Conclusion

This study investigates the profitability of three tillage practices in winter wheat production and evaluates the economic benefits of water quality improvement in the SHP. Wheat yield simulations were based on a 12-year field experiment after converting native grassland to cropping systems. An enterprise budget was developed to calculate crop income and account for the economic values of nutrient discharge in both runoff and groundwater. Economic risk analysis was performed using the SERF approach to compare the risk-adjusted profits under moldboard, sweeps, and no-till production.

Consistent with wheat yield results, sweeps tillage has the greatest mean net return, and no-till has the lowest net return. Previous research shows no-till and reduced tillage had greater agronomic and economic advantages (DeLaune et al., 2020; Pendell et al., 2007; Watkins et al., 2008). However, while the cropping systems in this study were in their initial 12 years after converting from native grassland, no-till was just used for 6 years. This may suggest the advantage of intensive tillage after land conversion, while no-till typically performs better than intensive tillage after it has been practiced for decades or longer.

SERF analysis suggests that sweeps tillage is the most preferred system regardless of risk aversion level without considering any environmental benefits. The difference in RP between sweeps and no-till becomes smaller as producers become more risk-averse, and this indicates that producers would be more easily to shift from sweeps tillage to no-till as they become more risk-averse. Moreover, after combining crop income with economic benefits associated with water quality improvement, the RP of no-till is greatly improved. This indicates that substantial water quality benefits associated with no-till would likely encourage wheat producers to utilize conservation tillage (Owens et al., 2002). Higher nutrient prices further make no-till more likely to be preferred for producers with moderate or even less risk aversion attitudes.

Appendix A

Figs. 3.A1 and 3.A2

Table 3.A1 Validation tests for wheat yields under three alternative tillage methods and wheat price.

	Wheat yield (kg ha^{-1})		No-till		Price ($ kg^{-1})
	Moldboard	Sweeps			
Experimental data					
Mean	2385	2690	1895		0.2139
Std. dev.	419	473	540		0.0556
Coef. vari.	17.6	17.6	28.5		26
Simulated data					
Mean	2386	2691	1896		0.2139
Std. dev.	419	473	540		0.0556
Coef. vari.	18	18	28		26
	Test value	Critical value	P-value		P-value
Two-sample Hotelling T^2 test	0.00	7.90	1.000	Two-sample t-test	0.998
Complete homogeneity test	0.00	16.92	1.000	F-test	0.470

Table 3.A2 Operating costs ($ ha^{-1}) under three alternative tillage methods.

Operating inputs	Moldboard	Sweeps	No-till
Wheat seed	42.63	42.63	42.63
Fertilizer	71.76	55.62	55.62
Pesticide	47.74	47.74	71.73
Interest	18.65	18.65	9.87
Custom hire	145.47	145.47	70.50
Custom harvest	86.64	86.64	86.64
Total operating costs	412.89	396.75	336.98

Notes: We did not include crop insurance, land rent, and machinery investment in the cost estimates.

Table 3.A3 Price assumptions for soil erosion reduction and soil nutrients.

Scenarios	Runoff $ cm^{-1}	Sediment $ kg^{-1}	P	N
(A) Baseline price	2	0.006	0.66	0.84
(B) Low price	1.6	0.005	0.55	0.7
(C) High price	2.4	0.0072	0.8	1
(D) Very high price	3	0.0086	1	1.2

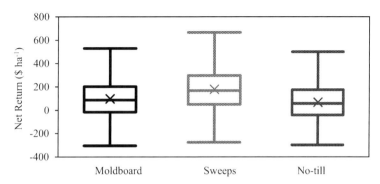

FIGURE 3.A1

Boxplot of the net returns under alternative tillage methods.

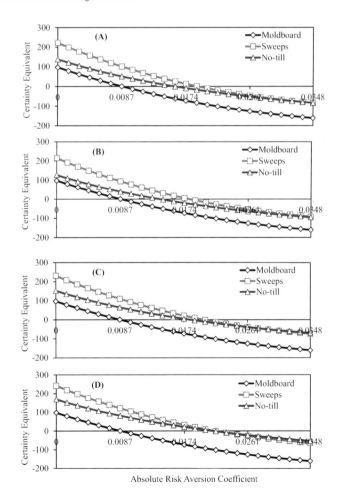

FIGURE 3.A2

Certainty equivalents ($ ha^{-1}) after combining with the estimated economic benefits of water quality improvement in four price scenarios: (A) baseline price, (B) low price, (C) higher price, and (D) very high price.

References

Anderson, J. R., & Dillon, J. L. (1992). *Risk analysis in dryland farming systems*. Rome, Italy: Food and Agriculture Organization of the United Nations. Available from https://agris.fao.org/agris-search/search.do?recordID = XF9326159.

Anderson, J. R., & Hardaker, J. B. (2003). *Risk aversion in economic decision making: Pragmatic guides for consistent choice by natural resource managers. In risk and uncertainty in environmental economics*. Cheltenham, UK: Edward Elgar Publishing Ltd.

Benham, B. L., Vaughan, D. H., Laird, M. K., Ross, B. B., & Peek, D. R. (2007). Surface water quality impacts of conservation tillage practices on burley tobacco production systems in southwest Virginia. *Water, Air, and Soil Pollution, 179*(1), 159−166.

BLS. 2021. Bureau of Labor Statistics. *CPI Inflation Calculator*. Available at: http://www.bls.gov/data/inflation_calculator.htm (accessed January 15, 2021).

Bosch, D. D., Potter, T. L., Truman, C. C., Bednarz, C., & Strickland, T. (2005). Surface runoff and lateral subsurface flow as a response to conservation tillage and soil-water conditions. *Transactions of the ASAE, 48*(6), 2137−2144.

Cartwright, L., & Kirwan, B. 2015. *Economic analysis of cover crops, version 2.1*, Released October 20, 2014. Available at: http://www.sites.google.com/site/kscovercrops/economics. (accessed December 10, 2020).

Claassen, R., Bowman, M., McFadden, J., Smith, D., & Wallander, S. 2018. *Tillage intensity and conservation cropping in the United States*. Available at: http://www.ers.usda.gov/publications/pub-details/?pubid = 90200 (accessed December 10, 2020).

Daryanto, S., Wang, L., & Jacinthe, P. A. (2017). Meta-analysis of phosphorus loss from no-till soils. *Journal of Environmental Quality, 46*(5), 1028−1037.

Decker, J. E., Epplin, F. M., Morley, D. L., & Peeper, T. F. (2009). Economics of five wheat production systems with no-till and conventional tillage. *Agronomy Journal, 101*(2), 364−372.

DeLaune, P. B., Keeling, J. W., Kelly, M., & Provin, T. 2015. Soil properties in long-term conservation tillage cotton cropping systems. *Beltwide Cotton Conference*. San Antonio, TX. January 5−7, 2015.

DeLaune, P. B., Mubvumba, P., Fan, Y., & Bevers, S. (2020). Agronomic and economic impacts of cover crops in Texas Rolling Plains cotton. *Agrosystems, Geosciences & Environment, 3*(1), e20027.

DeLaune, P. B., & Sij, J. W. (2012). Impact of tillage on runoff in long term no-till wheat systems. *Soil and Tillage Research, 124*, 32−35.

Djodjic, F., Bergström, L., & Ulén, B. (2002). Phosphorus losses from a structured clay soil in relation to tillage practices. *Soil Use and Management, 18*(2), 79−83.

Endale, D. M., Schomberg, H. H., Jenkins, M. B., Franklin, D. H., & Fisher, D. S. (2010). Management implications of conservation tillage and poultry litter use for Southern Piedmont USA cropping systems. *Nutrient Cycling in Agroecosystems, 88*(2), 299−313.

Fan, Y., DeLaune, P. B., Mubvumba, P., Kimura, E., & Park, S. (2019). *Economics of no-till and cover crops in Texas Rolling Plains dryland cotton*. Texas A&M AgriLife Extension Publication. Available from http://www.agrilifebookstore.org/product-p/esc-062.htm.

Fan, Y., Liu, Y., DeLaune, P. B., Mubvumba, P., Park, S. C., & Bevers, S. J. (2020). Economic analysis of adopting no-till and cover crops in irrigated cotton production under risk. *Agronomy Journal, 112*(1), 395−405.

Fathelrahman, E. M., Ascough, J. C., II, Hoag, D. L., Malone, R. W., Heilman, P., Wiles, L. J., & Kanwar, R. S. (2011). Economic and stochastic efficiency comparison of experimental tillage systems in corn and soybean under risk. *Experimental Agriculture, 47*(17), 111.

Franzluebbers, A., Hons, F., & Zuberer, D. (1994). Long-term changes in soil carbon and nitrogen pools in wheat management systems. *Soil Science Society of America Journal, 58*(6), 1639–1645.

Gaynor, J., & Findlay, W. (1995). Soil and phosphorus loss from conservation and conventional tillage in corn production. *Journal of Environmental Quality, 24*(4), 734–741.

Hardaker, J. B., Lien, G., Anderson, J. R., & Huirne, R. B. M. (2015). *Coping with risk in agriculture: Applied decision analysis* (3rd (ed.)). Cambridge, MA: CABI Publication.

Hardaker, J. B., Richardson, J. W., Lien, G., & Schumann, K. D. (2004). Stochastic efficiency analysis with risk aversion bounds: A simplified approach. *Australian Journal of Agricultural and Resource Economics, 48*(2), 379–383.

King, K. W., Williams, M. R., Macrae, M. L., Fausey, N. R., Frankenberger, J., Smith, D. R., Kleinman, P. J., & Brown, L. C. (2015). Phosphorus transport in agricultural subsurface drainage: A review. *Journal of Environmental Quality, 44*(2), 467–485.

Kleinman, P. J., Sharpley, A. N., Saporito, L. S., Buda, A. R., & Bryant, R. B. (2009). Application of manure to no-till soils: phosphorus losses by sub-surface and surface pathways. *Nutrient Cycling in Agroecosystems, 84*(3), 215–227.

Lien, G., Brian Hardaker, J., & Flaten, O. (2007). Risk and economic sustainability of crop farming systems. *Agricultural Systems, 94*(2), 541–552.

Locke, M. A., Krutz, L. J., Steinriede, R. W., Jr, & Testa, S., III (2015). Conservation management improves runoff water quality: Implications for environmental sustainability in a glyphosate-resistant cotton production system. *Soil Science Society of America Journal, 79*(2), 660–671.

Lozier, T., Macrae, M., Brunke, R., & Van Eerd, L. (2017). Release of phosphorus from crop residue and cover crops over the non-growing season in a cool temperate region. *Agricultural Water Management, 189*, 39–51.

Mueller, D., Wendt, R., & Daniel, T. (1984). Phosphorus losses as affected by tillage and manure application. *Soil Science Society of America Journal, 48*(4), 901–905.

Nail, E. L., Young, D. L., & Schillinger, W. F. (2007). Diesel and glyphosate price changes benefit the economics of conservation tillage vs traditional tillage. *Soil and Tillage Research, 94*(2), 321–327.

NASS. 2021. *U.S. Department of Agriculture, National Agricultural Statistics Service. Quick Stats.* Available at: https://quickstats.nass.usda.gov/ (accessed Febuary 12, 2021).

OSU. 2021. *Oklahoma Cooperative Extension Service. Enterprise budgets.* Available at: http://www.agecon.okstate.edu/budgets/ (accessed Febuary 8, 2021).

Owens, L., Malone, R., Hothem, D., Starr, G., & Lal, R. (2002). Sediment carbon concentration and transport from small watersheds under various conservation tillage practices. *Soil and Tillage Research, 67*(1), 65–73.

Patrignani, A., Godsey, C., Ochsner, T., & Edwards, J. (2012). Soil water dynamics of conventional and no-till wheat in the Southern Great Plains. *Soil Science Society of America Journal, 76*(5), 1768–1775.

Pendell, D. L., Williams, J. R., Boyles, S. B., Rice, C. W., & Nelson, R. G. (2007). Soil carbon sequestration strategies with alternative tillage and nitrogen sources under risk. *Review of Agricultural Economics, 29*(2), 247–268.

Richardson, J. W. (2010). *Simulation for applied risk management with an introduction to SIMETAR©.* College Station, Texas: Department of Agricultural Economics, Agricultural and Food Policy Center, Texas A&M University. Available from http://www.afpc.tamu.edu/courses/643/2017/simulation/Textbook-Simulation-Applied-Risk-Manual.pdf.

Richardson, J. W., Klose, S. L., & Gray, A. W. (2000). An applied procedure for estimating and simulating multivariate empirical (MVE) probability distributions in farm-level risk assessment and policy analysis. *Journal of Agricultural and Applied Economics, 32*(2), 299–315.

Richardson, J. W., Schumann, K., & Feldman, P. (2008). *Simetar: Simulation for excel to analyze risk*. Department of Agricultural Economics, Texas A&M University.

Rusu, T. (2014). Energy efficiency and soil conservation in conventional, minimum tillage and no-tillage. *International Soil and Water Conservation Research*, *2*(4), 42−49.

Schelde, K., de Jonge, L. W., Kjaergaard, C., Laegdsmand, M., & Rubæk, G. H. (2006). Effects of manure application and plowing on transport of colloids and phosphorus to tile drains. *Vadose Zone Journal*, *5*(1), 445−458.

Schumann, K. D., Richardson, J. W., Lien, G. D., & Hardaker, J. B. 2004. Stochastic efficiency analysis using multiple utility functions. In: *Selected Presentation at the American Agricultural Economics Association Annual Meeting*, Denver, Colorado, August 1−4, 2004. Available at: https://ideas.repec.org/p/ags/aaea04/19957.html (accessed December 10, 2020).

Sharpley, A. N., & Smith, S. (1994). Wheat tillage and water quality in the Southern Plains. *Soil and Tillage Research*, *30*(1), 33−48.

Smith, S., & Cassel, D. (1991). Estimating nitrate leaching in soil materials. In R. F. Follett, D. R. Keeney, & R. M. Cruse (Eds.), *Managing nitrogen for groundwater quality and farm profitability* (pp. 165−188). Madison, WI: Soil Science Society of America.

Tiessen, K., Elliott, J. A., Yarotski, J., Lobb, D., Flaten, D., & Glozier, N. (2010). Conventional and conservation tillage: Influence on seasonal runoff, sediment, and nutrient losses in the Canadian prairies. *Journal of Environmental Quality*, *39*(3), 964−980.

Triplett, G., & Dick, W. A. (2008). No-tillage crop production: A revolution in agriculture!. *Agronomy Journal*, *100*(S3), S153−S165.

Watkins, K., Hill, J., & Anders, M. (2008). An economic risk analysis of no-till management and rental arrangements in Arkansas rice production. *Journal of Soil and Water Conservation*, *63*(4), 242−250.

Williams, J. R., Pachta, M. J., Roozeboom, K. L., Llewelyn, R. V., Claassen, M. M., & Bergtold, J. S. (2012). Risk analysis of tillage and crop rotation alternatives with winter wheat. *Journal of Agricultural and Applied Economics*, *44*(4), 561−576.

Zhao, S. L., Gupta, S. C., Huggins, D. R., & Moncrief, J. F. (2001). Tillage and nutrient source effects on surface and subsurface water quality at corn planting. *Journal of Environmental Quality*, *30*(3), 998−1008.

CHAPTER 4

Bioremediation of contaminated soils by bacterial biosurfactants

Sabah Fatima[1], Muzafar Zaman[1], Basharat Hamid[1], Faheem Bashir[2], Zahoor Ahmad Baba[3] and Tahir Ahmad Sheikh[4]

[1]Department of Environmental Science, University of Kashmir, Srinagar, India, [2]Centre of Research for Development/Department of Environmental Science, University of Kashmir, Srinagar, India, [3]Division of Basic Science and Humanities, FOA, Sher-e-Kashmir University of Agricultural Sciences and Technology, Wadura, India, [4]Division of Agronomy, FOA, Sher-e-Kashmir University of Agricultural Sciences and Technology, Wadura, India

4.1 Introduction

Soil contamination inflicts global challenge since it causes ecosystem imbalances, thereby impeding environmental and public health (Gupta, 2020; Mao et al., 2015). Such soil pollution is mostly caused by the unhinged disposal of mining, industrial wastes, modern agricultural practices and by accidents caused during the transportation of these toxic substances (Martinez-Costa & Leyva-Ramos, 2017; Pacwa-Płociniczak et al., 2011). The increased amounts of various toxic compounds in the soil result mainly because of the use of pesticides, heavy metals (HMs), and petroleum derivatives (Palansooriya et al., 2020). The main concern of these compounds is their ease of binding with the soil particles as they possess high interfacial tension and reduced water solubility that make their remediation difficult from the contaminated sites (de Soza et al., 2013). Thus, efficient remediation approaches are needed to prevent the damage of the soils caused by these contaminants. A potential approach for the biological remediation of soil contaminated with organic and inorganic pollutants is the use of microbial biosurfactants (Luna et al., 2016; Karlapudi et al., 2018; Mekwichai et al., 2020). Biosurfactants produced by various microorganisms, including bacteria, yeast, and fungi, are surface-active metabolites having potential with amphipathic nature, that is, containing both hydrophilic and hydrophobic domains (John et al., 2021; Kour et al., 2021). The hydrophilic moiety may be amphoteric, negatively or positively charged ions or anionic as the hydrophobic moiety is a hydrocarbon chain. These form an interesting class of

67

Advances in Remediation Techniques for Polluted Soils and Groundwater. DOI: https://doi.org/10.1016/B978-0-12-823830-1.00011-0

alternatives to chemically synthesized surfactants because of their environment-friendly character (Islam & Sarma, 2021).

Biosurfactants increase the bioavailability of hydrophobic compounds and HMs through mobilization, solubilization, and micelle formation that increase their bioavailability to the bacterial community in the soil (Dell'Anno et al., 2018; Lee et al., 2018; Maia et al., 2019; Mnif et al., 2017; Pourfadakari et al., 2021; Santos et al., 2016). They play key roles in the interaction between the hydrocarbons, HMs, and the soil particles, by stimulating the mass transport of contaminants from the soil to aqueous phase at a particular concentration level called critical micelle concentration (CMC) where the surfactants assemble into an organized molecular arrangement (Chakraborty & Das, 2014). Biosurfactants also have some important properties such as biodegradability, bioavailability, lower toxicity, high foaming, high specific activity and selectivity at extreme pH, temperature, and salinity (Meena et al., 2021). All these benefits make biosurfactants desirable molecules for the bioremediation of soils and thus prompt us to find novel biosurfactants. Since, in nature, many bacterial species produce biosurfactants and have a wide range of commercial applications, they are considered an excellent option. Hence, this chapter aims to describe the bacterial biosurfactants (BBSs) and their role in the bioremediation of the environmental contaminants.

4.2   Bacterial biosurfactants and their classification

BBSs are a cluster of surface-active compounds synthesized by a diverse array of microorganisms (Nurfarahin et al., 2018; Sharma et al., 2015; Varjani & Upasani, 2017). Structurally, all BBSs are amphiphilic, that is, they comprise a hydrophilic head and a hydrophobic tail. The hydrophilic moiety is generally composed of mono/dipolysaccharides, peptides, amino acids, cations, and anions, while the hydrophobic part comprises saturated or unsaturated fatty acids (Fig. 4.1). Among kingdom bacteria, the genera, *Bacillus* and

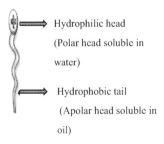

Hydrophilic head
(Polar head soluble in
water)

Hydrophobic tail
(Apolar head soluble in
oil)

FIGURE 4.1
Structure of a biosurfactant molecule.

Pseudomonas, represent the group of bacteria that are the major producers of BBSs (Jha et al., 2016; Loiseau et al., 2018; Purwasena et al., 2019). Among bacterial species, *Pseudomonas* (mostly *Pseudomonas aeruginosa*) have earned the name of best producers of glycolipid principally rhamnolipid consisting of rhamnose and 3-hydroxy fatty acids (Hassan et al., 2016; Rikalović et al., 2015). The other *Pseudomonas* species that produce rhamnolipids include *P. putida, P. rhizophila, P. fluorescens, P. alcaligenes, P. cepacia, P. teessidea, P. luteola, P. stutzeri, P. clemencea, P. chlororaphis, P. collierea* (Hassen et al., 2018).

The BBSs have a plethora of properties such as low toxicity, digestibility, biodegradability, and biocompatibility and significantly survive under extreme conditions, including salinity, heat, and alkalinity (Mulligan, 2021; Sarubbo, Lunaa, et al., 2015; Sarubbo, Rocha, et al., 2015). Such properties make them a vital choice for bioremediation purposes in addition to other applications.

Based on chemical composition and microbial origin, they are categorized into lipopeptides, glycolipids, lipoproteins, phospholipids, fatty acids, polymeric biosurfactants, neutral lipids, and particulate BBSs. The molecular weight of BBSs normally ranges from 500 to 1500 Da (Akbari et al., 2018). Based on their molecular mass, the bacterial BBSs are categorized into two major types: low molecular weight (LMW) and high molecular weight (HMW).

4.2.1   High molecular weight biosurfactants

HMW BBSs are synthesized by a diverse group of bacterial species and in general are called polymeric BBSs, which are mainly composed of proteins, lipoproteins, lipopolysaccharides, and polysaccharides (Fenibo et al., 2019; Santos et al., 2016). HMW BBSs can robustly adhere to different surfaces and behave as bioemulsifiers (McClements & Gumus, 2016). Among bacterial species, *Acinetobacter* has been identified to produce the majority of HMW BBSs. A few HMW BBSs and their components are shown in Fig. 4.2.

4.2.2   Low molecular weight biosurfactants

LMW BBSs include lipopeptides, lipoproteins, glycolipids (rhamnolipids, trehalolipids, and sophorolipids), phospholipids, fatty acids, and neutral lipids (Henkel & Hausmann, 2019; Mnif & Ghribi, 2016; Yuliani et al., 2018). The key role of these BBSs is to boost the surface area of hydrophobic substrates and in this way to improve the bioavailability of these substances via solubilization or desorption that regulates the microbe interaction (attachment and removal) from the surface (Vijayakumar & Saravanan, 2015). The bacterial genera such as *Pseudomonas* and *Bacillus* are the best known producers of LMW BBSs. The LMW BBSs and their components are shown in Fig. 4.3.

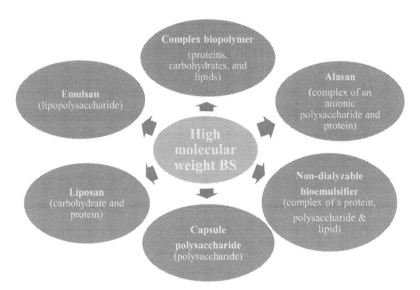

FIGURE 4.2
High molecular weight BBSs and their chemical composition. *BBSs*, Bacterial biosurfactants.

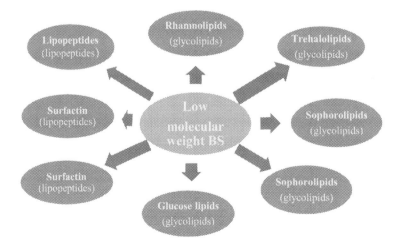

FIGURE 4.3
Low molecular weight BBSs and their chemical composition. *BBSs*, Bacterial biosurfactants.

4.3   Role of bacterial biosurfactants in the bioremediation of contaminated soils

In recent times, bacterial BBSs have been used for the bioremediation of various contaminants responsible for soil pollution. Their role in the biodegradation of various contaminants is discussed later.

4.3.1 Biosurfactants in the bioremediation of pesticides

The accelerated use of pesticides causes a harmful impact on the environment and primarily degrades soil ecosystem (Nicolopoulou-Stamati et al., 2016). Several studies were conducted to isolate and characterize novel BBS-producing bacteria to enhance the bioremediation process for contaminated soils (Carolin et al., 2020; Gaur et al., 2019; Uqab et al., 2016). Simultaneously the focus has been given to the use of BBS products for the solubilization and mobilization of pesticide molecules on the soil constituents (Ortiz-Hernández et al., 2013). The most studied BBSs in bioremediation are lipopeptides rhamnolipids, sophorolipids, and surfactin (Ortiz-Hernández et al., 2013). Among the bacterial communities, the bacterial species are best known producers of different types of BBSs for pesticide biodegradation. This is supported by the voluminous research on the isolation of BBS-producing bacteria such as *Pseudomonas*, *Klebsiella*, *Stenotrophomonas*, *Ochrobactrum*, and *Bacillus* sp. and their potential for the degradation of different types of pesticides such as organophosphate, chlorpyrifos, and beta-cypermethrin (Hassen et al., 2018; Lamilla et al., 2021; Nair et al., 2015; Singh & Cameotra, 2004). Some of the important bacterial species that have pesticide-degrading potential are presented in Table 4.1.

Table 4.1 Degradation of pesticides by different classes/types of biosurfactants synthesized by different bacteria.

Class/types	Source	Type of pesticide	References
Glycolipids; glycopeptides	*Pseudomonas rhodesiae* C4; *Pseudomonas marginalis* C9	Chlorpyrifos	Lamilla et al. (2021)
Glycolipid	*Pseudomonas* sp. B0406	Endosulfan and methyl parathion	García-Reyes et al. (2018)
Rhamnolipid	*Pseudomonas rhizophila* S211	Pentachlorophenol	Hassen et al. (2018)
Rhamnolipid	*Lysinibacillus sphaericus* IITR51	Endosulfan and hexachlorocyclohexane (HCH)	Gaur et al. (2019)
Rhamnolipids	*Pseudomonas* sp., *Serratia* sp., and *Pseudomonas aeruginosa*	Hexachlorocyclohexane (HCH)	Niar et al. (2015)
Rhamnolipid	*Pseudomonas*, *Klebsiella*, *Stenotrophomonas*, *Ochrobactrum*, and *Bacillus*	Chlorpyrifos	Singh and Cameotra (2004)
Rhamnolipids	*Arthrobacter globiformis*	DDT	Wang et al. (2018)
Rhamnolipid	*P. aeruginosa*; *Sphingomonas* sp. NM05	Hexachlorocyclohexane (HCH)	Manickam et al. (2012)
Rhamnolipid	*P. aeruginosa* CH7	Beta-cypermethrin	Zhang et al. (2011)

4.3.1.1 Mechanism

BBSs possess the potential to increase the solubility of hydrophobic or low—-water soluble organic pesticides, which in turn enhances their bioavailability to the bacteria that cleave them down to mineralized or nontoxic forms. However, when the concentration of BBS in water is above a standard level known as the CMC, the self-aggregation of surfactant molecules to form a cluster known as micelles that leads to the solubilization of hydrophobic organic pollutants (Zeng et al., 2018). The micelle formation in solution imparts detergent and solubilizing properties of the BBSs. Based on BBS concentration and their molecular weight, the BBS action for pollutant removal occurs through three basic modes, that is, mobilization, solubilization, and emulsification (Vijayakumar & Saravanan, 2015). The mobilization of the pollutant occurs when BBS concentration is below CMC that causes a desorption of the adsorbed organic compounds from a matrix with the help of biosurfactants making them available as free molecules in the aqueous phase. Oppositely when BBS concentration is above their CMC level, the solubilization of pollutants in the soil is enhanced through the entrapment of hydrophobic pesticides into the micelles (as microphase). Here, the hydrophobic contaminants are guarded against the water-based chemicals present in the aqueous phase and present an enhanced formal concentration of pollutants in water over its typical equilibrium value (so-called pseudo-solubility). Thus, BBSs through the process of mobilization and solubilization accelerate the mobility and bioavailability of pesticides in soils. Furthermore, BBS increases the biodegradation of pesticides in the soil, by lowering the surface tension of the aqueous phase and leads to a more efficient contact between microbial cells and pollutants, thus accelerating their bioavailability to the microbes (Silva et al., 2014).

4.3.2 Biosurfactants in bioremediation of hydrocarbons

Environmental contaminations, caused by hydrocarbons, have become a global concern because of their detrimental impacts on living beings (Diallo et al., 2019). Oil spills may cause serious environmental changes influencing quality soil, air, water, flora, and fauna as well (Duran & Cravo-Laureau, 2016; McWatters et al., 2016). In soil, the petroleum hydrocarbons exist as a complex mixture of several hydrocarbons that include aliphatic, asphaltenes, cycloalkanes, resins, mono and polyaromatics that are well known for their high toxicity, carcinogenicity, and persistence (da Silva et al., 2020; Kuppusamy et al., 2020). In addition, petroleum-derived hydrocarbon compounds are also reported as genotoxic, teratogenic, and immunotoxic (Balseiro-Romero et al., 2017). The polyaromatic hydrocarbons (PAHs) that are well known for their toxicity and carcinogenicity are considered very dangerous contaminants of soil, posing threat to living beings, including humans. The concerns about the PAHs are increasing because they are very difficult to get removed from contaminated soils (Parthipan et al., 2021). Bioremediation is a good technique for the recovery of areas contaminated with PAHs. A large number of hydrocarbon-degrading bacteria have been explored and applied for the bioremediation of hydrocarbons (Hentati et al., 2021). It has been reported that more than 79 genera of bacteria have the ability to degrade hydrocarbons (Xu et al., 2018). Many of them, including *Acinetobacter, Achromobacter, Alkanindiges, Arthrobacter, Alteromonas, Burkholderia, Enterobacter, Dietzia, Kocuria, Mycobacterium, Marinobacter, Pseudomonas, Pandorea, Streptobacillus, Staphylococcus, Streptococcus,* and *Rhodococcus* are known as key petroleum hydrocarbon degraders (Xu et al., 2018). Moreover, the recent developments in microbial biotechnology and high-throughput sequencing technologies have revealed that hydrocarbonclastic microbes are present abundantly and diversely in the oil-rich environments, signifying their key role in hydrocarbon biodegradation (Xu et al., 2018). In soil environments, the process of biodegradation is complex since the physicochemical characteristics of the soil and the oily contaminants might affect their bioavailability in the environment. Additionally, due to the lipophilic nature of such contaminants, only limited interaction occurs between the hydrocarbons and microbes because of impeded cell membrane crossing, causing the contaminants to remain intact (sorption) with the soil particles for longer periods (Decesaro et al., 2021). Consequently, the adsorbed compounds may become unavailable to microbes, which makes such contaminants difficult for biodegradation (Oualha et al., 2019). The environmental impacts of soil contamination due to oily compounds can be reversed by biosurfactants as supplements in the process of biodegradation (Decesaro et al., 2021). Several microbes bypass this obstruction by secreting biosurfactants into the soil, thereby increasing the surface area of hydrocarbons that enables these nonpolar molecules to enter into the cells and facilitate microbial access (Machado et al., 2020; Mao et al., 2015; Pereira et al., 2019). Thus, the bioremediation of soil through biosurfactant-producing microbes is a very promising, eco-friendly and cost-effective approach for removing the hydrocarbons by employing potential microorganisms. The genera, *Pseudomonas* and *Bacillus*, are the most widespread and important oil-degrading bacteria that excrete BBSs. The most used BBSs in the oil industry belong to glycolipids and their most excellent representatives are rhamnolipids produced mainly by the genus *Pseudomonas* (Kaczorek et al., 2018; Kiran et al., 2014). The second important group of oil-degrading biosurfactants belongs to lipopeptides, mostly produced by genus *Bacillus*. Sufactin is a well-known representative of this particular group (Patel et al., 2019). The other BBSs include polymeric BBSs such as alasan, emulsan, liposan, and polysaccharide−protein complexes, phospholipids, free fatty acids, and neutral lipids (Kaczorek et al., 2018; Santos et al., 2016). Several bacterial BBSs are known to enhance biodegradation process of hydrocarbons in various soil environments and are represented in Table 4.2.

Table 4.2 Degradation of hydrocarbons in soils by different biosurfactants produced by bacteria.

Biosurfactant type	Source	Hydrocarbons	Medium	References
Lipopeptide	*Pseudomonas aeruginosa* Lbp 3	PAHs	Soil	Bezza and Chirwa (2016)
Rhamnolipids	*P. aeruginosa* R21, *P. aeruginosa* R25, and *P. aeruginosa* R7	Crude oil	Soil	Tahseen et al. (2016)
Surfactin	*Bacillus nealsonii* S2MT	Heavy engine oil	Soil	Phulpoto et al. (2020)
Rhamnolipid	*P. aeruginosa* SR17	Crude oil	Soil	Patowary et al. (2018)
Rhamnolipid	*Shewanella* sp. BS4	Hydrocarbons (crude oil)	Soil	Joe et al. (2019)
Syringafactin	*Pseudomonas* sp. E311, E313, and E39	Diesel	Sand	Zouari et al. (2019)
Rhamnolipid	*P. aeruginosa* (*DKB1*)	Crude oil	Soil	Deivakumari et al. (2020)
Surfactin	*Bacillus flexus* S1I26 and *Paenibacillus* sp. S1I8	Benzo(a)pyrene	Rhizosphere soil	Kotoky and Pandey (2020)
Monorhamnolipid	*Klebsiella* sp. KOD36	Phenanthrene	Soil, soil slurry	Ahmad et al. (2021)
Rhamnolipid	*Pseudomonas* sp. SA3	Petroleum	Soil	Ambust et al. (2021)
Surfactin	*Bacillus methylotrophicus*	Biodiesel	Clayey soil	Decesaro et al. (2021)
Rhamnolipid	*P. aeruginosa* AHV-KH10	Diesel	Sea-shore sediment	Pourfadakari et al. (2021)

PAHs, *Polyaromatic hydrocarbons.*

It has been often observed that the best suited microbes for the bioremediation process are the native species from a contaminated habitat. Indigenous microbes have a greater ability to survive and multiply despite the toxic substances that present in such sites, exhibiting broader potential for producing BBSs and biodegradation (Jeanbille et al., 2016). Although other microbial strains have also been recognized for their potential for bioremediation, they are usually less efficient than the indigenous populations. Therefore, natural environments, both marine and terrestrial, that are highly contaminated by hydrocarbons are considered the potential biotopes for the isolation of microorganisms having the capability to degrade such contaminants (Jeanbille et al., 2016).

4.3.2.1 Mechanism

BBSs play a key role in the interaction between the hydrocarbons and the soil particles, stimulating the mass transport of contaminants from the soil to an aqueous phase. Two mechanisms are known to be involved through which the bioavailability of hydrophobic compounds is increased. First, the biosurfactants solubilize the hydrophobic compounds, mainly alkaline hydrophobic organic components, within their micellar structures. This increases the aqueous solubility of the hydrocarbon contaminants, enhancing their mobility and desorption from solid matrixes and making them available for the microorganisms (Lee et al., 2018; Maia et al., 2019; Pourfadakari et al., 2019). Second, the BBSs can cause changes in the surface cell structure of bacteria, improving their hydrophobicity that allows the bacterial cell to adsorb the oily hydrophobic contaminants easily (Dell'Anno et al., 2018; Maia et al., 2019; Mnif et al., 2017; Pourfadakari et al., 2019). These chemical changes occur mainly because of the loss of lipopolysaccharides to the medium (Maia et al., 2019; Singh et al., 2019). Therefore, the interaction of BBSs with the hydrophobic organics improves their solubility by reducing the surface and interfacial tensions.

Therefore, exploring the bacterial strains having the capability to produce BBSs suitable for bioremediation of hydrocarbons remains an unending quest. Generally, several factors of BBSs are responsible for the effective degradation of hydrocarbons. These include increased mobility, micelle formation, and enhanced bioavailability to bacteria-degrading microorganisms (Parthipan et al., 2021). For such reasons, intensive research studies have been conducted focused on selecting novel microbes that can produce BBSs with lower CMC values using cost-effective substrates (Borah et al., 2019). Among BBS classes, lipopeptides attract most attention due to their higher surface activities. Surfactin, a cyclic lipopeptide, produced by various species of genus *Bacillus* can lower the surface tension of water as much as from 72 to $27 \, \text{mN m}^{-1}$ and that of interfacial tension to less than $1.0 \, \text{mN m}^{-1}$ at even low concentrations as $20 \, \mu\text{M}$ (Kubicki et al., 2019). Though the most dominant producer of surfactin is *Bacillus subtilis*, other species, including *B. amyloliquefaciens*, *B. licheniformis*, *B. pumilus*, *B. mojavensis*, and some other members of *Bacillus* are also known to produce surfactin (Liu, Lin, et al., 2015; Liu, Mbadinga, et al., 2015; Patel et al., 2019).

4.3.3 Biosurfactants in bioremediation of heavy metals

HM contamination of soils is one of the world's major environmental problems. The contamination of soil by HMs occurs as a result of different industrial activities, such as mining, vehicle emissions, industrial waste deposits, and the dispersal of ash due to incineration processes (Sarubbo, Lunaa, et al., 2015; Sarubbo, Rocha, et al., 2015). Exposure to HMs poses significant risks to human health due to their immune toxicity, genotoxicity, carcinogenicity, and mutagenicity (Jin et al., 2019). BBS use in HM remediation has indisputable advantages such as microorganisms capable of producing surfactant compounds that do not need to have survival capacity in contaminated soil with HMs, although BBSs require continuous addition of new portions of these compounds (Pacwa-Płociniczak et al., 2011). BBSs facilitate the solubilization, dispersal, and desorption of contaminants in soil, thereby allowing the reuse of the land. The viability of BBSs for the removal of metals has been demonstrated and various studies have described the potential of these natural compounds of bacterial origin, such as rhamnolipids (Diaz et al., 2015; Santos et al., 2016; Saravanan et al., 2019; Sarubbo, Lunaa, et al., 2015; Sarubbo, Rocha, et al., 2015). BBSs produced by different microorganisms have unique metal-binding capacities and selectivity (Arjoon et al., 2013). It has been shown that cadmium and lead have higher affinities for rhamnolipids from *P. aeruginosa* than any other soil components to which they are bound (Mulligan & Wang, 2006).

The removal of HMs from the soil by BBSs is explained by the fact that the sorption of the BBS on the soil surface occurs, followed by the formation of metal−biosurfactant complex and then the desorption of the metal from the soil, which gets associated with BBS micelles (Mao et al., 2015). BBS is applied to contaminated soil for metal removal in a small quantity by using a cement mixer in the polluted sites. The complex formed between the BBS and metal ion is then flushed out and the soil is deposited back on site. BBS is then precipitated out of the complex formed leaving behind the metal ion. The bond formed by the BBS-metal ion is so strong that flushing with water eliminates the complex from the soil matrix. Similarly, the process can be repeated for deep subsurface (in situ) contamination, but with increased flushing activities. Studies have revealed the potential of lipopeptides, rhamnolipids, and surfactin in metal removal (Diaz et al., 2015; Singh & Cameotra, 2013). Table 4.3 presents the main potential application of bacterial BBSs in the remediation of HM contaminated soil.

4.3.3.1   Mechanism

The removal of HMs by BBSs consists of the following three main steps: sorption and binding of the BBS to the soil surface and complexation with metal; detachment of metal from the soil to the solution; and lastly association of the HM with micelles. The possible mechanisms for the extraction of HMs by BBSs are electrostatic interactions, ion exchange, precipitation−dissolution, and counter ion binding. HMs are trapped within the micelles through electrostatic interactions and can be easily recovered through precipitation or membrane separation techniques as illustrated in Fig. 4.4 (Santos et al., 2016). The surfactant concentration at which the aggregation of ions or molecules (micelles) first begins to form is called the CMC, where the surfactants assemble into an organized molecular arrangement (Chakraborty & Das, 2014). Above the CMC, the surfactant molecules

Table 4.3 Remediation of heavy metal contaminated soil by biosurfactants synthesized by different bacteria.

Class/ Subclass	Producer strain	Heavy metal remediation										Reference
		Mn	Cr	Fe	As	Ni	Co	Cu	Zn	Cd	Pb	
Lipopeptide	*Bacillus subtilis A21*					Ni	Co	Cu	Zn	Cd	Pb	Singh and Cameotra (2013)
Glycolipid	*Burkholderia sp. Z-90*	Mn			As			Cu	Zn	Cd	Pb	Yang et al. (2016)
Rhamnolipid	*P. aeruginosa*		Cr			Ni		Cu		Cd	Pb	Mulligan and Wang (2006)
Rhamnolipid (JBR 425)	*P. aeruginosa JBR425*				As			Cu	Zn		Pb	Wang and Mulligan (2009)
Trehalolipids	*Rhodococcus sp.*						Co					Narimannejad et al. (2019)
Lipopeptide	*Bacillus sp.*							Cu			Pb	Saleem et al. (2019)
Rhamnolipid	*Pseudomonas aeruginosa CVCM 411*			Fe					Zn			Diaz (2015)
Surfactin	*Bacillus subtilis*HIP3		Cr	Fe		Ni		Cu	Zn	Cd	Pb	Md Badrul Hisham et al. (2019)
Rhamnolipids	*Psuedomonas sp.*					Ni		Cu		Cd		Lee and Kim (2019)

assemble, forming aggregates and a range of micellar and vesicular configurations that depend on the pH of the solution.

Nonionic metals form complex with BBSs, which in turn decreases the solution phase activity of the metal and, therefore, promotes desorption. Under conditions of reduced interfacial tension, BBSs can bind to adsorbed metals directly and can accumulate metals at a solid−solution interface (Singh & Cameotra, 2013). Anionic BBSs create complexes with metals through ionic bonds, which are stronger than the bond between the metal and soil. The metal−BBS complex is then desorbed from the soil matrix to the soil solution due to the reduction in interfacial tension (Wu et al., 2017). Cationic BBSs can replace the same charged ions through competition with some, but not all, negatively charged surfaces (ion exchange). Metal ions can also be removed from the soil by BBS micelles (Diaz et al., 2015; Pacwa-Płociniczak et al., 2011).

The BBSs that have been effective soil washing agents for metal removal include rhamnolipids and lipopeptides. The potential of BBS-rhamnolipid in the washing of HM ions from river sediment has been reported (Chen et al., 2017). Chen et al. (2017) observed that 0.8% rhamnolipid removed Cu

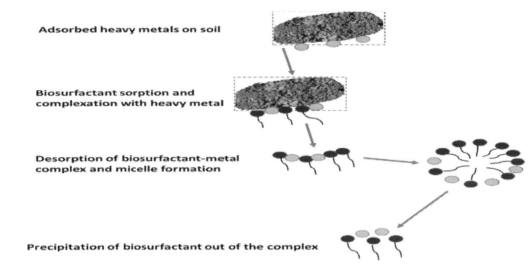

Adsorbed heavy metals on soil

Biosurfactant sorption and complexation with heavy metal

Desorption of biosurfactant-metal complex and micelle formation

Precipitation of biosurfactant out of the complex

FIGURE 4.4

Mechanism of HM removal by bacterial biosurfactants. *HM*, Heavy metal.

(80.21%), Cd (86.87%), Pb (63.54%), and Cr (47.85%) after 12 h at pH 7.0. They also emphasized that the efficiency of washing depended on initial HM ion speciation, rhamnolipid concentration, washing time, liquid/solid ratio, and pH. Increases in metal mobility, binding ability, and removal efficiency of Cu, Zn, Cr, Ni, and Mn up to 62%, 74%, 60%, 68%, and 64%, respectively, more than Pb (only 15%), using rhamnolipids and saponins in multiple washing steps have been reported (Tang et al., 2019). A recent study investigated the BBS production by a novel strain, *Pseudomonas* sp. CQ2, and the bioleaching effects of the BBSs on HMs, including Cd, Cu, and Pb in soil. They concluded that the use of BBSs attained the HM removal efficiencies of 78.7%, 65.7%, 56.9%, for Cd, Cu, and Pb, respectively (Sun et al., 2021). Therefore, BBSs can also be harnessed as an effective approach for the recovery of HMs from the contaminated sites.

4.4 Conclusion and future prospectus

The facts regarding the hazardous impacts of various pesticides, petroleum products, and HMs on the soil health have aimed at improving in the research demonstrations to formulate the methods of remediation that could be applied for the elimination of pollutants. Among the treatment techniques available, bioremediation has proved to be an effective approach in the remediation of contaminated soils. This chapter has summarized the bioremediation of contaminated soils by using effective BBs produced by bacteria. The BBs have earned respectable acceptance in the field of bioremediation due to their biodegradability, low toxicity, and effectiveness. Undoubtedly, biosurfactants perform a number of functions that will restore environmental sustainability not only in the 21st century but also beyond. A diverse group of bacterial species produce a variety of biosurfactants that detoxify the soil contaminants by using different modes of action. The present chapter concludes that this approach can open up new ways for the remediation of inorganic and organic pollutants from the environment. Therefore, more research is needed to isolate and characterize efficient biosurfactant-synthesizing microbial species, to identify novel biosurfactants and to scale up biosurfactant production for a pollutant-free planet and environmental sustainability.

References

Ahmad, Z., Zhang, X., Imran, M., Zhong, H., Andleeb, S., Zulekha, R., & Coulon, F. (2021). Production, functional stability, and effect of rhamnolipid biosurfactant from Klebsiella sp. on phenanthrene degradation in various medium systems. *Ecotoxicology and Environmental Safety*, *207*, 111514.

Akbari, S., Abdurahman, N. H., Yunus, R. M., Fayaz, F., & Alara, O. R. (2018). Biosurfactants—A new frontier for social and environmental safety: A mini review. *Biotechnology Research and Innovation*, *2*(1), 81−90.

Ambust, S., Das, A. J., & Kumar, R. (2021). Bioremediation of petroleum contaminated soil through biosurfactant and *Pseudomonas* sp. SA3 amended design treatments. *Current Research in Microbial Sciences*, *2*, 100031.

Arjoon, A., Olaniran, A. O., & Pillay, B. (2013). Co-contamination of water with chlorinated hydrocarbons and heavy metals: Challenges and current bioremediation strategies. *International Journal of Environmental Science and Technology*, *10*(2), 395−412.

Balseiro-Romero, M., Gkorezis, P., Kidd, P. S., Van Hamme, J., Weyens, N., Monterroso, C., & Vangronsveld, J. (2017). Characterization and degradation potential of diesel-degrading bacterial strains for application in bioremediation. *International Journal of Phytoremediation*, *19*(10), 955−963.

Bezza, F. A., & Chirwa, E. M. N. (2016). Bioremediation of polycyclic aromatic hydrocarbon contaminated soil by a microbial consortium through supplementation of biosurfactant produced by *Pseudomonas aeruginosa* strain. *Polycyclic Aromatic Compounds*, *36*(5), 848−872.

Borah, S. N., Sen, S., Goswami, L., Bora, A., Pakshirajan, K., & Deka, S. (2019). Rice based distillers dried grains with solubles as a low cost substrate for the production of a novel rhamnolipid biosurfactant having anti-biofilm activity against *Candida tropicalis*. *Colloids and Surfaces B: Biointerfaces*, *182*, 110358.

Carolin, C. F., Kumar, P. S., & Ngueagni, P. T. (2020). A review on new aspects of lipopeptide biosurfactant: Types, production, properties and its application in the bioremediation process. *Journal of Hazardous Materials*, *407*, 124827.

Chakraborty, J., & Das, S. (2014). *Biosurfactant-based bioremediation of toxic metals*. *Microbial Biodegradation and Bioremediation* (pp. 167−201). Elsevier.

Chen, W., Qu, Y., Xu, Z., He, F., Chen, Z., Huang, S., & Li, Y. (2017). Heavy metal (Cu, Cd, Pb, Cr) washing from river sediment using biosurfactant rhamnolipid. *Environmental Science and Pollution Research, 24*(19), 16344–16350.

da Silva, S., Gonçalves, I., Gomes de Almeida, F. C., Padilha da Rocha e Silva, N. M., Casazza, A. A., Converti, A., & Asfora Sarubbo, L. (2020). Soil bioremediation: Overview of technologies and trends. *Energies, 13*(18), 4664.

de Souza, R. B., Maziviero, T. G., Christofoletti, C. A., Pinheiro, T. G., & Fontanetti, C. S. (2013). *Soil contamination with heavy metals and petroleum derivates: impact on Edaphic Fauna and remediation strategies.* Soil process and current trends in quality assessment, (pp. 175–203). InTech.

Decesaro, A., Rempel, A., Machado, T. S., Cappellaro, Â. C., Machado, B. S., Cechin, I., & Colla, L. M. (2021). Bacterial biosurfactant increases ex situ biodiesel bioremediation in clayey soil. *Biodegradation, 32*(4), 389–401.

Deivakumari, M., Sanjivkumar, M., Suganya, A. M., Prabakaran, J. R., Palavesam, A., & Immanuel, G. (2020). Studies on reclamation of crude oil polluted soil by biosurfactant producing *Pseudomonas aeruginosa* (DKB1). *Biocatalysis and Agricultural Biotechnology, 29*, 101773.

Dell'Anno, F., Sansone, C., Ianora, A., & Dell'Anno, A. (2018). Biosurfactant-induced remediation of contaminated marine sediments: Current knowledge and future perspectives. *Marine Environmental Research, 137*, 196–205.

Diallo, M. M., Vural, C., Şahar, U., & Ozdemir, G. (2019). Kurstakin molecules facilitate diesel oil assimilation by *Acinetobacter haemolyticus* strain 2SA through overexpression of alkane hydroxylase genes. *Environmental Technology, 42*, 2031–2045.

Diaz, M. A., De Ranson, I. U., Dorta, B., Banat, I. M., Blazquez, M. L., Gonzalez, F., & Ballester, A. (2015). Metal removal from contaminated soils through bioleaching with oxidizing bacteria and rhamnolipid biosurfactants. *Soil and Sediment Contamination: An International Journal, 24*(1), 16–29.

Duran, R., & Cravo-Laureau, C. (2016). Role of environmental factors and microorganisms in determining the fate of polycyclic aromatic hydrocarbons in the marine environment. *FEMS Microbiology Reviews, 40*(6), 814.

Fenibo, E. O., Ijoma, G. N., Selvarajan, R., & Chikere, C. B. (2019). Microbial surfactants: The next generation multifunctional biomolecules for applications in the petroleum industry and its associated environmental remediation. *Microorganisms, 7*(11), 581.

García-Reyes, S., Yáñez-Ocampo, G., Wong-Villarreal, A., Rajaretinam, R. K., Thavasimuthu, C., Patiño, R., & Ortiz-Hernández, M. L. (2018). Partial characterization of a biosurfactant extracted from *Pseudomonas* sp. B0406 that enhances the solubility of pesticides. *Environmental technology, 39*(20), 2622–2631.

Gaur, A., Bajaj, A., Regar, R. K., Kamthan, M., Jha, R. R., Srivastava, J. K., & Manickam, N. (2019). Rhamnolipid from a Lysininbacillus sphaericus IITR51 and its potential application for dissolution of hydrophobic pesticides. *Bioresource Technology, 272*, 19–25.

Gupta, P. K. (2020). Fate, transport, and bioremediation of biodiesel and blended biodiesel in subsurface environment: A review. *Journal of Environmental Engineering, 146*(1), 03119001.

Hassan, M., Essam, T., Yassin, A. S., & Salama, A. (2016). Optimization of rhamnolipid production by biodegrading bacterial isolates using Plackett–Burman design. *International Journal of Biological Macromolecules, 82*, 573–579.

Hassen, W., Neifar, M., Cherif, H., Najjari, A., Chouchane, H., Driouich, R. C., & Cherif, A. (2018). Pseudomonas rhizophila S211, a new plant growth-promoting rhizobacterium with potential in pesticide-bioremediation. *Frontiers in microbiology, 9*, 34.

Henkel, M., & Hausmann, R. (2019). *Diversity and classification of microbial surfactants. Biobased surfactants* (pp. 41–63). AOCS Press.

Hentati, D., Cheffi, M., Hadrich, F., Makhloufi, N., Rabanal, F., Manresa, A., & Chamkha, M. (2021). Investigation of halotolerant marine *Staphylococcus* sp. CO100, as a promising hydrocarbon-degrading and biosurfactant-producing bacterium, under saline conditions. *Journal of Environmental Management, 277*, 111480.

Islam, N. F., & Sarma, H. (2021). Metagenomics approach for selection of biosurfactant producing bacteria from oil contaminated soil: An insight into its technology. *Biosurfactants for a Sustainable Future: Production and Applications in the Environment and Biomedicine*, 43−58.

Jeanbille, M., Gury, J., Duran, R., Tronczynski, J., Agogué, H., Ben Said, O., & Auguet, J. C. (2016). Response of core microbial consortia to chronic hydrocarbon contaminations in coastal sediment habitats. *Frontiers in Microbiology, 7*, 1637.

Jha, S. S., Joshi, S. J., & Geetha, S. J. (2016). Lipopeptide production by *Bacillus subtilis* R1 and its possible applications. *Brazilian Journal of Microbiology, 47*, 955−964.

Jin, Z., Deng, S., Wen, Y., Jin, Y., Pan, L., Zhang, Y., & Zhang, D. (2019). Application of Simplicillium chinense for Cd and Pb biosorption and enhancing heavy metal phytoremediation of soils. *Science of The Total Environment, 697*, 134148.

Joe, M. M., Gomathi, R., Benson, A., Shalini, D., Rengasamy, P., Henry, A. J., & Sa, T. (2019). Simultaneous application of biosurfactant and bioaugmentation with rhamnolipid-producing shewanella for enhanced bioremediation of oil-polluted soil. *Applied Sciences, 9* (18), 3773.

John, W. C., Ogbonna, I. O., Gberikon, G. M., & Iheukwumere, C. C. (2021). Evaluation of biosurfactant production potential of *Lysinibacillus fusiformis* MK559526 isolated from automobile-mechanic-workshop soil. *Brazilian Journal of Microbiology, 52*, 663−674.

Kaczorek, E., Pacholak, A., Zdarta, A., & Smułek, W. (2018). The impact of biosurfactants on microbial cell properties leading to hydrocarbon bioavailability increase. *Colloids and Interfaces, 2*(3), 35.

Karlapudi, A. P., Venkateswarulu, T. C., Tammineedi, J., Kanumuri, L., & Ravuru, B. K. (2018). Role of biosurfactants in bioremediation of oil pollutiona review. *Petroleum*, 241−249.

Kiran, G. S., Sabarathnam, B., Thajuddin, N., & Selvin, J. (2014). Production of glycolipid biosurfactant from spongeassociated marine actinobacterium *Brachybacterium paraconglomeratum* MSA21. *Journal of Surfactants and Detergents, 17*, 531−542.

Kotoky, R., & Pandey, P. (2020). Rhizosphere mediated biodegradation of benzo (A) pyrene by surfactin producing soil bacilli applied through *Melia azedarach* rhizosphere. *International Journal of Phytoremediation, 22*(4), 363−372.

Kour, D., Kaur, T., Devi, R., Yadav, A., Singh, M., Joshi, D., & Saxena, A. K. (2021). Beneficial microbiomes for bioremediation of diverse contaminated environments for environmental sustainability: Present status and future challenges. *Environmental Science and Pollution Research, 28*(20), 24917−24939.

Kubicki, S., Bollinger, A., Katzke, N., Jaeger, K. E., Loeschcke, A., & Thies, S. (2019). Marine biosurfactants: Biosynthesis, structural diversity and biotechnological applications. *Marine Drugs, 17*(7), 408.

Kuppusamy, S., Naga Raju, M., Mallavarapu, M., & Kadiyala, V. (2020). *An overview of total petroleum hydrocarbons. Total petroleum hydrocarbons* (pp. 1−27). Springer.

Lamilla, C., Schalchli, H., Briceño, G., Leiva, B., Donoso-Piñol, P., Barrientos, L., & Diez, M. C. (2021). A Pesticide Biopurification System: A Source of Biosurfactant-Producing Bacteria with Environmental Biotechnology Applications. *Agronomy, 11*, 624.

Lee, A., & Kim, K. (2019). Removal of heavy metals using rhamnolipid biosurfactant on manganese nodules. *Water, Air, & Soil Pollution, 230*(11), 1−9.

Lee, D. W., Lee, H., Kwon, B. O., Khim, J. S., Yim, U. H., Kim, B. S., & Kim, J. J. (2018). Biosurfactant-assisted bioremediation of crude oil by indigenous bacteria isolated from Taean beach sediment. *Environmental Pollution, 241*, 254−264.

Liu, J. F., Mbadinga, S. M., Yang, S. Z., Gu, J. D., & Mu, B. Z. (2015). Chemical structure, property and potential applications of biosurfactants produced by *Bacillus subtilis* in petroleum recovery and spill mitigation. *International Journal of Molecular Sciences, 16*(3), 4814−4837.

Liu, Q., Lin, J., Wang, W., Huang, H., & Li, S. (2015). Production of surfactin isoforms by *Bacillus subtilis* BS-37 and its applicability to enhanced oil recovery under laboratory conditions. *Biochemical Engineering Journal, 93*, 31−37.

Loiseau, C., Portier, E., Corre, M. H., Schlusselhuber, M., Depayras, S., Berjeaud, J. M., & Verdon, J. (2018). Highlighting the potency of biosurfactants produced by *Pseudomonas* strains as anti-Legionella agents. *BioMed research international, 8194368.*

Luna, J. M., Rufino, R. D., & Sarubbo, L. A. (2016). Biosurfactant from *Candida sphaerica* UCP0995 exhibiting heavy metal remediation properties. *Process Safety and Environmental Protection, 102*, 558−566.

Machado, T. S., Decesaro, A., Cappellaro, Â. C., Machado, B. S., van Schaik Reginato, K., Reinehr, C. O., & Colla, L. M. (2020). Effects of homemade biosurfactant from *Bacillus methylotrophicus* on bioremediation efficiency of a clay soil contaminated with diesel oil. *Ecotoxicology and Environmental Safety, 201*, 110798.

Maia, M., Capão, A., & Procópio, L. (2019). Biosurfactant produced by oil-degrading *Pseudomonas putida* AM-b1 strain with potential for microbial enhanced oil recovery. *Bioremediation Journal, 23*(4), 302−310.

Manickam, N., Bajaj, A., Saini, H. S., & Shanker, R. (2012). Surfactant mediated enhanced biodegradation of hexachlorocyclohexane (HCH) isomers by *Sphingomonas* sp. NM05. *Biodegradation, 23*(5), 673−682.

Mao, X., Jiang, R., Xiao, W., & Yu, J. (2015). Use of surfactants for the remediation of contaminated soils: A review. *Journal of Hazardous Materials, 285*, 419−435.

Martinez-Costa, J. I., & Leyva-Ramos, R. (2017). Effect of surfactant loading and type upon the sorption capacity of organobentonite towards pyrogallol. *Colloids and Surfaces A: Physicochemical and Engineering Aspects, 520*, 676−685.

McClements, D. J., & Gumus, C. E. (2016). Natural emulsifiers—Biosurfactants, phospholipids, biopolymers, and colloidal particles: Molecular and physicochemical basis of functional performance. *Advances in Colloid and Interface Science, 234*, 3−26.

McWatters, R. S., Wilkins, D., Spedding, T., Hince, G., Raymond, B., Lagerewskij, G., & Snape, I. (2016). On site remediation of a fuel spill and soil reuse in Antarctica. *Science of the Total Environment, 571*, 963−973.

Md Badrul Hisham, N. H., Ibrahim, M. F., Ramli, N., & Abd-Aziz, S. (2019). Production of biosurfactant produced from used cooking oil by *Bacillus* sp. HIP3 for heavy metals removal. *Molecules (Basel, Switzerland), 24*(14), 2617.

Meena, K. R., Dhiman, R., Singh, K., Kumar, S., Sharma, A., Kanwar, S. S., & Mandal, A. K. (2021). Purification and identification of a surfactin biosurfactant and engine oil degradation by Bacillus velezensis KLP2016. *Microbial Cell Factories, 20*(1), 1−12.

Mekwichai, P., Tongcumpou, C., Kittipongvises, S., & Tuntiwiwattanapun, N. (2020). Simultaneous biosurfactant-assisted remediation and corn cultivation on cadmium-contaminated soil. *Ecotoxicology and environmental safety, 192*, 110298.

Mnif, I., & Ghribi, D. (2016). Glycolipid biosurfactants: Main properties and potential applications in agriculture and food industry. *Journal of the Science of Food and Agriculture, 96*(13), 4310−4320.

Mnif, I., Sahnoun, R., Ellouz-Chaabouni, S., & Ghribi, D. (2017). Application of bacterial biosurfactants for enhanced removal and biodegradation of diesel oil in soil using a newly isolated consortium. *Process Safety and Environmental Protection, 109*, 72−81.

Mulligan, C. N. (2021). Sustainable remediation of contaminated soil using biosurfactants. *Frontiers in Bioengineering and Biotechnology, 9*, 195.

Mulligan, C. N., & Wang, S. (2006). Remediation of a heavy metal-contaminated soil by a rhamnolipid foam. *Engineering Geology, 85*(1−2), 75−81.

Nair, A. M., Rebello, S., Rishad, K. S., Asok, A. K., & Jisha, M. S. (2015). Biosurfactant facilitated biodegradation of quinalphos at high concentrations by *Pseudomonas aeruginosa* Q10. *Soil and Sediment Contamination: An International Journal, 24*, 542−553.

Narimannejad, S., Zhang, B., & Lye, L. (2019). Adsorption behavior of cobalt onto saline soil with/without a biosurfactant: Kinetic and isotherm studies. *Water, Air, & Soil Pollution, 230*(2), 47.

Nicolopoulou-Stamati, P., Maipas, S., Kotampasi, C., Stamatis, P., & Hens, L. (2016). Chemical pesticides and human health: the urgent need for a new concept in agriculture. *Frontiers in public health, 4*, 148.

Nurfarahin, A. H., Mohamed, M. S., & Phang, L. Y. (2018). Culture medium development for microbial-derived surfactants production—An overview. *Molecules (Basel, Switzerland), 23*(5), 1049.

Ortiz-Hernández, M. L., Sánchez-salinas, E., Godínez, M. L. C., González, E. D., & Ursino, E. C. P. (2013). Mechanisms and strategies for pesticide biodegradation: opportunity for waste, soils and water cleaning. *Revista Internacional de Contaminacion Ambiental, 29*, 85−104.

Oualha, M., Al-Kaabi, N., Al-Ghouti, M., & Zouari, N. (2019). Identification and overcome of limitations of weathered oil hydrocarbons bioremediation by an adapted *Bacillus sorensis* strain. *Journal of Environmental Management, 250*, 109455.

Pacwa-Płociniczak, M., Płaza, G. A., Piotrowska-Seget, Z., & Cameotra, S. S. (2011). Environmental applications of biosurfactants: Recent advances. *International Journal of Molecular Sciences, 12*(1), 633−654.

Palansooriya, K. N., Shaheen, S. M., Chen, S. S., Tsang, D. C., Hashimoto, Y., Hou, D., & Ok, Y. S. (2020). Soil amendments for immobilization of potentially toxic elements in contaminated soils: A critical review. *Environment International, 134*, 105046.

Parthipan, P., Cheng, L., Rajasekar, A., & Angaiah, S. (2021). Microbial surfactants are next-generation biomolecules for sustainable remediation of polyaromatic hydrocarbons. *Biosurfactants for a Sustainable Future: Production and Applications in the Environment and Biomedicine, 7*, 139−158.

Patel, S., Homaei, A., Patil, S., & Daverey, A. (2019). Microbial biosurfactants for oil spill remediation: Pitfalls and potentials. *Applied Microbiology and Biotechnology, 103*(1), 27−37.

Patowary, R., Patowary, K., Kalita, M. C., & Deka, S. (2018). Application of biosurfactant for enhancement of bioremediation process of crude oil contaminated soil. *International Biodeterioration & Biodegradation, 129*, 50−60.

Pereira, E., Napp, A. P., Allebrandt, S., Barbosa, R., Reuwsaat, J., Lopes, W., & Vainstein, M. H. (2019). Biodegradation of aliphatic and polycyclic aromatic hydrocarbons in seawater by autochthonous microorganisms. *International Biodeterioration & Biodegradation, 145*, 104789.

Phulpoto, I. A., Yu, Z., Hu, B., Wang, Y., Ndayisenga, F., Li, J., & Qazi, M. A. (2020). Production and characterization of surfactin-like biosurfactant produced by novel strain *Bacillus nealsonii* S2MT and it's potential for oil contaminated soil remediation. *Microbial Cell Factories, 19*(1), 1−12.

Pourfadakari, S., Ahmadi, M., Jaafarzadeh, N., Takdastan, A., Ghafari, S., & Jorfi, S. (2019). Remediation of PAHs contaminated soil using a sequence of soil washing with biosurfactant produced by Pseudomonas aeruginosa strain PF2 and electrokinetic oxidation of desorbed solution, effect of electrode modification with Fe_3O_4 nanoparticles. *Journal of Hazardous Materials, 379*, 120839.

Pourfadakari, S., Ghafari, S., Takdastan, A., & Jorfi, S. (2021). A salt resistant biosurfactant produced by moderately halotolerant *Pseudomonas aeruginosa* (AHV-KH10) and its application for bioremediation of diesel-contaminated sediment in saline environment. *Biodegradation, 32*(3), 327−341.

Purwasena, I. A., Astuti, D. I., Fauziyyah, N. A., Putri, D. A. S., & Sugai, Y. (2019). Inhibition of microbial influenced corrosion on carbon steel ST37 using biosurfactant produced by Bacillus sp. *Materials Research Express, 6*(11).

Rikalović, M. G., Vrvić, M. M., & Karadžić, I. M. (2015). Rhamnolipid biosurfactant from Pseudomonas aeruginosa: from discovery to application in contemporary technology. *Journal of the Serbian Chemical Society, 80*, 279−304.

Saleem, H., Pal, P., Haija, M. A., & Banat, F. (2019). Regeneration and reuse of bio-surfactant to produce colloidal gas aphrons for heavy metal ions removal using single and multistage cascade flotation. *Journal of Cleaner Production, 217*, 493−502.

Santos, D. K. F., Rufino, R. D., Luna, J. M., Santos, V. A., & Sarubbo, L. A. (2016). Biosurfactants: Multifunctional biomolecules of the 21st century. *International Journal of Molecular Sciences, 17*(3), 401.

Saravanan, A., Jayasree, R., Hemavathy, R. V., Jeevanantham, S., Hamsini, S., Yaashikaa, P. R., & Yuvaraj, D. (2019). Phytoremediation of Cr (VI) ion contaminated soil using Black gram (Vigna mungo): Assessment of removal capacity. *Journal of Environmental Chemical Engineering, 7*(3), 103052.

Sarubbo, L. A., Lunaa, J. M., & Rufinoa, R. D. (2015). Application of a biosurfactant produced in low-cost substrates in the removal of hydrophobic contaminants. *Chemical Engineering, 43*, 295−300.

Sarubbo, L. A., Rocha, R. B., Jr, Luna, J. M., Rufino, R. D., Santos, V. A., & Banat, I. M. (2015). Some aspects of heavy metals contamination remediation and role of biosurfactants. *Chemistry and Ecology, 31*(8), 707−723.

Sharma, D., Saharan, B. S., Chauhan, N., Procha, S., & Lal, S. (2015). Isolation and functional characterization of novel biosurfactant produced by *Enterococcus faecium*. *SpringerPlus, 4*(1), 1−14.

Singh, A. K., & Cameotra, S. S. (2013). Efficiency of lipopeptide biosurfactants in removal of petroleum hydrocarbons and heavy metals from contaminated soil. *Environmental Science and Pollution Research, 20*(10), 7367−7376.

Silva, E. J., Silva, N. M. P. R., Rufino, R. D., Luna, J. M., Silva, R. O., & Sarubbo, L. A. (2014). Characterization of a biosurfactant produced by Pseudomonas cepacia CCT6659 in the presence of industrial wastes and its application in the biodegradation of hydrophobic compounds in soil. *Colloids and Surfaces B: Biointerfaces, 117*, 36−41.

Singh, P., & Cameotra, S. S. (2004). Enhancement of metal bioremediation by use of microbial surfactants. *Biochemical and Biophysical Research Communications, 319*(2), 291−297.

Singh, P., Patil, Y., & Rale, V. (2019). Biosurfactant production: Emerging trends and promising strategies. *Journal of Applied Microbiology, 126*(1), 2−13.

Sun, W., Zhu, B., Yang, F., Dai, M., Sehar, S., Peng, C., & Naz, I. (2021). Optimization of biosurfactant production from *Pseudomonas* sp. CQ2 and its application for remediation of heavy metal contaminated soil. *Chemosphere, 265*, 129090.

Tahseen, R., Afzal, M., Iqbal, S., Shabir, G., Khan, Q. M., Khalid, Z. M., & Banat, I. M. (2016). Rhamnolipids and nutrients boost remediation of crude oil-contaminated soil by enhancing bacterial colonization and metabolic activities. *International Biodeterioration and Biodegradation, 115*, 192−198.

Tang, J., He, J., Qiu, Z., & Xin, X. (2019). Metal removal effectiveness, fractions, and binding intensity in the sludge during the multiple washing steps using the combined rhamnolipid and saponin. *Journal of Soils and Sediments, 19*(3), 1286−1296.

Uqab, B., Mudasir, S., & Nazir, R. (2016). Review on bioremediation of pesticides. *Journal of Bioremediation & Biodegradation, 7*, 1000343.

Varjani, S. J., & Upasani, V. N. (2017). Critical review on biosurfactant analysis, purification and characterization using rhamnolipid as a model biosurfactant. *Bioresource Technology, 232*, 389−397.

Vijayakumar, S., & Saravanan, V. (2015). Biosurfactants-types, sources and applications. *Research Journal of Microbiology, 10*(5), 181.

Wang, S., & Mulligan, C. N. (2009). Rhamnolipid biosurfactant-enhanced soil flushing for the removal of arsenic and heavy metals from mine tailings. *Process Biochemistry, 44*(3), 296−301.

Wang, X., Sun, L., Wang, H., Wu, H., Chen, S., & Zheng, X. (2018). Surfactant-enhanced bioremediation of DDTs and PAHs in contaminated farmland soil. *Environmental technology, 39*(13), 1733−1744.

Wu, J., Zhang, J., Wang, P., Zhu, L., Gao, M., Zheng, Z., & Zhan, X. (2017). Production of rhamnolipids by semi-solid-state fermentation with *Pseudomonas aeruginosa* RG18 for heavy metal desorption. *Bioprocess and Biosystems Engineering, 40*(11), 1611−1619.

Xu, X., Liu, W., Tian, S., Wang, W., Qi, Q., Jiang, P., & Yu, H. (2018). Petroleum hydrocarbon-degrading bacteria for the remediation of oil pollution under aerobic conditions: A perspective analysis. *Frontiers in Microbiology, 9*, 2885.

Yang, Z., Zhang, Z., Chai, L., Wang, Y., Liu, Y., & Xiao, R. (2016). Bioleaching remediation of heavy metal-contaminated soils using Burkholderia sp. Z-90. *Journal of Hazardous Materials, 301*, 145−152.

Yuliani, H., Perdani, M. S., Savitri, I., Manurung, M., Sahlan, M., Wijanarko, A., & Hermansyah, H. (2018). Antimicrobial activity of biosurfactant derived from *Bacillus subtilis* C19. *Energy Procedia, 153*, 274−278.

Zeng, Z., Liu, Y., Zhong, H., Xiao, R., Zeng, G., Liu, Z., & Qin, L. (2018). Mechanisms for rhamnolipids-mediated biodegradation of hydrophobic organic compounds. *Science of The Total Environment, 634*, 1−11.

Zhang, C., Wang, S., & Yan, Y. (2011). Isomerization and biodegradation of beta-cypermethrin by *Pseudomonas aeruginosa* CH7 with biosurfactant production. *Bioresource technology, 102* (14), 7139−7146.

Zouari, O., Lecouturier, D., Rochex, A., Chataigne, G., Dhulster, P., Jacques, P., & Ghribi, D. (2019). Bio-emulsifying and biodegradation activities of syringafactin producing Pseudomonas spp. strains isolated from oil contaminated soils. *Biodegradation, 30*(4), 259−272.

Evaluation of machine learning-based modeling approaches in groundwater quantity and quality prediction

Madhumita Sahoo[1,2]

[1]University of Alaska Fairbanks, Fairbanks, AK, United States, [2]Gandhi Institute for Technology, Biju Pattnaik University of Technology, India

5.1 Overview

Machine learning (ML) is a subset of artificial intelligence (AI) that gives computers the ability to learn without being explicitly programmed. To program most of the processes, ML techniques have found applications in various fields, including groundwater hydrology. The complexity of problems in groundwater hydrology is high. Therefore both computational approaches and experiments have been used for solving multiple issues in groundwater modeling. However, conducting experiments in subsurface hydrology is arduous, expensive, and requires extensive instrumentation and equipment.

However, computational approaches do not require any equipment and can make significant advances with acceptable errors (Tahmasebi et al., 2020). Beginning with application in optimal groundwater remediation (Ranjithan et al., 1993; Rogers et al., 1995; Wagner, 1995) and geophysical explorations (Constable et al., 1987; Zohdy, 1989) in the late 1980s and early 1990s, ML has become one of the most popular approaches in groundwater modeling studies. These techniques can handle huge datasets with ease and do predictive analytics faster than any human can (Mueller & Massaron, 2016). ML techniques' superiority in handling complex and nonlinear problems efficiently makes it an obvious choice for analysis in several studies.

Statistical methods also handle predictive analytics. These methods, indeed, have proved their prediction efficacy in various previous works, including quantitative and qualitative modeling and analysis in groundwater systems (Gibbons et al., 2009; Sahoo & Jha, 2015; Upton & Jackson, 2011). However, groundwater systems are known to have high levels of uncertainty and nonlinearity among parameters. ML proves its superiority in handling

Advances in Remediation Techniques for Polluted Soils and Groundwater. DOI: https://doi.org/10.1016/B978-0-12-823830-1.00016-X

Table 5.1 Comparison between machine learning (ML) techniques and statistical methods.

Technique	ML techniques	Statistical methods
Data handling	Works with big data; raw data from sensors or the web text is split into training and test data.	Models are used to create predictive power on small samples.
Data input	The data are sampled, randomized and transformed to maximize accuracy scoring in the prediction of out-of-sample (or completely new) examples.	Parameters interpret real-world phenomena and provide stress on magnitude.
Result	Probability is taken into account for comparing what could be the best guess or decision.	The output captures the variability and uncertainty of parameters.
Assumptions	The scientist learns from the data.	The scientist assumes a certain output and tries to prove it.
Distribution	The distribution is unknown or ignored before learning from data.	The scientist assumes a well-defined distribution.
Fitting	The scientist creates the best fit, but generalizable model.	The result is fit to the present data distribution.

Taken from Mueller, J.P. & Massaron, L. (2016). Machine learning for dummies. John Wiley & Sons.

these uncertainties compared to statistical methods in the following ways (Table 5.1).

With the availability of affordable and good quality computational facilities, more researchers prefer ML over statistical methods during the recent decade. As a result, the application of several ML techniques is gaining popularity compared to relying on traditional statistical methods. In the following subsections, we will discuss few popularly used ML techniques in groundwater modeling studies.

5.1.1 ML techniques—introduction

The machine learns by looking for a pattern among the input data. It keeps adjusting the pattern to improve upon the estimated output. Based on the type of learning algorithms the machines can use, ML techniques are classified into three groups: (1) supervised learning algorithm, (2) unsupervised learning algorithm, and (3) reinforcement learning algorithm. Supervised learning algorithm is trained to predict some observed dependent variables, either categorical or continuous, given some independent attributes (Shen, 2018). Unsupervised learning algorithms seek to learn how to represent independent variables to find meaningful structures. In reinforcement learning, the algorithm keeps on learning continuously from the output it obtains at each stage. The best solution is rewarded, and poor solutions are punished. Reinforcement learning algorithm aims at maximizing performance through continuous learning.

5.1.1.1 Supervised learning algorithm

Supervised learning algorithms (Fig. 5.1) are based on the patterns learned by the existing data or historical datasets to predict the estimated output. Datasets are labeled as dependent and independent datasets. Supervised learning algorithms are used to solve classification and regression problems. Classification problems require finding a pattern to group the input dataset into distinct groups based on a specific pattern. The variables involved are discrete-valued. Regression problems predict the outcomes based on the relationship pattern identified between the input and output datasets. The variables in the regression problem are continuous-valued. Examples of classification problems in groundwater modeling can be used in categorical prediction for water quality in wells (Messier et al., 2019), classification of areas vulnerable to groundwater contamination (Sajedi-Hosseini et al., 2018), and groundwater potential mapping (Pham et al., 2019). Regression modeling problems can be used to predict groundwater level variation within a study area (Maheswaran & Khosa, 2013; Shiri et al., 2013; Suryanarayana et al., 2014).

5.1.1.2 Unsupervised learning algorithm

Unsupervised learning algorithms (Fig. 5.2) do not have any labeled data. All data are independent of each other. The algorithm finds identification of patterns among the data points to group them distinctively. Clustering and association problems are known to use unsupervised learning algorithms. In groundwater modeling, clustering analyses are used in groundwater quality

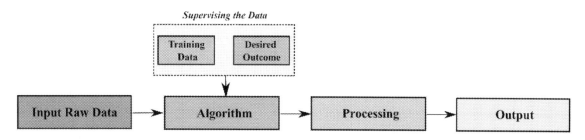

FIGURE 5.1

Schematic diagram of a supervised learning algorithm.

FIGURE 5.2

Schematic diagram of an unsupervised learning algorithm.

assessment based on various contaminants present in groundwater (Pacheco Castro et al., 2018; Marín Celestino et al., 2018; Kim et al., 2014). Association problems such as the apriori algorithm are used in handling missing well log data in lithology prediction (Hossain et al., 2020).

5.1.1.3   Reinforcement learning algorithm

Applications of these algorithms (Fig. 5.3) are not yet common in groundwater modeling. Examples of reinforcement algorithms are—Markov decision models, Q-learning, and deep Q-networks.

5.1.2   Generic steps involved in using an ML technique

Most ML problems involve the following four steps to solve the problem:

1. Data collection: Raw input data are collected in this step. Data which can represent a wide variation in the variables are preferable. A reasonably large dataset with a relevant representation of variations can be considered good for learning. On the other hand, a huge dataset with little to no variation among variables can be of no use. Additionally, large datasets may result in overtraining of the model and may result in poor performance.
2. Preprocessing of input: Input data need to be preprocessed for identifying any missing data or treating outliers. Imputation or interpolation methods can be used to fill in the missing data.
3. Training a model: This step is exclusively for problems using supervised or reinforcement learning algorithms. The dataset is split into two parts—train and test. A training set is used to develop and calibrate the model, and a testing set is used to validate the developed model.

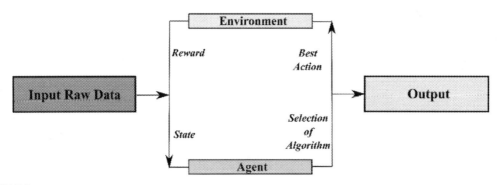

FIGURE 5.3
Schematic diagram of a reinforcement learning algorithm.

4. Validating the model: A testing set is used to evaluate the accuracy of the developed model. The various error statistics can determine accuracy. The difference between the input and output variables gives the error. The smaller the error, the better the model.

Usually, researchers use more than one ML technique in a single study to compare the methods. The method which gives the lower error is preferred for further modeling in the study. In recent years, many review articles on the application of ML techniques in groundwater were published. Each of the articles covered a different aspect of groundwater modeling and enlisted the ML methods used for quality and quantity modeling. This chapter lists popularly used methods with a brief detail on their methodologies, and examples of where it has been previously used are provided in the following sections. This chapter aims to give a brief overview of methods and their suitability in different groundwater modeling problems.

5.2 Popular Ml techniques used in groundwater modeling

The ML techniques can be generalized with the generic features they have. This section enlists the popular techniques along with a brief description of their various features.

5.2.1 Supervised ML techniques

Every supervised ML method has hypothesis, cost function, and gradient descent function. A hypothesis is a function that models the problem. Cost function measures the error between predicted and expected values. Gradient descent function optimizes the model parameters. Under the supervised ML family, linear algorithms are very widely used. In this family of hypothesis classes, linear regression and logistic regression predictors are popularly used in groundwater modeling. Linear regression has been used in determining water quality indices (Kadam et al., 2019; Yazdanpanah, 2016), contamination source identification (Aguilar et al., 2007; Hofmann et al., 2015), and predicting groundwater level variations temporally (Ebrahimi & Rajaee, 2017; Huang et ai., 2019). Logistic regression is used for solving similar groundwater problems (Mair & El-Kadi, 2013; Nguyen, Ha, Avand et al., 2020; Nguyen, Ha, Nguyen et al., 2020; Park et al., 2017; Rizeei et al., 2018). In high dimension feature space, the complexity of learning for linear predictors increases. This complexity by tackled by using support vector machines (SVMs). SVMs can perform classification and regression in high-dimensional spaces using kernels. Kernels are functions that transform the input data into the required form to be used in the feature space. Commonly used kernel functions in SVMs are—linear, polynomial, nonlinear, sigmoid, and radial basis functions. SVMs have been used in groundwater modeling to predict groundwater level variation (Yoon et al., 2016; Zhou et al., 2017), potential zone/recharge zone mapping (Huang et al., 2017; Naghibi et al., 2017) predicting groundwater quality (Arabgol et al., 2016; Isazadeh et al., 2017). Boosting algorithm is used to enhance the learning ability of weak learners. One of the most commonly used boosting algorithms is AdaBoost (abbreviated for Adaptive Boosting). The AdaBoost algorithm can be used in conjunction with other learning algorithms to improve the learner's performance. In groundwater modeling, AdaBoost has been used in conjunction with multiple neural networks or ML methods to predict water quality (Chen, Zhao et al., 2020) and potential zone mapping (Mosavi et al., 2021). Bootstrap aggregating or "bagging" is another effective method for enhancing the performance of the ML methods. The bagging method is commonly used in decision trees (e.g., Random Forests). The decision trees are constructed by randomly picking a subset of the training dataset with replacement. Each decision tree is produced independently. A forest is grown up to a predefined number of trees with low bias and high variance (Breiman, 1996; Tahmasebi et al., 2020). The random forest method has been used in estimating water quantity/source of groundwater (Baudron et al., 2013; Parveen et al., 2020; Sihag et al., 2020) and water quality prediction/vulnerability mapping (Lahjouj et al., 2020; Norouz et al., 2016; Ouedraogo et al., 2019). Nearest neighbor algorithms are another set of popularly used ML methods for classification and regression. These methods predict based on labels of the training data points present in the neighborhood. Search in the neighborhood depends on the distances between the data points. The nearest neighbor algorithm in the form of the k-nearest neighbor algorithm has been used in predicting groundwater quality and quantity mapping (Motevalli et al., 2019; Rahmati, Choubin et al., 2019; Rahmati, Ebrahimi et al., 2019). Naïve Bayes method is a probabilistic ML model under supervised class. This method assumes features of each data point as independent. The naïve Bayes model first gets trained and then generates data using Baye's rule. It has been used in groundwater modeling for mapping groundwater potential zones (Chen, Tsangaratos et al., 2019; Martínez-Santos & Renard, 2020; Miraki et al., 2019).

5.2.2 Unsupervised ML techniques

Unsupervised methods can identify hidden patterns among unlabeled data. In unsupervised learning methods, distance-based clustering methods can predict patterns in unlabeled data. K-means clustering is one of the very popular distance-based cluster algorithms. It has been used in groundwater modeling to classify areas based on groundwater quality (Marín Celestino et al., 2018;

Javadi et al., 2017) and in classifying areas based on water level variation (Naranjo-Fernández et al., 2020; Everest & Özcan, 2019). Hierarchical clustering is another kind of clustering that builds trees (called dendograms). Dendograms are formed by continuous partition of data by merging two nearest clusters. Hierarchical clustering has been used for groundwater quality analysis and classification (Egbueri, 2020; El-Hames et al., 2013; Rahbar et al., 2020). Another set of unsupervised algorithms used for dimensionality reduction can provide a meaningful structure of the data points. One of the popular methods of this class—Principal Component Analysis (PCA)—linearly transforms the original data and maps it onto a smaller dimensional space. PCA has successfully demonstrated its efficacy in characterizing groundwater contamination zones (Hu et al., 2013; Rao, 2014; Ravikumar & Somashekar, 2017). Factor analysis is another example of the dimensionality reduction class used in groundwater modeling. This method has been used for hydrogeochemical analysis (Biglari et al., 2018; El-Rawy et al., 2019; Guo et al., 2017). There are several other examples under both supervised and unsupervised classes. It is beyond the ability of this chapter to discuss all methods. Readers and ML enthusiasts can find more examples in standard books on ML.

5.3   Efficacy of ML-based modeling

Examples of popular methods and their applications identified the prevalence of ML methods in groundwater quality and quantity monitoring. In this section, we enlisted studies (Table 5.2) and identified if the analysis done by ML methods was compared with any physical model.

The groundwater modeling problems are usually solved using ML techniques predicted water level (temporally or spatially), water quality, vulnerability to contamination, or estimated subsurface hydraulic parameters. Most studies rely heavily on the comparative analysis of multiple ML methods. These methods, however, have drawbacks such as large data requirements, low generalization ability, overtraining, using irrelevant data as input, using improper modeling methods, and tuning parameters (Malakar et al., 2021; Rajaee et al., 2019; Willard et al., 2020).

Regardless of these weaknesses, these data-driven methods have been chosen more often by researchers for groundwater resource modeling and quality estimation due to their simplistic approach, promising results, and satisfactory application in scientific problems that are not completely understood (Malakar et al., 2021; Willard et al., 2020). Comparison of spatial/temporal analysis with the physical flow or transport models is not common. These analyses identified the dependent and most relevant parameters affecting the

Table 5.2 List of groundwater studies in which machine learning (ML) techniques were used.

Author(s) and year	Quality/ quantity modeling	Spatial/ temporal analysis	ML method(s) used	Comparison with physical model
Al-Abadi and Alsamaani (2020)	Quantity	Spatial	SVM, RFM, CART, SGB, kNN	–
Al-Fugara et al. (2020)	Quantity	Spatial	SVM, BRT, MDA, RF, MARS	–
Arabameri et al. (2019)	Quantity	Spatial	RF, SVM, Binary LR	-
Barzegar et al. (2017)	Quality	Spatiotemporal	ELM, SVM	–
Baudron et al. (2013)	Quality	Spatial	RFM, CART, LDA	–
Chakraborty et al. (2020)	Quality	Spatial	RF, BRT, LR	Yes
Chen, Pradhan et al. (2019)	Quantity	Spatial	LDA	–
Chen, He et al. (2020)	Quantity	Spatial	SVM	Yes
Chen, Li et al. (2020)	Quantity	Spatial	RF, LR, decision tree	–
Chen, Zhao et al. (2020)	Quantity	Spatial	J48, RF	–
DeSimone et al. (2020)	Quality	Spatial	BRT	–
Díaz-Alcaide and Martínez-Santos (2019)	Quantity	Spatial	RF, LR, SVM, QDA, kNN, LDA, Naïve Bayes, AdaBoost classifier, GBM, decision tree	–
Ebrahimi and Rajaee (2017)	Quantity	Temporal	MLR, SVM, linear regression	–
Friedel et al. (2020)	Quality	Temporal	LDA, BRT, RF	–
Golkarian et al. (2018)	Quantity	Spatial	RF, MARS, C5.0	–
Guzman et al. (2019)	Quantity	Temporal	SVM	–
Huang et al. (2017)	Quantity	Temporal	SVM	–
Kalantar et al. (2019)	Quantity	Spatial	MDA, LDA, RF	–
Kenda et al. (2018)	Quantity	Temporal	RF, GBM, regression trees	–
Khosravi et al. (2020)	Quantity	Spatial	kNN, M5P, LWL, RBD, KStar	–
Knoll et al. (2019)	Quality	Spatial	RF, MLR, CART, BRT	–
Knoll et al. (2020)	Quality	Spatial	RF	–
Kordestani et al. (2019)	Quantity	Spatial	BRT	–
Lal and Datta (2018)	Quality	Spatial	SVM	–
Lee et al. (2020)	Quantity	Spatial	BCT	–
Malekzadeh et al. (2019)	Quantity	Temporal	ELM, wavelet-ELM	Yes
Martínez-Santos and Renard (2020)	Quantity	Spatial	SVM, LR, Naives Bayes, AdaBoost classifier, GBM, QDA, LR, kNN, LDA	–
Messier et al. (2019)	Quality	Spatial	RF, GBM, SVM, maximum likelihood-linear model	–
Miraki et al. (2019)	Quantity	Spatial	SVM, RF	–
Mosavi et al. (2021)	Quantity	Spatial	AdaBoost classification trees, RF, CART	–

Continued

Table 5.2 List of groundwater studies in which machine learning (ML) techniques were used. *Continued*

Author(s) and year	Quality/ quantity modeling	Spatial/ temporal analysis	ML method(s) used	Comparison with physical model
Naghibi and Pourghasemi (2015)	Quantity	Spatial	CART, RF, BRT, GLM	–
Naghibi and Dashtpagerdi (2017)	Quantity	Spatial	kNN, LDA, MARS, QDA	–
Naghibi et al. (2016)	Quantity	Spatial	BRT, CART, RF	–
Naghibi et al. (2017)	Quantity	Spatial	SVM, RF	–
Naghibi et al. (2018)	Quantity	Spatial	LDA, QDA, FDA, PDA, BRT, RF, kNN, SVM, MARS	–
Nguyen, Ha, Avand et al. (2020)	Quantity	Spatial	CDT	–
Nguyen, Ha, Nguyen et al. (2020)	Quantity	Spatial	LR and ensembles	–
Nolan et al. (2015)	Quality	Temporal	BRT	–
Pal et al. (2020)	Quantity	Spatial	RF	–
Pham et al. (2019)	Quantity	Spatial	Rotation forest with boosting	–
Podgorski et al. (2018)	Quality	Spatiotemporal	RF, MLR, LR	–
Podgorski and Berg (2020)	Quality	Spatial	RF	–
Rahman et al. (2020)	Quantity	Temporal	EGB, RF	–
Rahmati, Choubin et al. (2019) and Rahmati, Ebrahimi et al. (2019)	Quality	Spatial	RF, SVM, kNN	–
Rizeei et al. (2019)	Quantity	Spatial	LR, SVM	–
Rodriguez-Galiano et al. (2018)	Quality	Spatial	CART, RF, SVM	–
Sahoo et al. (2018)	Quantity	Temporal	SVM, RF, GBM	–
Sahour et al. (2020)	Quality	Spatial	MLR, EGB	–
Sajedi-Hosseini et al. (2018)	Quality	Spatial	MDA, SVM, BRT	–
Sameen et al. (2019)	Quantity	Spatial	RF, SVM, decision trees	–
Seyoum et al. (2019)	Quantity	Spatial	BRT	–
Singh et al. (2014)	Quality	Temporal	SVM, decision tree	–
Suryanarayana et al. (2014)	Quantity	Temporal	SVM	–
Tan et al. (2020)	Quality	Spatial	BRT, RF, LR	–
Tang et al. (2019)	Quantity	Spatiotemporal	SVM	–
Tien Bui et al. (2019)	Quantity	Spatial	ADTree with AdaBoost, LMT, LR, RF, SVM, SGD	–
Wei et al. (2019)	Quantity	Temporal	SVM	–
Xu et al. (2014)	Quantity	Spatial and Temporal	SVM	Yes

Continued

Table 5.2 List of groundwater studies in which machine learning (ML) techniques were used. *Continued*

Author(s) and year	Quality/ quantity modeling	Spatial/ temporal analysis	ML method(s) used	Comparison with physical model
Yadav et al. (2017)	Quantity	Temporal	SVM, ELM	–
Yadav et al. (2020)	Quantity	Temporal	SVM	–
Yang et al. (2019)	Quantity and Quality	Spatial	RF	–
Yazdanpanah (2016)	Quality	Spatiotemporal	Linear regression	–
Yousefi et al. (2020)	Quantity	Spatial	MARS-SVM, RF-BRT, FDA-GLM-MDA	–

BRT, *Boosted regression tree*; CART, *classification and regression tree*; CDT, *credal decision tree*; EGB, *extreme gradient boosting*; ELM, *extreme learning machine*; FDA, *flexible discriminant analysis*; GBM, *gradient boosting mechanism*; GLM, *generalized linear model*; LDA, *linear discriminant analysis*; LR, *logistic regression*; LWL, *locally weighted learning*; MARS, *multivariate adaptive regression spline*; MDA, *mixture discriminant analysis*; MLR, *multiple linear regression*; QDA, *quadratic discriminant analysis*; RBD, *regression by discretization*; RF, *random forest*; SGB, *stochastic gradient boosting*; SVM, *support vector machine*.

process. However, the physics of most processes is still not known. For developing physics-guided ML techniques, the groundwater system's governing equations are of paramount importance. In recent years, physics-guided ML techniques are developed where scientific knowledge and data are integrated together. Physics-guided ML techniques require the determination of hidden physical parameters affecting the processes. Existing ML techniques are improving their predictability and integrating scientific domain knowledge. However, physics-guided ML methods are in their developing stages and need further research.

5.4   Conclusion

The application of neural networks and deep learning algorithms are prevalent in groundwater modeling. Neural networks, deep learning, and the numerous amalgamations of neural networks and different optimization algorithms are not covered in this chapter and need a distinct discussion of their own. This chapter evaluated the popular ML algorithms used in groundwater modeling studies. ML techniques have been demonstrating their power where high levels of uncertainty are encountered. Efficient groundwater models in both spatial and temporal analyses have been obtained using a plethora of ML techniques. The drawbacks of ML-based modeling can be overcome through the use of physics-guided ML methods. Existing ML methods can be sensibly used to understand the causal process.

References

Aguilar, J. B., Orban, P., Dassargues, A., & Brouyère, S. (2007). Identification of groundwater quality trends in a chalk aquifer threatened by intensive agriculture in Belgium. *Hydrogeology Journal, 15*(8), 1615.

Al-Abadi, A. M., & Alsamaani, J. J. (2020). Spatial analysis of groundwater flowing artesian condition using machine learning techniques. *Groundwater for Sustainable Development, 11*, 100418.

Al-Fugara, A. K., Pourghasemi, H. R., Al-Shabeeb, A. R., Habib, M., Al-Adamat, R., Al-Amoush, H., & Collins, A. L. (2020). A comparison of machine learning models for the mapping of groundwater spring potential. *Environmental Earth Sciences, 79*, 1–19.

Arabameri, A., Roy, J., Saha, S., Blaschke, T., Ghorbanzadeh, O., & Tien Bui, D. (2019). Application of probabilistic and machine learning models for groundwater potentiality mapping in Damghan Sedimentary Plain, Iran. *Remote Sensing, 11*(24), 3015.

Arabgol, R., Sartaj, M., & Asghari, K. (2016). Predicting nitrate concentration and its spatial distribution in groundwater resources using support vector machines (SVMs) model. *Environmental Modeling & Assessment, 21*(1), 71–82.

Barzegar, R., Moghaddam, A. A., Adamowski, J., & Fijani, E. (2017). Comparison of machine learning models for predicting fluoride contamination in groundwater. *Stochastic Environmental Research and Risk Assessment, 31*(10), 2705–2718.

Baudron, P., Alonso-Sarría, F., García-Aróstegui, J. L., Cánovas-García, F., Martínez-Vicente, D., & Moreno-Brotóns, J. (2013). Identifying the origin of groundwater samples in a multi-layer aquifer system with Random Forest classification. *Journal of Hydrology, 499*, 303–315.

Biglari, H., Saeidi, M., Karimyan, K., Narooie, M. R., & Sharafi, H. (2018). Data for factor analysis of hydro-geochemical characteristics of groundwater resources in Iranshahr. *Data in Brief, 19*, 548–563.

Breiman, L. (1996). Bagging predictors. *Machine Learning, 24*(2), 123–140.

Chakraborty, M., Sarkar, S., Mukherjee, A., Shamsudduha, M., Ahmed, K. M., Bhattacharya, A., & Mitra, A. (2020). Modeling regional-scale groundwater arsenic hazard in the transboundary Ganges River Delta, India and Bangladesh: Infusing physically-based model with machine learning. *Science of The Total Environment, 748*, 141107.

Chen, C., He, W., Zhou, H., Xue, Y., & Zhu, M. (2020). A comparative study among machine learning and numerical models for simulating groundwater dynamics in the Heihe River Basin, northwestern China. *Scientific Reports, 10*(1), 1–13.

Chen, W., Li, Y., Tsangaratos, P., Shahabi, H., Ilia, I., Xue, W., & Bian, H. (2020). Groundwater spring potential mapping using artificial intelligence approach based on kernel logistic regression, random forest, and alternating decision tree models. *Applied Sciences, 10*(2), 425.

Chen, W., Pradhan, B., Li, S., Shahabi, H., Rizeei, H. M., Hou, E., & Wang, S. (2019). Novel hybrid integration approach of bagging-based fisher's linear discriminant function for groundwater potential analysis. *Natural Resources Research, 28*(4), 1239–1258.

Chen, W., Tsangaratos, P., Ilia, I., Duan, Z., & Chen, X. (2019). Groundwater spring potential mapping using population-based evolutionary algorithms and data mining methods. *Science of the Total Environment, 684*, 31–49.

Chen, W., Zhao, X., Tsangaratos, P., Shahabi, H., Ilia, I., Xue, W., Wang, X., & Ahmad, B. B. (2020). Evaluating the usage of tree-based ensemble methods in groundwater spring potential mapping. *Journal of Hydrology, 583*, 124602.

Constable, S. C., Parker, R. L., & Constable, C. G. (1987). Occam's inversion: A practical algorithm for generating smooth models from electromagnetic sounding data. *Geophysics, 52*(3), 289–300.

DeSimone, L. A., Pope, J. P., & Ransom, K. M. (2020). Machine-learning models to map pH and redox conditions in groundwater in a layered aquifer system, Northern Atlantic Coastal Plain, eastern USA. *Journal of Hydrology: Regional Studies, 30*, 100697.

Díaz-Alcaide, S., & Martínez-Santos, P. (2019). Mapping fecal pollution in rural groundwater supplies by means of artificial intelligence classifiers. *Journal of Hydrology, 577*, 124006.

Ebrahimi, H., & Rajaee, T. (2017). Simulation of groundwater level variations using wavelet combined with neural network, linear regression and support vector machine. *Global and Planetary Change, 148*, 181−191.

Egbueri, J. C. (2020). Groundwater quality assessment using pollution index of groundwater (PIG), ecological risk index (ERI) and hierarchical cluster analysis (HCA): a case study. *Groundwater for Sustainable Development, 10*, 100292.

El-Hames, A. S., Hannachi, A., Al-Ahmadi, M., & Al-Amri, N. (2013). Groundwater quality zonation assessment using GIS, EOFs and hierarchical clustering. *Water Resources Management, 27*(7), 2465−2481.

El-Rawy, M., Ismail, E., & Abdalla, O. (2019). Assessment of groundwater quality using GIS, hydrogeochemistry, and factor statistical analysis in Qena Governorate, Egypt. *Desalination and Water Treatment, 162*, 14−29.

Everest, T., & Özcan, H. (2019). Applying multivariate statistics for identification of groundwater resources and qualities in NW Turkey. *Environmental Monitoring and Assessment, 191*(2), 47.

Friedel, M. J., Wilson, S. R., Close, M. E., Buscema, M., Abraham, P., & Banasiak, L. (2020). Comparison of four learning-based methods for predicting groundwater redox status. *Journal of Hydrology, 580*, 124200.

Gibbons, R. D., Bhaumik, D., & Aryal, S. (2009). *Statistical methods for groundwater monitoring* (Vol. 2). New York: Wiley.

Golkarian, A., Naghibi, S. A., Kalantar, B., & Pradhan, B. (2018). Groundwater potential mapping using C5. 0, random forest, and multivariate adaptive regression spline models in GIS. *Environmental Monitoring and Assessment, 190*(3), 1−16.

Guo, X., Zuo, R., Shan, D., Cao, Y., Wang, J., Teng, Y., Fu, Q., & Zheng, B. (2017). Source apportionment of pollution in groundwater source area using factor analysis and positive matrix factorization methods. *Human and Ecological Risk Assessment, 23*(6), 1417−1436.

Guzman, S. M., Paz, J. O., Tagert, M. L. M., & Mercer, A. E. (2019). Evaluation of seasonally classified inputs for the prediction of daily groundwater levels: NARX networks vs support vector machines. *Environmental Modeling & Assessment, 24*(2), 223−234.

Hofmann, J., Watson, V., & Scharaw, B. (2015). Groundwater quality under stress: Contaminants in the Kharaa River basin (Mongolia). *Environmental Earth Sciences, 73*(2), 629−648.

Hossain, T. M., Watada, J., Jian, Z., Sakai, H., Rahman, S., & Aziz, I. A. (2020). Missing well log data handling in complex lithology prediction: An NIS apriori algorithm approach. *Int. J. of Innovative Computing, Information and Control, 16*(3), 1077−1091.

Hu, S., Luo, T., & Jing, C. (2013). Principal component analysis of fluoride geochemistry of groundwater in Shanxi and Inner Mongolia, China. *Journal of Geochemical Exploration, 135*, 124−129.

Huang, F., Huang, J., Jiang, S. H., & Zhou, C. (2017). Prediction of groundwater levels using evidence of chaos and support vector machine. *Journal of Hydroinformatics, 19*(4), 586−606.

Huang, X., Gao, L., Crosbie, R. S., Zhang, N., Fu, G., & Doble, R. (2019). Groundwater recharge prediction using linear regression, multi-layer perception network, and deep learning. *Water, 11*(9), 1879.

Isazadeh, M., Biazar, S. M., & Ashrafzadeh, A. (2017). Support vector machines and feed-forward neural networks for spatial modeling of groundwater qualitative parameters. *Environmental Earth Sciences, 76*(17), 1−14.

Javadi, S., Hashemy, S. M., Mohammadi, K., Howard, K. W. F., & Neshat, A. (2017). Classification of aquifer vulnerability using K-means cluster analysis. *Journal of Hydrology, 549*, 27−37.

Kadam, A. K., Wagh, V. M., Muley, A. A., Umrikar, B. N., & Sankhua, R. N. (2019). Prediction of water quality index using artificial neural network and multiple linear regression modelling approach in Shivganga River basin, India. *Modeling Earth Systems and Environment, 5*(3), 951−962.

Kalantar, B., Al-Najjar, H. A., Pradhan, B., Saeidi, V., Halin, A. A., Ueda, N., & Naghibi, S. A. (2019). Optimized conditioning factors using machine learning techniques for groundwater potential mapping. *Water, 11*(9), 1909.

Kenda, K., Čerin, M., Bogataj, M., Senožetnik, M., Klemen, K., Pergar, P., ... & Mladenić, D. (2018). Groundwater modeling with machine learning techniques: Ljubljana polje aquifer. In Multidisciplinary Digital Publishing Institute Proceedings (Vol. 2, No. 11, p. 697).

Khosravi, K., Barzegar, R., Miraki, S., Adamowski, J., Daggupati, P., Alizadeh, M. R., Pham, B. T., & Alami, M. T. (2020). Stochastic modeling of groundwater fluoride contamination: Introducing lazy learners. *Groundwater, 58*(5), 723−734.

Kim, K. H., Yun, S. T., Park, S. S., Joo, Y., & Kim, T. S. (2014). Model-based clustering of hydrochemical data to demarcate natural vs human impacts on bedrock groundwater quality in rural areas, South Korea. *Journal of Hydrology, 519*, 626−636.

Knoll, L., Breuer, L., & Bach, M. (2019). Large scale prediction of groundwater nitrate concentrations from spatial data using machine learning. *Science of the Total Environment, 668*, 1317−1327.

Knoll, L., Breuer, L., & Bach, M. (2020). Nation-wide estimation of groundwater redox conditions and nitrate concentrations through machine learning. *Environmental Research Letters, 15*(6), 064004.

Kordestani, M. D., Naghibi, S. A., Hashemi, H., Ahmadi, K., Kalantar, B., & Pradhan, B. (2019). Groundwater potential mapping using a novel data-mining ensemble model. *Hydrogeology Journal, 27*(1), 211−224.

Lahjouj, A., El Hmaidi, A., Bouhafa, K., & Boufala, M. H. (2020). Mapping specific groundwater vulnerability to nitrate using random forest: Case of Sais basin, Morocco. *Modeling Earth Systems and Environment, 6*(3), 1451−1466.

Lal, A., & Datta, B. (2018). Development and implementation of support vector machine regression surrogate models for predicting groundwater pumping-induced saltwater intrusion into coastal aquifers. *Water Resources Management, 32*(7), 2405−2419.

Lee, S., Hyun, Y., Lee, S., & Lee, M. J. (2020). Groundwater potential mapping using remote sensing and GIS-based machine learning techniques. *Remote Sensing, 12*(7), 1200.

Maheswaran, R., & Khosa, R. (2013). Long term forecasting of groundwater levels with evidence of non-stationary and nonlinear characteristics. *Computers & Geosciences, 52*, 422−436.

Mair, A., & El-Kadi, A. I. (2013). Logistic regression modeling to assess groundwater vulnerability to contamination in Hawaii, USA. *Journal of Contaminant Hydrology, 153*, 1−23.

Malakar, P., Sarkar, S., Mukherjee, A., Bhanja, S., & Sun, A. Y. (2021). *Use of machine learning and deep learning methods in groundwater. Global groundwater* (pp. 545−557). Elsevier.

Malekzadeh, M., Kardar, S., & Shabanlou, S. (2019). Simulation of groundwater level using MODFLOW, extreme learning machine and Wavelet-Extreme Learning Machine models. *Groundwater for Sustainable Development, 9*, 100279.

Marín Celestino, A. E., Martínez Cruz, D. A., Otazo Sánchez, E. M., Gavi Reyes, F., & Vásquez Soto, D. (2018). Groundwater quality assessment: An improved approach to k-means clustering, principal component analysis and spatial analysis: A case study. *Water, 10*(4), 437.

Martínez-Santos, P., & Renard, P. (2020). Mapping groundwater potential through an ensemble of big data methods. *Groundwater, 58*(4), 583–597.

Messier, K. P., Wheeler, D. C., Flory, A. R., Jones, R. R., Patel, D., Nolan, B. T., & Ward, M. H. (2019). Modeling groundwater nitrate exposure in private wells of North Carolina for the Agricultural Health Study. *Science of the Total Environment, 655*, 512–519.

Miraki, S., Zanganeh, S. H., Chapi, K., Singh, V. P., Shirzadi, A., Shahabi, H., & Pham, B. T. (2019). Mapping groundwater potential using a novel hybrid intelligence approach. *Water resources Management, 33*(1), 281–302.

Mosavi, A., Hosseini, F. S., Choubin, B., Goodarzi, M., Dineva, A. A., & Sardooi, E. R. (2021). Ensemble boosting and bagging based machine learning models for groundwater potential prediction. *Water Resources Management, 35*(1), 23–37.

Motevalli, A., Naghibi, S. A., Hashemi, H., Berndtsson, R., Pradhan, B., & Gholami, V. (2019). Inverse method using boosted regression tree and k-nearest neighbor to quantify effects of point and non-point source nitrate pollution in groundwater. *Journal of Cleaner Production, 228*, 1248–1263.

Mueller, J. P., & Massaron, L. (2016). *Machine learning for dummies.* John Wiley & Sons.

Naghibi, S. A., & Dashtpagerdi, M. M. (2017). Evaluation of four supervised learning methods for groundwater spring potential mapping in Khalkhal region (Iran) using GIS-based features. *Hydrogeology Journal, 25*(1), 169–189.

Naghibi, S. A., & Pourghasemi, H. R. (2015). A comparative assessment between three machine learning models and their performance comparison by bivariate and multivariate statistical methods in groundwater potential mapping. *Water Resources Management, 29*(14), 5217–5236.

Naghibi, S. A., Ahmadi, K., & Daneshi, A. (2017). Application of support vector machine, random forest, and genetic algorithm optimized random forest models in groundwater potential mapping. *Water Resources Management, 31*(9), 2761–2775.

Naghibi, S. A., Pourghasemi, H. R., & Abbaspour, K. (2018). A comparison between ten advanced and soft computing models for groundwater qanat potential assessment in Iran using R and GIS. *Theoretical and Applied Climatology, 131*(3), 967–984.

Naghibi, S. A., Pourghasemi, H. R., & Dixon, B. (2016). GIS-based groundwater potential mapping using boosted regression tree, classification and regression tree, and random forest machine learning models in Iran. *Environmental Monitoring and Assessment, 188*(1), 1–27.

Naranjo-Fernández, N., Guardiola-Albert, C., Aguilera, H., Serrano-Hidalgo, C., & Montero-González, E. (2020). Clustering groundwater level time series of the exploited Almonte-Marismas aquifer in Southwest Spain. *Water, 12*(4), 1063.

Nguyen, P. T., Ha, D. H., Avand, M., Jaafari, A., Nguyen, H. D., Al-Ansari, N., Phong, T. V., Sharma, R., Kumar, R., Le, H. V., & Ho, L. S. (2020). Soft computing ensemble models based on logistic regression for groundwater potential mapping. *Applied Sciences, 10*(7), 2469.

Nguyen, P. T., Ha, D. H., Nguyen, H. D., Van Phong, T., Trinh, P. T., Al-Ansari, N., Le, H. V., Pham, B. T., Ho, L. S., & Prakash, I. (2020). Improvement of credal decision trees using ensemble frameworks for groundwater potential modeling. *Sustainability, 12*(7), 2622.

Nolan, B. T., Fienen, M. N., & Lorenz, D. L. (2015). A statistical learning framework for groundwater nitrate models of the Central Valley, California, USA. *Journal of Hydrology, 531*, 902–911.

Norouz, H., Asghari Moghaddam, A., & Nadiri, A. (2016). Determining vulnerable areas of Malekan plain Aquifer for Nitrate, Using random forest method. *Journal of Environmental Studies, 41*(4), 923–942.

Ouedraogo, I., Defourny, P., & Vanclooster, M. (2019). Application of random forest regression and comparison of its performance to multiple linear regression in modeling groundwater nitrate concentration at the African continent scale. *Hydrogeology Journal, 27*(3), 1081–1098.

Pacheco Castro, R., Pacheco Ávila, J., Ye, M., & Cabrera Sansores, A. (2018). Groundwater quality: Analysis of its temporal and spatial variability in a karst aquifer. *Groundwater*, *56*(1), 62–72.

Pal, S., Kundu, S., & Mahato, S. (2020). Groundwater potential zones for sustainable management plans in a river basin of India and Bangladesh. *Journal of Cleaner Production*, *257*, 120311.

Park, S., Hamm, S. Y., Jeon, H. T., & Kim, J. (2017). Evaluation of logistic regression and multivariate adaptive regression spline models for groundwater potential mapping using R and GIS. *Sustainability*, *9*(7), 1157.

Parveen, S., Anastasia, A., & Barkha, C. (2020). Estimation of the recharging rate of groundwater using random forest technique. *Applied Water Science*, *10*(7).

Pham, B. T., Jaafari, A., Prakash, I., Singh, S. K., Quoc, N. K., & Bui, D. T. (2019). Hybrid computational intelligence models for groundwater potential mapping. *Catena*, *182*, 104101.

Podgorski, J., & Berg, M. (2020). Global threat of arsenic in groundwater. *Science (New York, N.Y.)*, *368*(6493), 845–850.

Podgorski, J. E., Labhasetwar, P., Saha, D., & Berg, M. (2018). Prediction modeling and mapping of groundwater fluoride contamination throughout India. *Environmental Science & Technology*, *52*(17), 9889–9898.

Rahbar, A., Vadiati, M., Talkhabi, M., Nadiri, A. A., Nakhaei, M., & Rahimian, M. (2020). A hydrogeochemical analysis of groundwater using hierarchical clustering analysis and fuzzy C-mean clustering methods in Arak plain, Iran. *Environmental Earth Sciences*, *79*(13), 1–17.

Rahman, A. S., Hosono, T., Quilty, J. M., Das, J., & Basak, A. (2020). Multiscale groundwater level forecasting: Coupling new machine learning approaches with wavelet transforms. *Advances in Water Resources*, *141*, 103595.

Rahmati, O., Choubin, B., Fathabadi, A., Coulon, F., Soltani, E., Shahabi, H., Mollaefar, E., Tiefenbacher, J., Cipullo, S., Ahmad, B. B., & Bui, D. T. (2019). Predicting uncertainty of machine learning models for modelling nitrate pollution of groundwater using quantile regression and uneec methods. *Science of the Total Environment*, *688*, 855–866.

Rahmati, O., Moghaddam, D. D., Moosavi, V., Kalantari, Z., Samadi, M., Lee, S., & Tien Bui, D. (2019). An automated python language-based tool for creating absence samples in groundwater potential mapping. *Remote Sensing*, *11*(11), 1375.

Rajaee, T., Ebrahimi, H., & Nourani, V. (2019). A review of the artificial intelligence methods in groundwater level modeling. *Journal of Hydrology*, *572*, 336–351.

Ranjithan, S., Eheart, J. W., & Garrett, J. H., Jr (1993). Neural network-based screening for groundwater reclamation under uncertainty. *Water Resources Research*, *29*(3), 563–574.

Rao, N. S. (2014). Spatial control of groundwater contamination, using principal component analysis. *Journal of Earth System Science*, *123*(4), 715–728.

Ravikumar, P., & Somashekar, R. K. (2017). Principal component analysis and hydrochemical facies characterization to evaluate groundwater quality in Varahi river basin, Karnataka state, India. *Applied Water Science*, *7*(2), 745–755.

Rizeei, H. M., Azeez, O. S., Pradhan, B., & Khamees, H. H. (2018). Assessment of groundwater nitrate contamination hazard in a semi-arid region by using integrated parametric IPNOA and data-driven logistic regression models. *Environmental Monitoring and Assessment*, *190*(11), 1–17.

Rizeei, H. M., Pradhan, B., Saharkhiz, M. A., & Lee, S. (2019). Groundwater aquifer potential modeling using an ensemble multi-adoptive boosting logistic regression technique. *Journal of Hydrology*, *579*, 124172.

Rodriguez-Galiano, V. F., Luque-Espinar, J. A., Chica-Olmo, M., & Mendes, M. P. (2018). Feature selection approaches for predictive modelling of groundwater nitrate pollution: An evaluation of filters, embedded and wrapper methods. *Science of the Total Environment, 624*, 661−672.

Rogers, L. L., Dowla, F. U., & Johnson, V. M. (1995). Optimal field-scale groundwater remediation using neural networks and the genetic algorithm. *Environmental Science & Technology, 29*(5), 1145−1155.

Sahoo, M., Kasot, A., Dhar, A., & Kar, A. (2018). On predictability of groundwater level in shallow wells using satellite observations. *Water Resources Management, 32*(4), 1225−1244.

Sahoo, S., & Jha, M. K. (2015). On the statistical forecasting of groundwater levels in unconfined aquifer systems. *Environmental Earth Sciences, 73*(7), 3119−3136.

Sahour, H., Gholami, V., & Vazifedan, M. (2020). A comparative analysis of statistical and machine learning techniques for mapping the spatial distribution of groundwater salinity in a coastal aquifer. *Journal of Hydrology, 591*, 125321.

Sajedi-Hosseini, F., Malekian, A., Choubin, B., Rahmati, O., Cipullo, S., Coulon, F., & Pradhan, B. (2018). A novel machine learning-based approach for the risk assessment of nitrate groundwater contamination. *Science of the Total Environment, 644*, 954−962.

Sameen, M. I., Pradhan, B., & Lee, S. (2019). Self-learning random forests model for mapping groundwater yield in data-scarce areas. *Natural Resources Research, 28*(3), 757−775.

Seyoum, W. M., Kwon, D., & Milewski, A. M. (2019). Downscaling GRACE TWSA data into high-resolution groundwater level anomaly using machine learning-based models in a glacial aquifer system. *Remote Sensing, 11*(7), 824.

Shen, C. (2018). A transdisciplinary review of deep learning research and its relevance for water resources scientists. *Water Resources Research, 54*(11), 8558−8593.

Shiri, J., Kisi, O., Yoon, H., Lee, K. K., & Nazemi, A. H. (2013). Predicting groundwater level fluctuations with meteorological effect implications—A comparative study among soft computing techniques. *Computers & Geosciences, 56*, 32−44.

Sihag, P., Angelaki, A., & Chaplot, B. (2020). Estimation of the recharging rate of groundwater using random forest technique. *Applied Water Science, 10*(7), 1−11.

Singh, K. P., Gupta, S., & Mohan, D. (2014). Evaluating influences of seasonal variations and anthropogenic activities on alluvial groundwater hydrochemistry using ensemble learning approaches. *Journal of Hydrology, 511*, 254−266.

Suryanarayana, C., Sudheer, C., Mahammood, V., & Panigrahi, B. K. (2014). An integrated wavelet-support vector machine for groundwater level prediction in Visakhapatnam, India. *Neurocomputing, 145*, 324−335.

Tahmasebi, P., Kamrava, S., Bai, T., & Sahimi, M. (2020). Machine learning in geo-and environmental sciences: From small to large scale. *Advances in Water Resources*, 103619.

Tan, Z., Yang, Q., & Zheng, Y. (2020). Machine learning models of groundwater arsenic spatial distribution in Bangladesh: Influence of holocene sediment depositional history. *Environmental Science & Technology, 54*(15), 9454−9463.

Tang, Y., Zang, C., Wei, Y., & Jiang, M. (2019). Data-driven modeling of groundwater level with Least-Square support vector machine and spatial−temporal analysis. *Geotechnical and Geological Engineering, 37*(3), 1661−1670.

Tien Bui, D., Shirzadi, A., Chapi, K., Shahabi, H., Pradhan, B., Pham, B. T., Singh, V. P., Chen, W., Khosravi, K., Bin Ahmad, B., & Lee, S. (2019). A hybrid computational intelligence approach to groundwater spring potential mapping. *Water, 11*(10), 2013.

Upton, K. A., & Jackson, C. R. (2011). Simulation of the spatio-temporal extent of groundwater flooding using statistical methods of hydrograph classification and lumped parameter models. *Hydrological Processes, 25*(12), 1949−1963.

Wei, Z. L., Lü, Q., Sun, H. Y., & Shang, Y. Q. (2019). Estimating the rainfall threshold of a deep-seated landslide by integrating models for predicting the groundwater level and stability analysis of the slope. *Engineering Geology, 253*, 14−26.

Wagner, B. J. (1995). Recent advances in simulation-optimization groundwater management modeling. *Reviews of Geophysics, 33*(S2), 1021−1028.

Willard, J., Jia, X., Xu, S., Steinbach, M., & Kumar, V. (2020). Integrating physics-based modeling with machine learning: A survey. arXiv preprint arXiv:2003.04919.

Xu, T., Valocchi, A. J., Choi, J., & Amir, E. (2014). Use of machine learning methods to reduce predictive error of groundwater models. *Groundwater, 52*(3), 448−460.

Yadav, B., Ch, S., Mathur, S., & Adamowski, J. (2017). Assessing the suitability of extreme learning machines (ELM) for groundwater level prediction. *Journal of Water and Land Development, 32*(1), 103−112.

Yadav, B., Gupta, P. K., Patidar, N., & Himanshu, S. K. (2020). Ensemble modelling framework for groundwater level prediction in urban areas of India. *Science of the Total Environment, 712*, 135539.

Yang, J., Griffiths, J., & Zammit, C. (2019). National classification of surface−groundwater interaction using random forest machine learning technique. *River Research and Applications, 35*(7), 932−943.

Yazdanpanah, N. (2016). Spatiotemporal mapping of groundwater quality for irrigation using geostatistical analysis combined with a linear regression method. *Modeling Earth Systems and Environment, 2*(1), 18.

Yoon, H., Hyun, Y., Ha, K., Lee, K. K., & Kim, G. B. (2016). A method to improve the stability and accuracy of ANN-and SVM-based time series models for long-term groundwater level predictions. *Computers & Geosciences, 90*, 144−155.

Yousefi, S., Sadhasivam, N., Pourghasemi, H. R., Nazarlou, H. G., Golkar, F., Tavangar, S., & Santosh, M. (2020). Groundwater spring potential assessment using new ensemble data mining techniques. *Measurement, 157*, 107652.

Zhou, T., Wang, F., & Yang, Z. (2017). Comparative analysis of ANN and SVM models combined with wavelet preprocess for groundwater depth prediction. *Water, 9*(10), 781.

Zohdy, A. A. (1989). A new method for the automatic interpretation of Schlumberger and Wenner sounding curves. *Geophysics, 54*(2), 245−253.

Microbial consortium for bioremediation of polycyclic aromatic hydrocarbons polluted sites

Pankaj Kumar Gupta

Wetland Hydrology Research Laboratory, Faculty of Environment, University of Waterloo, Waterloo, ON, Canada

6.1 Introduction

One of the fastest emerging needs nationally and internationally in the 21st century is the need for clean water. The available freshwater reserves are being fast depleted globally due to rapid urbanization and population increase. On the other hand, due to extensive industrialization, novel contaminants are being introduced into the environmental system. Environmental pollution has become a global concern due to rapid growth of industrialization, urbanization, and modern development under climate change conditions. Technological innovations in industries have given rise to new products and new pollutants in abundant level, which are above the self-cleaning capacity in the environment.

A large number of industrial chemicals such as total petroleum hydrocarbons (TPH), polychlorinated biphenyl, polycyclic aromatic hydrocarbons (PAHs), heavy metals, and pesticides are released in the environment without proper treatment. Therefore it is reported that the most common organic contaminants in groundwater include aromatic hydrocarbons, pesticides, and chlorinated compounds. These chemicals pose potential threats to human as well ecological health. These pollutants are released by various anthropogenic activities such as accidental spillage, industries, and leaks from underground storage tanks, agricultural practices, and poor waste disposal in landfills (Yadav & Hassanizadeh, 2011). This may cause ultimate sinks for such toxic pollutants into major components of the soil−water system. Therefore the biogeochemical properties of these natural resources have been transformed and resulting in the continuous loss of soil−water functions in sustaining living organisms. Large-scale production, transport, use, and disposal of petroleum have made it as one of the leading contaminants in the subsurface.

Advances in Remediation Techniques for Polluted Soils and Groundwater. DOI: https://doi.org/10.1016/B978-0-12-823830-1.00015-8

Naturally occurring microorganisms play a significant role in the degradation of such pollutants from soil—water resources. Box 6.1 presents some potential microbes generally used to remediate PAHs-polluted sites. These native microbes need more (micro)-nutrients, electron acceptors, and favoring environmental conditions to achieve complete degradation of pollutants. Therefore the modification in polluted sites by providing such essential components stimulates the microbial growth and ultimately the removal of pollutants. The main focus of this chapter is to present a review on microbial degradation of PAHs compounds using different remediation techniques.

6.2   PAHs pollutants: source, toxicity, and metabolic pathways

PAHs are a group of hydrocarbons having fused aromatic rings and are major concern due to their ability to bioaccumulate, which may cause carcinogenic effects on humans. These are compounds having high melting/boiling points along with high lipid solubility but they are known to have low vapor pressure and low water solubility (Ghosal et al., 2016). However, the release of small quantity of such pollutants in the soil—water system significantly degrades the quality of subsurface water (Yadav & Hassanizadeh, 2011). PAHs contaminants include a wide range of industrial compounds such as gasoline, fuel oils, chlorinated and fluorinated hydrocarbons, creosote, and transformer oils. Many PAHs, such as gasoline, creosote, and waste liquids, are multicomponent, with the constituents having a broad distribution of chemical properties (Geller & Hunt, 1993). According to the Toxic Release Inventory (EPA) report (2005), the oil refining industry is one of the 10 major sources releasing/emitting toxic chemicals into the environment. In the past few years, PAHs pollution has become one of the most serious global concerns due to its toxicity to microorganisms as well as to higher

---

BOX 6.1 Potential microorganisms used to remediate PAHs-polluted sites (Zhou et al., 2008, Bautista et al., 2009).

Bacteria: *Sphingomonas paucimobilis*, *Pseudomonas*, *Agrobacterium*, *Bacillus*, *Burkholderia*, *Sphingomonas Rhodococcus sp.*, *Mycobacterium*, mixed culture of *Pseudomonas*, *Flavobacterium species*, *Pseudomonas aeruginosa*, *Mycobacterium flavescens*, and *Rhodococcus*

Fungi: *Cunninghamella echinulata var. elegans*, *Ligninolytic fungi*, *Aspergillus*, *Trichocladium canadense*, *Fusarium*

oxysporum y, *T. canadense*, *Aspergillus sp.*, *Verticillium*, and *Achremonium*.

Algae: *Oscillatoria*, *Agmenellum*, *Selenastrum capricornutum* (freshwater green algae), *Pseudomonas migulae*, and *Sphingomonas yanoikuyae*.

forms of life including humans (Varjani & Upasani, 2016c). Four types of toxicities have been reported in the literature, that is, genotoxicity, carcinogenicity, immunotoxicity, teratogenicity. PAHs cause the genotoxic which have been confirmed in vitro tests for rodents and human cells (Thomas et al., 2002). While PAHs bind with cellular protein and DNA, which may cause development of tumors and cancer (Ghosal et al., 2016). Similarly, the intake of PAHs contaminated water/food may cause toxic effects to the embryo or fetus (Perera et al., 2012).

Natural attenuation of PAHs followed two metabolic pathways via aerobic and anaerobic metabolic biodegradation. Aerobic biodegradation is faster than anaerobic biodegradation of PAHs under ideal environmental conditions (Gupta et al., 2013). In aerobic degradation, microbial oxidative process, activation and incorporation of oxygen is the enzymatic key reaction catalyzed by oxygenases and peroxidases (Abbasian et al., 2015). Monooxygenases transfer one oxygen atom to the substrate and reduce the other oxygen atom to water (Das & Chandran, 2011). The low water solubility and high sorption capacity of PAHs are often found to greatly influence biodegradation (Abbasian et al., 2015; Van Hamme et al., 2003). Peripheral degradation pathways convert organic compounds step by step into intermediates of central intermediary metabolism, such as, the tricarboxylic acid cycle. Microbial degrading hydrocarbons need oxygen at two metabolic sites, during the initial attack of the substrate and at the end of the respiratory chain. Initial substrate conditions play a significant role in the degradations of PAHs. Gupta et al. (2013) investigated the impact of different substrate concentrations on the degradation of toluene. He found that the biodegradation rate increases with increasing initial substrate concentration up to 100 ppm and remains optimal till 150 ppm and then decreases with high concentrations.

This fortifies that the low initial substrate concentration may cause less carbon sources and high concentration may cause toxic effects to native microbes, both conditions may lead to decreased biodegradation rate of PAHs. Table 6.1 highlighted studies performed reported in the literature to investigate the PAHs biodegradation under varying subsurface conditions.

6.3 Bioremediation of PAHs

Technology to treat polluted water has advanced as a result of the contribution from growing concern over wastewater discharge from petroleum refineries in recent decades (Allen, 2008; Shpiner et al., 2009). In the need of addressing this issue, a number of conventional methods have developed over the years such as reverse osmosis (Mant et al., 2005; Murray-Gulde

Table 6.1 Laboratory and numerical investigation on biodegradation of polycyclic aromatic hydrocarbon (PAH) compounds under varying subsurface conditions.

Types	Reference	Contaminant source	Concentration level	Temp.	Water saturation	Experimental setups	Governing processes	Highlight
L & N	Da Silva and Alvarez (2002)	Methyl tertiary-butyl ether (MTBE) and benzene, toluene, ethylbenzene and xylene (BTEX)	BTEX:1 ppm MTBE:25 ppm	C (22°C)	F	Glass: 5 cm diameter and 20 cm length	Retardation; biodegradation (impact of ethanol)	MTBE did not degrade in the biologically active column and did not affect the degradation of BTEX, while ethanol significantly affect biodegradation.
L & N	Sovik et al. (2002)	Toluene and o-xylene	Mix: 50 ppm of both	C (12°C)	14%—15% at top	Glass: 0.5 m length and 10 cm diameters	Volatilization and biodegradation	First-order degradation coefficient was estimated in the range of 0.10—0.11 d^{-1} for o-xylene and 0.19—0.21 d^{-1} for toluene.
L & N	Zheng et al. (2002)	Toluene and 1,2,4-trimethylbenzene	Pure phase	C (8°C—10°C)	F	Stainless steel: 25 cm length and 2.5 cm diameter	Biodegradation rates and sorption	Zero-order biodegradation rates due to Fe(III) reduction.
L & N	Davis et al. (2003)	Diesel	Dissolved 337 ppm	Room Temperature (RT)	Top unsaturated	Glass: 10 cm length Aluminum: 50 cm Plexiglas: 120 cm	Biodegradation rates	An average biodegradation rate was 0.20 mg (kg d^{-1}.
L & N	Ranck et al. (2005)	Toluene	Mix BTEX: 15 ppm	RT	F	Glass: 17.8 cm diameter and 122 cm length	Biodegradation (using modified zeolites)	High removal efficiency of BTEX by modified zeolites.
L	Tindall et al. (2005)	Toluene	Pure phase	RT	70%	Polyvinyl chloride (PVC): 30 cm diameter and 150 cm length	Biodegradation (using nitrate and H_2O_2)	Nutrient-enhanced columns degraded significantly more toluene than the control columns.
L	Vogt et al. (2007)	Benzene	Dissolved	V (12°C—20°C)	F	Stainless: 6 m length and 25 cm diameter	Biodegradation under sulfate-reducing conditions	Estimated biodegradation rate was 8—36 $\mu M\ d^{-1}$.
L	Nerantzis and Dyer (2010)	BTEX	Mixed pure phase	C (21°C)	0%—12% from top	PVC: 50 cm length and	Vapor transport and	A thin soil layer of high moisture content can significantly obstruct the

	Reference	Contaminant	Concentration	Temperature	Moisture	Column dimensions	Process	Findings
L	Filho et al. (2013)	Gasoline	Mixed with ethanol	C (19.9 C ± 2.3 C)	F	10 cm diameter Glass: 110 mm high	biodegradation rates Advection; volatilization; biodegradation	gaseous transport of BTEX vapor. Ethylbenzene and toluene showed the highest volatilizations in the gasoline-ethanol column.
L & N	Picone et al. (2013)	Dissolved Toluene	12–21 ppm	C (20ºC)	9%–27%	Glass: 3.5 cm diameter and 19.5 cm length	Gas advection and biodegradation	Maximum mass removal rates were 0.69 ± 0.09, 0.73 ± 0.04, and 1.06 ± 0.22 mg h^{-1} at 27%, 14%, and 9% water saturation, respectively.
L	Zhao et al. (2015)	BTEX	B: 8.7 ppm; T: 5.44 ppm; E:0.37 ppm and X:0.38 ppm	RT	F	Glass: 98 cm length and 14 cm diameter	Biodegradation rates under nitrate, sulfate, and Fe (II) reducing conditions	Degradation of BTEX with four electron acceptors was in order as: nitrate > sulfate > chelated Fe(III) > DO.
L & N	Khan et al. (2016)	Toluene	Pure phase	RT	12% of pore space	Chromoflax glass column: 35 cm length and 4.1 cm diameter	Biodegradation of toluene vapors	High biodegradation rates of toluene were observed within few centimeters pathway of vapor in column.
L	Yang et al. (2017)	Toluene	Pure phase	RT	Unsaturated	Glass: 5 cm length and 5 cm diameter	Natural attenuation under fluctuating groundwater conditions	Toluene degradation is significantly affected by groundwater table fluctuation.

et al., 2003), membrane filtration (ultra/nano/microfiltration) (Allen, 2008; Ravanchi et al., 2009; Saien, 2010), etc. However, all these methods carry an array of unignorable disadvantages such as incomplete degradation (Das & Chandran, 2011), formation of toxic by-products (Baskar, 2011; Ojumu et al., 2005; Wuyep et al., 2007) along with the operational downside which limits their application in reality.

Bioremediation has been used to eliminate PAHs without affecting further environmental damage (Yadav & Hassanizadeh, 2011). Although the natural bioremediation is safe and less disruptive, it takes considerably long time to restore the contaminated site (Rifai et al., 2000). Therefore engineered bioremediation techniques are gaining popularity due to their faster-remediating rates particularly for sites facing extreme environmental conditions. The engineered bioremediation is achieved by maintaining favorable environmental conditions such as soil moisture level, pH, salinity, and or temperature of the target site (Yadav & Hassanizadeh, 2011). This can be performed by applying surface recharge, providing oxygen/ nutrients, using plants at the contaminated sites (Dzantor, 2007) which help in maintaining favorable bioremediation conditions and in enhancing metabolism of the microorganisms (Basu et al., 2015). This advance bioremediation influences the microbial activities and their neighboring environmental condition for accelerating the practice of biodegradation and is categorized as bioaugmentation and biostimulation. Bioaugmentation is a microorganism seeding practice for cultivating the volume of a PAH degrader by adding potential microbial cultures which are grown independently in well-defined conditions.

Biostimulation is enhanced by the adding of nutrients, electron acceptors, oxygen, and other relevant compounds to the polluted sites which enhanced the (co)-metabolic actions of the microflora. Generally, the oxygen releasing compounds such as H_2O_2. MgO_2, O_2, NO_3, SO_4, $Mn(IV)$, and $Fe(III)$ are used to stimulate the PAH-polluted sites. These electron acceptors significantly increased the oxygen level, which help in maintaining the aerobic condition. Alvarez and Vogel (1995) used nitrate as an electron acceptor as well as nutrient to incubate PAH degrader in batch system. The addition of nitrate to soil enhances the denitrification which causes increased oxygen level and significantly degrades the PAHs mass. Similarly, Röling et al. (2002) investigated the impact of nutrient amendment on dynamics of bacterial communities along with the PAHs biodegradation and found that the nutrient amendment significantly increased the bacterial population and improved biodegradation up to 92% of pollutant mass removal. Yadav et al. (2013) conducted a series of microcosm experiments, in which initially the natural biodegradation has been investigated and subsequently added the domestic wastewater as nutrients.

6.4 Plant—microbes interactions

Plant-assisted bioremediation refers to the use of selective plant species for the targeted pollutant to mitigate the toxic effects and removal of pollutant mass from (sub)-surface. This technique used the plant—geochemical interaction to modify the polluted site and also supply (micro)-nutrient, oxygen, etc. into subsurface for better performance of degrader on targeted pollutants (Susarla et al., 2002). PAHs are mostly removed by degradation, rhizoremediation, stabilization, and volatilization, with mineralization being possible when some plants such as *Canna generalis* are used (Yadav et al., 2013). The plant—geochemical interaction enhances the (1) physical and chemical properties of sites, (2) nutrient supply by releasing root exudates (Shimp et al., 1993), (3) aeration by transfer the oxygen, (4) intercepting and retarding the movements of chemicals, (5) the plant enzymatic transformation, and (6) resistant to the vertical and lateral migration of pollutants (Narayanan et al., 1998a, 1998b). Similarly, the plant—microbe interaction increased (1) mineralization in rhizosphere and (2) numbers of degraders and shorter the lag phase until disappearance of the compound. Some key factors to consider when choosing a plant include root system, which may be fibrous or tap subject to the depth of contaminant, toxicity of pollutant to plant, plant survival and its adaptability to prevailing environmental conditions, plant growth rate, and resistance to diseases and pests.

Constructed wetland (CW) works on phytoremediation technique to treat polluted soil—water system at low cost, with low energy consumption, low sludge production, and low maintenance. This economically viable option is nothing but a complex ecosystem with vegetation, hydric soils, microbes, and prevailing flow patterns assisting in the quality improvement of inflowed water. The wetland can be of either surface flow (SF) wetland type or subsurface flow (SSF) type. Considering the favorable climatic conditions, the application of CW in tropical areas is high, and as a matter of fact, these areas also contain 80% of the global oil resources. So, the treatment of large-polluted sites near petroleum refineries/oil wells can be well treated using CW. Such industries, in recent times, are also showing interest in adopting the method as the posttreated water can also be reused and is comparatively less expensive.

The different types of CWs can be used for the removal of PAHs from polluted sites as per the objective. Where SSFCWs are considered as the main treatment systems, tertiary treatment can be achieved through both SF and SSF wetlands. The SSF wetland type can be performed through either the horizontal flow method (HSSF) or the vertical flow method (VSSF). The major difference comes into the picture due to limited oxygen in HSSF type, where favorable conditions are created for organic matter and denitrification processes, however, in VSSF, ammonia removal and chemical oxygen demand (COD) are high. SF systems can filter, absorb, and retain particulate matter, nutrients, and so on from wastewater, and that is why they are good in the treatment of PAH-polluted resources. Horner et al. (2012) showed through a study that free-water surface flow, under prevailing conditions and suitable setup, can effectively reduce heavy metals concentration such as Fe, Mn, Ni, and Zn and along with reduced PAHs concentrations. For biological treatment of petroleum-polluted sites, SSF-engineered wetlands have been found more effective than SF wetlands, most likely due to the higher specific area present in gravel-bed (Davis et al., 2009). Both horizontal and vertical SSF wetlands are effective in lowering petroleum hydrocarbon concentration at a polluted subsurface zone. Forced Aeration Bed system built in either VSSF or HSSF CWs can improve the removal efficiency of BTEX, that is, Benzene, Toluene, Ethylbenzene, and Xylene (major PAHs constituents) in petroleum discharge to up to nondetectable levels. The proof studies were conducted by Wallace (2001) and Davis et al. (2009) in VSSF and HSSF wetland setup, respectively. Results from both the studies suggested that the forced aeration system enhanced the volatilization rate, which is mainly responsible for BTEX removal. A similar study by Ferro et al. (2002) at BP Amoco Refinery in Wyoming (USA) was worked out, which operated in an upward vertical flow mode, and achieved a decrease in benzene concentration from $0.2-0.6\,\mathrm{mg\,L^{-1}}$ (influent) to $0.05\,\mathrm{mg\,L^{-1}}$ (effluent). An engineered wetland system with aeration, constructed by Wallace and Kadlec (2005), was initially built at pilot scale, and later, at full scale of $6000\,\mathrm{m^3\,day^{-1}}$. This too presented alike results where enhanced volatilization and aerobic biodegradation due to additional aeration system provided desirable effluent contaminant concentration. A comparative experiment by Bedessem et al. (2007) in treating refinery-affected groundwater with and without aeration justified the previous arguments. Results showed that a decrease in concentration of petroleum hydrocarbon-diesel range organics was attained up to 77% without, and >95% with aeration, along with the removal of total benzene; toluene; ethylbenzene; and o-, m-, and p-xylenes. An important study that investigated the significance of hydraulic retention time (HRT) on TPH removal was conducted by Al-Baldawi et al (2014). It showed that for effective removal of TPH from water, a longer retention time is required. An evaluation of the performance of HSSF CWs under planted, unplanted, and plant root mat was done by Chen et al. (2012) and found that the rate of elimination of benzene was same for plant root mat and planted systems. There are certain conditions that need to be optimized for hydrocarbon (PAHs) removal through CWs. These conditions are optimum aeration rate, temperature, appropriate HRT, etc. Eke and Scholz (2008) after studying the removal of benzene through 12 vertical-flow microcosm wetlands with varied compositions concluded that the high removal efficiency could be achieved due to volatilization and also due to enhanced biodegradation rates because of high temperature. They, along with Davis et al. (2009), confirmed that aeration improves volatilization and hydrocarbon degradation. Plant/macrophyte selection in SF CWs is crucial as poor plant establishment may result in poor contaminant removal rate (Simi, 2000). Root-microbial interaction plays significant role in the degradation of PAHs in root zone (Gupta and Yadav, 2017). The deep root systems of plant improve the aeration in subsurface, which maintain the oxygen level in deep vadose zone. The root exudates, dead root hair and fine root, serve as an important source of the carbon for microbial growths (Shimp et al., 1993). Also, the root exudates also accelerate the enzyme synthesis of microbial metabolisms.

6.5 Microbes and their consortium to degrade PAHs

6.5.1 Sulfate-reducing bacteria

Microorganism that uses sulfate $\left(SO_4^{2-}\right)$ as terminal electron acceptor for their anaerobic respiration, converting it to hydrogen sulfide (H_2S), is known as sulfate-reducing bacteria (SRB). Although named after their ability to respire using sulfate, sulfur reducers are extremely flexible regarding electron donors and electron acceptors that are used for their growth. They can be found everywhere in the anoxic habitat, where they play a vital role in carbon and sulfur cycle (Muyzer & Stams, 2008). Although they are useful in removing sulfate and heavy metals from waste streams, they cause a nuisance in offshore oil industries due to production of sulfide, which is corrosive and toxic in nature. SRB that are isolated and categorized till now can be divided into four bacterial phyla and two Archaea. The bacterial phyla are the following: (1) *Proteobacteria* (Okoro et al., 2017), (2) *Firmicutes* (Nazina et al., 1988), (3) *Nitrospira*, and (4) *Thermodesulfobacterium*. Phyla are *Euryarchaeota* (Archaeoglobus genus) and *Crenarchaeota* (Thermocladium and Caldivirga genera). They can thrive in extremely varying environmental conditions with wide range of pH, temperature, and salinity (Varjani, 2017). Table 6.2 highlights the sulfate-reducing bacteria capable to reduce the PAHs contaminants from soil−water resources.

6.5.2 Methanogens

Microorganisms that produce methane as their respiratory by-product in hypoxic conditions are categorized as methanogens. They are abundant in wetlands, digestive tract of animals, and ecosystem of petroleum oil reservoirs (Liang et al., 2016). They are coccoid (spherical shaped) or bacilli (rod shaped). Archaea of these microorganism belongs to phyla *Euryarchaeota*, which further divided into five category: *Methanobacteriales, Methanococcales, Methanomicrobiales, Methanosarcinales, and Mehanopyrales* (Madigan et al., 2015). Process by which biogenic methane forms in marine sediments are called as methanogenesis plays a crucial role in anaerobic wastewater treatments. Methanogens can be capable of metabolizing H_2, CO_2, methylamines,

Table 6.2 Potential sulfate-reducing bacteria found in soil—water system to degrade the polycyclic aromatic hydrocarbons.

Microbes category	Name of the microorganisms	Optimal conditions	References
Sulfate-reducing bacteria	Sp. of genus *Marinobacterium, Oceanobacter, Holomonas, Desulfuromonas,* and *Desulfobolus*	Temperature: 30—60 (°C)	Okoro et al. (2017)
	Desulfonauticus autotrophicus	Salinity: 3% (w/v); temperature: 58 (°C)	Mayilraj et al. (2009)
	Desulfotignum toluenicum	Salinity; 1.5% (w/v); temperature: 20—40 (°C)	Ommedal and Torsvik (2007)
	Desulfovermiculus halophilus	Salinity: 10% (w/v); temperature: 25—47 (°C)	Belyakova et al. (2006)
	Desulfovibrio gracillis, Desulfovibrio bastinii	Salinity: 5%—6% (w/v); temperature: 20°C—50°C	Magot et al. (2004)
	Thermodesulfobacterium hydrogeniphilum	Salinity: 3% (w/v); Temperature: 75 (°C)	Jeanthon et al. (2002)
	Desulfotomaculum kuznetsovii	Salinity condition does not favor the growth of such organism. temperature: 50—90 (°C)	Nazina et al. (1988)
	Desulfovibrio gabonensis	Salinity: 5%—6% (w/v); temperature: 15—40 (°C)	Tardy-Jacquenod et al. (1996)
	Desulfovibrio vietnamensis	Salinity: 5% (w/v); temperature: 35 (°C).	Dang et al. (1996)
	Desulfacinum infernum	Temperature: 40—65 (°C). salinity: 1% (w/v)	Rees et al. (1995)
	Thermodesulforhabdus norvegicus	Salinity: 1.6% (w/v); temperature: 44—74 (°C)	Beeder et al. (1995)

acetate, and dimethylsulfides. Depending on the substrate utilized during biodegradation, they are classified into following three main groups: (1) hydrogenotrophic methanogens, (2) methylotrophic methanogens (Cheng et al., 2007), and (3) acetoclastic methanogens (Okoro et al., 2017). Table 6.3 highlights the methanogens capable to reduce the PAHs contaminants from soil—water resources (Table 6.4).

6.5.3 Iron-, nitrate-, and manganese-reducing bacteria

Microorganisms capable of reducing iron, nitrate and manganese from any environment, especially oilfield ecosystem, have proved to be beneficial for degradation of PAHs (Nealsonl & Myres, 1992). Most of such bacteria are capable of using various terminal acceptors, but their unique characteristic is their ability to grow under anaerobic conditions. Microbial sp. of the

Table 6.3 Potential methanogens bacteria found in soil–water system to degrade the polycyclic aromatic hydrocarbons.

Microbes category	Name of the microorganism	Optimal conditions	References
Methanogens	Sp. of genus *Methanosaeta* and *Methanolobus*	Temperature: 30–60 (°C).	Okoro et al. (2017)
	Sp. of genus *Methanothermobacter* and *Methanosaeta*	Temperature: 55 (°C)	Mayumi et al. (2011)
	Methanocalculus halotolerans	Salinity: 5%–6% (w/v); Temperature: 40 (°C).	Ollivier et al. (1998)
	Methanoplanus petrolearius	Salinity: 1%–3% (w/v); Temperature: 20–40 (°C)	Ollivier et al. (1997)
	Methanobacterium thermoalcaliphilum and *Methanobacterium bryantii*	Saline environmental conditions do not support its growth. Temperature: 30–80 (°C)	-
	Methanosarcina siciliae	Salinity: 2.5%–3.5% (w/v); temperature: 20–50 (°C)	Ni and Boone (1991)
	Methanobacterium ivanovii	Temperature: 10–55 (°C)	Belyaev et al. (1986)

Table 6.4 Potential fermentative bacteria found in soil–water system to degrade the polycyclic aromatic hydrocarbons.

Microbes category	Name of the microorganism	Optimal conditions	References
Fermentative bacteria	Sp. of genus *Marinobacterium* and *Halomonas*	Temperature: 30–60 (°C)	Okoro et al. (2017)
	Dethiosulfovibrio Peptidovorans and *Spirochaeta smaragdinae*	Salinity: 3%–5% (w/v); Temperature: 20–40 (°C)	Magot, Fardeau, et al. (1997); Magot, Ravot, et al. (1997)
	Thermotoga hypogea	Salinity does not favor the growth. Temperature: 70–80 (°C)	Fardeau et al. (1997)
	Thermotoga subterranea	Salinity: 1%–2% (w/v); temperature: 75 (°C)	Jeanthon et al. (1995)
	Geotoga subterranea	Salinity: 4% (w/v); temperature: 50–60 (°C)	Davey et al. (1993)
	Geotoga petraea	Salinity; 3% (w/v); temperature: <55 (°C)	Davey et al. (1993)

genera *Deferribacter* and *Geobacillus*. Such microorganisms can use peptone, yeast extract, hydrogen, casamino, and various acids such as lactate and acetate as energy sources in presence of iron, nitrate, and/or manganese (Greene et al., 1997). Table 6.5 highlights the iron, nitrate, and manganese

Table 6.5 Potential iron, nitrate, and manganese bacteria to degrade the polycyclic aromatic hydrocarbons.

Microbes category	Name of the microorganism	Optimal	References
Iron-, nitrate-, and manganese-reducing bacteria	*Thermovibrio ammonificans*	Salinity: 2% (w/v); pH:5.5; temperature: 75°C.	Vetriani et al. (2004)
	Geobacillus	They are thermophilic microaerophiles having ability to degrade alkanes only under aerobic conditions and some could reduce nitrate anaerobically.	Nazina et al. (2001)
	Deferribacter	This organism can use hydrogen, peptone, yeast extract, tryptone, casamino acids, and various acids such as acetate, lactate, and valerate as energy sources in the presence of iron, nitrate, and manganese.	Greene et al. (1997)

bacteria proficient to reduce the PAHs contaminants from soil—water resources.

6.6   Microbial degradation kinetics models

Microbial degradation kinetics is one of the major decision-making factors for the remediation and better management of PAHs-polluted site. The kinetic model helps in selection of different best techniques for polluted sites under hostile environmental conditions. Furthermore, the understanding of microbial degradation kinetics will promote the multimicrobial/electron acceptors strategy (Powell et al., 2014). Therefore the focus of this section is to present the overview of bioremediation kinetics models associated with PAHs-polluted sites.

The relationship between the specific growth rate (μ) of the microbial population and the subtract concentration (S) is called microbial growth kinetics (Kovarova-kovar & Egli, 1998). Microbial growth kinetics meet to mass transfer kinetics and enzymatic kinetics which results as the ultimate biodegradability's of substrates. Generally, the PAHs degradation is expressed as:

$$-\frac{\partial C}{\partial t} = \mu_{\max} C \left(\frac{C_o + X_o - C}{K_s + C} \right) \tag{6.1}$$

where $\mu_{\max}$ is the maximum specific growth rate, C is the pollutant concentration at time t, C_o is the initial concentration and X_o corresponding to contaminate required to produce initial microbial density, K_s is the substrate affinities constant (Yadav & Hassanizadeh, 2011). The growth and degradation phenomenon can be described satisfactorily with four parameters; the two kinetics parameters are μ_{max} and K_s and the other two stoichiometric parameters $Y_{s/x}$ and S_{min}.

Monod's kinetics, in which degradation rates resulting from zero-order to first orders kinetics to target substrate concentration, is broadly used to define biodegradation rate of NAPLs during remediation of soil−water system (Alvarez et al., 1994; Kovarova-kovar & Egli, 1998; Littlegohns & Daugulis, 2008; Trigueros et al., 2010; Yadav & Hassanizadeh, 2011). The equation is written as $-\frac{\partial C}{\partial t} = kmC/(Ks + C)$ where $k_m = \mu_{max}X_o$. Alvarez et al., 1994 calculated Monod's coefficient and K_s for benzene and toluene during aerobic biodegradation in sandy aquifer system. The Monod coefficients were calculated as $k = 8.3$ g benzene/g-cells/day and $K_s = 12.2$ mg L^{-1} for benzene and $k = 9.9$ g benzene g-cells/day and $Ks = 17.4$ mg L^{-1} for toluene.

The zero-order kinetics represents the oversimplification in PAHs degradation and mostly used low concentration degradation (Datta et al., 2014; Yadav & Hassanizadeh, 2011). In zero-order kinetics, the rate of depletion of pollutant is taken as constant under condition is $X_o >> C_o$; and $C_o >> K_s$. In the first-order model necessary condition are $X_o >> C_o$ and $K_s >> C_o$ and model represented as $-\frac{\partial C}{\partial t} = k1C$ where $k_1 = \mu_{max}X_o/K_s$. Gupta et al. (2013) investigated the first-order model for the toluene degradation up to 100 ppm substrate concentration. Similarly, Alvarez et al. (1994) investigated the first-order kinetics model during the biodegradation of NAPLs and calculated the values kinetic parameters K_s is equal to 0.01 for BTEX compounds in sandy aquifer. In logistic model, the half saturation constant is greater than the initial contaminate concentration. The differential form of equation is written as $-\frac{\partial C}{\partial t} = k1C/(Co + Xo - C)$, where $k_1 = X_o/K_s$. Similarly, the logarithmic model has lacked a horizontal asymptote as time become larger. In logarithmic model the rate constant K is equal to the maximum growth rate μ_{max}. These models are relevant to soils having nonlimiting nutrients, the absence of mass transfer limitations, optimal soil moisture content, and constant physicochemical factors (i.e., temperature, salinity, and pH). Out of these factors, some may affect the rate of substrate uptake by microbial assemblage and others may alter the rate of contaminant transport/supply to the microorganisms.

6.7 Conclusion and future recommendations

Effective implementation of bioremediation requires a thorough understanding of the soil−plant−microbes−atmospheric continuum processes which are currently poorly understood and make this technology expensive and inefficient despite the tremendous potential mentioned above. This lack of understanding hinders the efforts of researchers in their quest to develop concurrent treatments from contaminated soils—water system. In this direction, an effort has been made to evaluate the performance of microbial interaction with geo-environment to decontaminate the PAHs-polluted sites. Microorganisms capable to enhance the biodegradation by sulfate-reducing Fe/Mg/nitrate-N utilization as electron acceptors are discussed well in this chapter. Biodegradation kinetics is elaborated next to understand the biodegradation kinetics models. Further, there is urgent need to investigate the following aspect of research:

1. Impact of different level of pollutant load on microbial action and performances under stress conditions.
2. Microbial kinetics under different environmental conditions, especially salinity, temperature, pH, and so on.
3. Impact of cometabolisms and other background pollutants such as arsenic/nitrate on the microbial performances during a bioremediation project.
4. Consideration of subsurface heterogeneity and microbial distributions with space and depth is equally important to understand for better performance of a bioremediation project and management of polluted sites.

Acknowledgments

Author of the chapter is thankful to Dr. Basant Yadav, Director, Remwasol Remediation Technologies Pvt. Ltd. for presubmission review and constructive comment to improve the text and contents.

References

Abbasian, F., Lockington, R., Mallavarapu, M., & Naidu, R. (2015). A comprehensive review of aliphatic hydrocarbon biodegradation by bacteria. *Applied Biochemistry and Biotechnology*, *176*(3), 670−699.

Al-Baldawi, I. A., Abdullah, S. R., Abu Hasan, H., Suja, F., Anuar, N., & Mushrifah, I. (2014). Optimized conditions for phytoremediation of diesel by Scirpus grossus in horizontal subsurface flow constructed wetlands (HSFCWs) using response surface methodology. *Journal of Environmental Management*, *140*, 152−159. Available from https://doi.org/10.1016/j.jenvman.2014.03.007.

Allen, W. E. (2008). Process water treatment in Canada's oil sands industry: II. A review of emerging technologies. *Journal of Environmental Engineering and Science*, *7*, 499−524. Available from https://doi.org/10.1139/S08-020.

Alvarez, P. J. J., Anid, P. J., & Vogel, T. M. (1994). Kinetics of toluene degradation by denitrifying aquifer microorganisms. *Journal of Environmental Engineering*, *120*, 1327−1336.

Alvarez, P. J. J., & Vogel, T. M. (1995). Biodegradation of BTEX and their aerobic metabolise by indigenous microorganisms under nitrate reducing conditions. *Water Sciences Technology*, *3*(1), 15−28.

Baskar, G. (2011). *Studies on application of subsurface flow constructed wetland for wastewater treatment*. Kattankulathur: SRM University, 603 203, Department of Civil Engineering. Chennai, Tamil Nadu, India: SRM University, Kattankulathur - 603 203.

Basu, S., Yadav, B. K., & Mathur, S. (2015). Enhanced bioremediation of BTEX contaminated groundwater in pot-scale wetlands. *Environmental Science and Pollution Research, 22*(24), 20041−20049.

Bautista, L. F., et al. (2009). Effect of different non-ionic surfactants on the biodegradation of PAHs by diverse aerobic bacteria. *International Biodeterioration & Biodegradation, 63*(7), 913−922.

Bedessem, M. E., Ferro, A. M., & Hiegel, T. (2007). Pilot-scale constructed wetlands for petroleum contaminated groundwater. *Water Environment Research, 79*, 581−586. Available from https://doi.org/10.2175/106143006X111943.

Beeder, J., Torsvik, T., & Lien, T. (1995). Thermodesulforhabdus norvegicus gen. nov., sp. nov., a novel thermophilic sulfate-reducing bacterium from oil field water. *Archives of Microbiology, 164*, 331−336.

Belyaev, S. S., Obraztsova, A. Y., Laurinavichus, K. S., & Bezrukova, L. V. (1986). Characteristics of rod-shaped methane-producing bacteria from oil pool and description of Methanobacterium ivanovii. *Microbiology (Reading, England), 55*, 821−826.

Belyakova, E. V., Rozanova, E. P., Borzenkov, I. A., Tourova, T. P., Pusheva, M. A., Lysenko, A. M., & Kolganova, T. V. (2006). The new facultatively chemolithoautotrophic, moderately halophilic, sulfate reducing bacterium Desulfovermiculus halophilus gen. nov., sp. nov., isolated from an oil field. *Microbiology (Reading, England), 75*, 161−171.

Chen, Z., Kuschk, P., Reiche, N., Borsdorf, H., Kästner, M., & Köser, H. (2012). Comparative evaluation of pilot scale horizontal subsurface-flow constructed wetlands and plant root mats for treating groundwater contaminated with benzene and MTBE. *Journal of Hazardous Materials, 209−210*, 510−515. Available from https://doi.org/10.1016/j.jhazmat.2012.01.067.

Cheng, L., Qiu, T. L., Yin, X. B., Wu, X. L., Hu, G. Q., Deng, Y., & Zhang, H. (2007). Methermicoccus shengliensis gen. nov., sp. nov., a thermophilic, methylotrophic methanogen isolated from oil-production water, and proposal of Methermicoccaceae fam. nov. *International Journal of Systematic and Evolutionary Microbiology, 57*, 2964−2969.

Da Silva, M. L., & Alvarez, P. J. (2002). Effects of ethanol vs MTBE on benzene, toluene, ethylbenzene, and xylene natural attenuation in aquifer columns. *Journal of Environmental Engineering, 128*(9), 862−867.

Dang, P. N., Dang, T. C. H., Lai, T. H., & Stan-Lotter, H. (1996). Desulfovibrio vietnamensis sp. nov., a halophilic sulfate-reducing bacterium from vietnamese oil fields. *Anaerobe, 2*, 385−392.

Das, N., & Chandran, P. (2011). Microbial degradation of petroleum hydrocarbon contaminants: An overview. *Biotechnology Research International, 2011*, 1−13. Available from https://doi.org/10.4061/2011/941810, Article ID 941810.

Datta, A., Philip, L., & Murty Bhallamudi, S. (2014). Modeling the biodegradation kinetics of aromatic and aliphatic volatile pollutant mixture in liquid phase. *Chemical Engineering Journal, 241*, 288−300. Available from https://doi.org/10.1016/j.cej.2013.10.039.

Davey, M. E., Wood, W. A., Key, R., Nakamura, K., & Stahl, D. A. (1993). Isolation of three species of Geotoga and Petrotoga: Two new genera, representing a new lineage in the bacterial line of descent distantly related to the "Thermotogales". *Systematic and Applied Microbiology, 16*, 191−200.

Davis, B. M., Wallace, S., & Willison, R. (2009). Pilot-scale engineered wetlands for produced water treatment. *Society of Petroluem Engineers, 4*(3), 75−79. Available from https://doi.org/10.2118/120257-PA, Retrieved December 22, 2016, from.

Davis, C., Cort, T., Dai, D., Illangasekare, T. H., & Munakata-Marr, J. (2003). Effects of heterogeneity and experimental scale on the biodegradation of diesel. *Biodegradation, 14*(6), 373−384.

Dzantor, E. K. (2007). Phytoremediation: The state of rhizosphere 'engineering' for accelerated rhizodegradation of xenobiotic contaminants. *Journal of Chemical Technology & Biotechnology, 82*(3), 228−232.

Eke, P. E., & Scholz, M. (2008). Benzene removal with vertical - flow constructed treatment wetlands. *Journal of Chemical Technology and Biotechnology (Oxford, Oxfordshire: 1986), 83*, 55−63. Available from https://doi.org/10.1002/jctb.1778.

Fardeau, M. L., Ollivier, B., Patel, B. K. C., Magot, M., Thomas, P., Rimbault, A., Rocchiccioli, F., & Garcia, J. L. (1997). Thermotoga hypogea sp. nov., a xylanolytic, thermophilic bacterium from an oil-producing well. *International Journal of Systematic Bacteriology, 47*, 1013−1019.

Ferro A. M., Utah N. L., Kadlec R. H., Deschamp J., & Wyoming C. (2002) Constructed wetland system to treat wastewater at the BP Amoco former Casper refinery: Pilot scale project. In *Proceedings of the 9th international petroleum envoronmental conference*. Albuquerque, NM: IPECConsortium. Retrieved from http://ipec.utulsa.edu/Ipec/Conf2002/tech_sessions.html.

Filho, N. I., Vieceli, N. C., Cardoso, E. M., & Lovatel, E. R. (2013). Analysis of BTEX in experimental columns containing neat gasoline and gasoline-ethanol. *Journal of the Brazilian Chemical Society, 24*(3), 410−417.

Geller, J. T., & Hunt, J. R. (1993). Mass transfer from nonaqueous phase organic liquids in water-saturated porous media. *Water Resources Research, 29*(4), 833−845.

Ghosal, D., Ghosh, S., Dutta, T. K., & Ahn, Y. (2016). Current state of knowledge in microbial degradation of polycyclic aromatic hydrocarbons (PAHs): A review. *Frontiers in Microbiology, 7*, 1369.

Greene, A. C., Patel, B. K. C., & Sheehy, A. (1997). Deferribacter thermophilus gen. nov., sp. nov. a novel thermophilic manganese- and iron-reducing bacterium isolated from a petroleum reservoir. *International Journal of Systematic Bacteriology, 47*, 505−509.

Gupta, P. K., Ranjan, S., & Yadav, B. K. (2013). BTEX biodegradation in soil-water system having different substrate concentrations. *International Journal of Engineering Research & Technology, 2* (12), 1765−1772, ISSN: 2278-0181.

Horner, J. E., Castle, J. W., Rodgers, J. H., Jr, Murray Gulde, C., & Myers, J. E. (2012). Design and performance of pilot-scale constructed wetland treatment systems for treating oilfield produced water from sub-saharan Africa. *Water, Air, and Soil Pollution, 223*, 1945−1957. Available from https://doi.org/10.1007/s11270-011-0996-1.

Jeanthon, C., L'Haridon, S., Cueff, V., Banta, A., Reysenbach, A. L., & Prieur, D. (2002). Thermodesulfobacterium hydrogeniphilum sp. nov., a thermophilic, chemolithoautotrophic, sulfate-reducing bacterium isolated from a deep-sea hydrothermal vent at Guaymas Basin, and emendation of the genus Thermodesulfobacterium. *International Journal of Systematic and Evolutionary Microbiology, 52*, 765−772.

Jeanthon, C., Reysenbach, A. L., L'Haridon, S., Gambacorta, A., Pace, N. R., Glenat, P., & Prieur, D. (1995). Thermotoga subterranea sp. nov., a new thermophilic bacterium isolated from a continental oil reservoir. *Archives of Microbiology, 164*, 91−97.

Khan, A. M., Wick, L. Y., Harms, H., & Thullner, M. (2016). Biodegradation of vapor-phase toluene in unsaturated porous media: Column experiments. *Environmental Pollution, 211*, 325−331.

Kovarova-kovar, K., & Egli, T. (1998). Growth kinetics of suspended microbial cells: From single-substrate-controlled growth to mixed-substrate kinetics. *Microbiology and Molecular Biology Reviews, 62*(3), 646, 1998.

Liang, B., Wang, L. Y., Zhou, Z., Mbadinga, S. M., Zhou, L., Liu, J. F., Yang, S. Z., Gu, J. D., & Mu, B. Z. (2016). High frequency of Thermodesulfovibrio spp. and Anaerolineaceae in association with Methanoculleus spp. in a long-term incubation of n-alkanes-degrading methanogenic enrichment culture. *Frontiers in Microbiology, 7*(1431), 1−12.

Littlegohns, J. V., & Daugulis, A. J. (2008). Kinetics and interactions of BTEX compounds during degradation by a bacterial consortium. *Process Biochemistry*, *43*(10), 1068–1076. Available from https://doi.org/10.1016/j.procbio.2008.05.010.

Madigan, M. T., Martinko, J. M., Bender, K. S., Buckley, D. H., Stahl, D. A., & Brock, T. (2015). *Brock biology of microorganisms* (14th (ed.)). Harlow, United Kingdom: Pearson Higher Education Limited.

Magot, M., Basso, O., Tardy-Jacquenod, C., & Caumette, P. (2004). Desulfovibrio bastinii sp. nov. and Desulfovibrio gracilis sp. nov., moderately halophilic, sulfate-reducing bacteria isolated from deep subsurface oilfield water. *International Journal of Systematic and Evolutionary Microbiology*, *54*, 1693–1697.

Magot, M., Fardeau, M. L., Arnauld, O., Lanau, C., Ollivier, B., Thomas, P., & Patel, B. K. C. (1997). Spirochaeta smaragdinae sp. nov., a new mesophilic strictly anaerobic spirochete from an oil field. *FEMS Microbiology Letters*, *155*, 185–191.

Magot, M., Ravot, G., Campaignolle, X., Ollivier, B., Patel, B. K. C., Fardeau, M. L., Thomas, P., Crolet, J. L., & Garcia, J. L. (1997). Dethiosulfovibrio peptidovorans gen. nov., sp. nov., a new anaerobic, slightly halophilic, thiosulfate-reducing bacterium from corroding offshore oil wells. *International Journal of Systematic Bacteriology*, *47*, 818–824.

Mant, C., Coasta, S., Williams, J., & Tambourgi, E. (2005). Studies of removal of chromium by model constructed wetlands. *The Brazilian Journal of Chemical Engineering*, *22*(3), 381–387. Available from https://doi.org/10.1590/S0104-66322005000300007.

Mayilraj, S., Kaksonen, A. H., Cord-Ruwisch, R., Schumann, P., Sproer, C., Tindall, B. J., & spring, S. (2009). Desulfonauticus autotrophicus sp. nov., a novel thermophilic sulfatereducing bacterium isolated from oil-production water and emended description of the genus Desulfonauticus. *Extremophiles: Life Under Extreme Conditions*, *13*, 247–255.

Murray-Gulde, C., Heatley, J. E., Karanfil, T., Rodgers, J. R., Jr, & Myers, J. E. (2003). Performance of a hybrid reverse osmosis-constructed wetland treatment system for brackish oil field produced water. *Water Research*, *37*(3), 705–713. Available from https://doi.org/10.1016/S0043-1354(02)00353-6.

Muyzer, G., & Stams, A. J. M. (2008). The ecology and biotechnology of sulphate-reducing bacteria. *Nature Reviews Microbiology*, *6*, 441–454.

Narayanan, M., Tracy, J. C., Davis, L. C., & Erickson, L. E. (1998a). Modeling the fate of toluene in a chamber with alfalfa plants 1. Theory and modeling concepts. *Journal of Hazardous Substance Research*, *1*, 1–30.

Narayanan, M., Tracy, J. C., Davis, L. C., & Erickson, L. E. (1998b). Modeling the fate of toluene in a chamber with alfalfa plants 2. Numerical results and comparison study. *Journal of Hazardous Substance Research.*, *1*(1994).

Nazina, T. N., Ivanova, A. E., Kanchaveli, L. P., & Rozanova, E. P. (1988). A new thermophilic methylotrophic sulfate-reducing bacterium Desulfotomaculum kuznetsovii sp. nov. *Microbiology (Reading, England)*, *57*, 823–827.

Nazina, T. N., Tourova, T. P., Poltaraus, A. B., Novikova, E. V., Grigoryan, A. A., Ivanova, A. E., Lysenko, A. M., Petrunyaka, V. V., Osipov, G. A., Belyaev, S. S., & Ivanov, M. V. (2001). Taxonomic study of aerobic thermophilic Bacilli: Descriptions of Geobacillus subterraneus gen. nov., sp. nov. and Geobacillus uzenensis sp. nov. from petroleum reservoirs and transfer of Bacillus stearothermophilus, Bacillus thermocatenulatus, Bacillus thermoleovorans, Bacillus kaustophilus, Bacillus thermodenitrificans to Geobacillus as the new combinations G. stearothermophilus, G. th. *International Journal of Systematic and Evolutionary Microbiology*, *51*, 433–446.

Nealsonl, K. H., & Myres, R. C. (1992). Microbial reduction of manganese and iron: New approaches to carbon cycling. *Applied and Environmental Microbiology*, *58*(2), 439–443.

Nerantzis, P. C., & Dyer, M. R. (2010). Transport of BTEX vapours through granular soils with different moisture contents in the vadose zone. *Geotechnical and Geological Engineering, 28*(1), 1—13.

Ni, S., & Boone, D. R. (1991). Isolation and characterization of a di- methyl sulfide-degrading methanogen, Methanolobus siciliae H1350, from an oil well. *International Journal of Systematic Bacteriology, 41*, 410—416.

Ojumu, T. V., Beelo, O. O., & Solomon, B. O. (2005). Evaluation of microbial systems for bioremediation of petroleum refinery effluents in Nigeria. *African Journal of. Biotechnology*, 10.4314%2Fajb.v4i1.15048.

Okoro, C. C., Ekun, O. A., & Nwume, M. I. (2017). Microbial community structures of an offshore and near-shore oil production facilities after biocide treatment and the potential to induce souring and corrosion. *African Journal of Microbiology Research, 11*(5), 171—184.

Ollivier, B., Cayol, J. L., Patel, B. K. C., Magot, M., Fardeau, M. L., & Garcia, J. L. (1997). Methanoplanus petrolearius sp. nov., a novel methanogenic bacterium from an oil producing well. *FEMS Microbiology Letters, 147*, 51—56.

Ollivier, B., Fardeau, M. L., Cayol, J. L., Magot, M., Patel, B. K. C., Prensier, G., & Garcia, J. L. (1998). Characterization of Methanocalculus halotolerans gen. nov., sp. nov., isolated from an oil-producing well. *International Journal of Systematic Bacteriology, 48*, 821—828.

Ommedal, H., & Torsvik, T. (2007). Desulfotignum toluenicum sp. nov., a novel toluene-degrading, sulphate-reducing bacterium isolated from an oil-reservoir model column. *International Journal of Systematic Bacteriology, 57*, 2865—2869.

Perera, F. P., Tang, D., Wang, S., Vishnevetsky, J., Zhang, B., Diaz, D., Camann, D., & Rauh, V. (2012). Prenatal polycyclic aromatic hydrocarbon (PAH) exposure and child behavior at age 6—7 years. *Environmental Health Perspectives, 120*, 921—926.

Picone, S., Grotenhuis, T., Van Gaans, P., Valstar, J., Langenhoff, A., & Rijnaarts, H. (2013). Toluene biodegradation rates in unsaturated soil systems vs liquid batches and their relevance to field conditions. *Applied Microbiology and Biotechnology, 97*, 7887—7898.

Powell, C. L., Goltz, M. N., & Agrawal, A. (2014). Degradation kinetics of chlorinated aliphatic hydrocarbons by methane oxidizers naturally- associated with wetland plant roots. *Journal of Contaminant Hydrology, 170*, 68—75.

Ranck, J. M., Bowman, R. S., Weeber, J. L., Katz, L. E., & Sullivan, E. J. (2005). BTEX removal from produced water using surfactant-modified zeolite. *Journal of Environmental Engineering, 131*(3), 434—442.

Ravanchi, M. T., Kaghazchi, T., & Kargari, A. (2009). Application of membrane seperation processes in petrochemical industry: A review. *Desalination, 235*(1—3), 199—244. Available from https://doi.org/10.1016/j.desal.2007.10.043.

Rees, G. N., Grassia, G. S., Sheehy, A. J., Dwivedi, P. P., & Patel, B. K. C. (1995). Desulfacinum infernum gen. nov., sp. nov., a thermophilic sulfate-reducing bacterium from a petroleum reservoir. *International Journal of Systematic Bacteriology, 45*, 85—89.

Rifai, H. S., Newell, C. J., Gonzales, J. R., & Wilson, J. T. (2000). Modeling natural attenuation of fuels with BIOPLUME III. *Journal of Environmental Engineering, 126*(5), 428—438.

Röling, W. F. M., et al. (2002). Robust hydrocarbon degradation and dynamics of bacterial communities during nutrient-enhanced oil spill bioremediation. *Applied and Environmental Microbiology, 68*(11), 5537—5548, PMC. Web. 9 Oct. 2016.

Saien J. (2010) *Treatment of the refinery wastewater by nano particles of TiO2*. USA Patent No. Pub. No.: US 20100200515 A1. Retrieved February 2011.

Shimp, J. F., Tracy, J. C., Davis, L. C., Lee, E., Huang, W., Erickson, L. E., & Schnoor, J. L. (1993). Beneficial effects of plants in the remediation of soil and groundwater contaminated with organic materials. *Critical Reviews in Environmental Science and Technology, 23*(1), 41—77.

Shpiner, R., Vathi, S., & Stuckey, D. C. (2009). Treatment of "produced water" by waste stabilization ponds: removal of heavy metals. *Water Research, 43*, 4258–4268. Available from https://doi.org/10.1016/j.watres.2009.06.004.

Simi A. (2000). Water quality assessmentof a surface flow constructed wetland treating oil refinery wastewater. In K.R. Reddy (Ed.), *7th international conference on wetlands system for water pollution control. 3*, pp. 1295–1304. Lake Buena Vista, Boca Raton, Florida, USA: IWA. Retrieved January 17, 2011.

Søvik, A. K., Alfnes, E., Breedveld, G. D., French, H. K., Pedersen, T. S., & Aagaard, P. (2002). Transport and degradation of toluene and o-xylene in an unsaturated soil with dipping sedimentary layers. *Journal of Environmental Quality, 31*(6), 1809–1823.

Susarla, S., Medina, V. F., & McCutcheon, S. C. (2002). Phytoremediation: An ecological solution to organic chemical contamination. *Ecological Engineering, 18*(5), 647–658. Available from https://doi.org/10.1016/S0925-8574(02)00026-5.

Tardy-Jacquenod, C., Magot, M., Laigret, F., Kaghad, M., Patel, B. K. C., Guezennec, J., Iwitheron, R., & Caumette, P. (1996). Desulfovibrio gabonensis sp. nov., a new moderately halophilic sulfate- reducing bacterium isolated from an oil pipeline. *International Journal of Systematic Bacteriology, 46*, 710–715.

Thomas, K. V., Balaam, J., Barnard, N., Dyer, R., Jones, C., Lavender, J., & McHugh, M. (2002). Characterisation of potentially genotoxic compounds in sediments collected from United Kingdom estuaries. *Chemosphere, 49*(3), 247–258.

Tindall, J. A., Friedel, M. J., Szmajter, R. J., & Cuffin, S. M. (2005). Part 1: Vadose-zone column studies of toluene (enhanced bioremediation) in a shallow unconfined aquifer. *Water, Air, and Soil Pollution, 168*(1–4), 325–357.

Trigueros, D. E. G., Modenes, A. N., Kroumov, A. D., & Espinoza-Quinones, F. R. (2010). Modeling of biodegradation process of BTEX compounds: Kinetic parameters estimation by using Particle Swarm Global Optimizer. *Process Biochemistry, 45*(8), 1355–1361. Available from https://doi.org/10.1016/j.procbio.2010.05.007.

Van Hamme, J. D., Singh, A., & Ward, O. P. (2003). Recent advances in petroleum microbiology. *Microbiology and Molecular Biology Reviews, 67*(4), 503–536.

Varjani, S. J., & Upasani, V. N. (2016c). Biodegradation of petroleum hydrocarbons by oleophilic strain of Pseudomonas aeruginosa NCIM 5514. *Bioresource Technology, 222*, 195–201.

Vetriani, C., Speck, M. D., Ellor, S. V., Lutz, R. A., & Starovoytov, V. (2004). Thermovibrio ammonificans sp. nov., a thermophilic, chemolithotrophic, nitrate-ammonifying bacterium from deep-sea hydrothermal vents. *International Journal of Systematic and Evolutionary Microbiology, 54*, 175–181.

Vogt, C., Gödeke, S., Treutler, H. C., Weiß, H., Schirmer, M., & Richnow, H. H. (2007). Benzene oxidation under sulfate-reducing conditions in columns simulating in situ conditions. *Biodegradation, 18*(5), 625–636.

Wallace, S., & Kadlec, R. (2005). BTEX degradation in a cold-climate wetland system. *Water Science and Technology, 51*(9), 165–171.

Wallace S.D. (2001) On-site remediation of petroleum contact wastes using surface flow wetlands. In *Proceedings of the 2nd international conference on wetlands and remediation*. Burlington.

Wuyep, P. A., Chuma, A. G., Awodi, S., & Nok, A. J. (2007). Biosorption of Cr, Mn, Fe, Ni, Cu and Pb metals from petroleum refinery effluent by calcium alginate immobilized mycelia of Polyporus squmosus. *Scientific Research and Essays, 2*(7), 217–221.

Yadav, B. K., & Hassanizadeh, S. M. (2011). An overview of biodegradation of LNAPLs in coastal (semi)-arid environment. *Water, Air, & Soil Pollution, 220*(1–4), 225–239.

Yadav, B. K., Ansari, F. A., Basu, S., & Mathur, A. (2013). Remediation of LNAPL contaminated groundwater using plant-assisted biostimulation and bioaugmentation methods. *Water, Air, & Soil Pollution, 225*(1), 1793. Available from https://doi.org/10.1007/s11270-013-1793-9.

Yang, Y., Li, J., Xi, B., Wang, Y., Tang, J., Wang, Y., & Zhao, C. (2017). Modeling BTEX migration with soil vapor extraction remediation under low-temperature conditions. *Journal of Environmental Management, 203*, 114−122.

Zhao, Y., Qu, D., Hou, Z., & Zhou, R. (2015). Enhanced natural attenuation of BTEX in the nitrate-reducing environment by different electron acceptors. *Environmental Technology, 36*(5), 615−621.

Zheng, Z., Aagaard, P., & Breedveld, G. D. (2002). Sorption and anaerobic biodegradation of soluble aromatic compounds during groundwater transport. 1. Laboratory column experiments. *Environmental Geology, 41*(8), 922−932.

Zhou, H. W., et al. (2008). Different bacterial groups for biodegradation of three-and four-ring PAHs isolated from a Hong Kong mangrove sediment. *Journal of Hazardous Materials, 152*(3), 1179−1185.

Fate, transport, and bioremediation of PAHs in experimental domain: an overview of current status and future prospects

Pankaj Kumar Gupta[1], Manik Goel[2], Sanjay K. Gupta[3] and Ram N. Bhargava[4]

[1]Wetland Hydrology Research Laboratory, Faculty of Environment, University of Waterloo, Waterloo, ON, Canada, [2]Department of Hydrology, Indian Institute of Technology Roorkee, Roorkee, India, [3]Environmental Engineering, Department of Civil Engineering, Indian Institute of Technology Delhi, New Delhi, India, [4]Department of Microbiology (DM), Babasaheb Bhimrao Ambedkar University, Lucknow, India

7.1 Introduction

Spill from the industries and accidental spillage of petroleum derivatives from tankers, pipelines, and storage tanks are unavoidable circumstances (Salleh et al., 2003). Polycyclic aromatic hydrocarbons (PAHs) are the most widespread contaminants of soil and groundwater systems (Margesin et al., 2000) that are classified as either light or dense non-aqueous phase liquids (LNAPLs and DNAPLs, respectively) depending on their relative density (Kamath et al., 2004). The most threats to groundwater pollution are due to the leakage of PAHs from the aboveground and the underground storage tanks. During the imports and exports petrochemical handling, transportation, pipeline, storage, and associated leakage cause the PAHs spill in the soil−water system. According to the Directorate General of Commercial Intelligence and Statistics, Government of India, there is increasing growth in imports of petroleum, crude oil products over last years, and the current imports percentage growth is 54.1. Similarly, exports percentage growth is 41.92 over previous years (Directorate General of Commercial Intelligence, and Statistics, Ministry of Commerce and Industry, Government of India, 2015). As per the Ministry of Petroleum and Natural Gas, Government of India, 13.70% growth in refining capacity in 2012−13. Such increasing capacity may make soil−water systems more vulnerable to PAHs pollution.

The PAHs movement in unsaturated porous media is sufficiently complex when only two fluid phases air and water, are present; flow becomes even

Advances in Remediation Techniques for Polluted Soils and Groundwater. DOI: https://doi.org/10.1016/B978-0-12-823830-1.00017-1

more complicated when a third fluid phase, such as an immiscible organic fluid, is involved. This third fluid phase (NAPL) arises when liquid hydrocarbon fuels or solvents are spilled accidentally on the ground surface or when they leak from underground storage tanks. The resulting subsurface flow problem then involves three fluids, air, water, and PAHs, each having different interfacial tensions with each other, different viscosities, and different capillary interactions with the soil. The PAH movement in the subsurface is dominated by advection and hydrodynamics mechanisms. Advection and dispersions mechanisms result in the translation or spreading of plumes in a porous medium and a large spatial area are covered. The sorption of dissolved constituents, on the other hand, results in the partition of species between the solid and aqueous phases thereby reducing the mass of contaminant that is in solution; however, sorption is responsible for soil phase soil pollution (Gupta et al., 2019). The biodegradation of organic contaminants by a native subsurface microbial population is the most likely process whereby mass can be removed without physically extracting the contaminant from the aquifer. Fig. 7.1 presents schematic diagram of PAHs fate and transport in subsurface.

Most of the current research deals with the effect of different physical and chemical mechanisms individually (Gupta et al., 2020; Kumar et al., 2021), where associated impacts of variable environmental conditions (Basu et al., 2020) and water-table dynamics, heterogeneous domains are poorly mentions (Gupta and Bhargava, 2020; Gupta and Yadav, 2020; Gupta and Yadav,

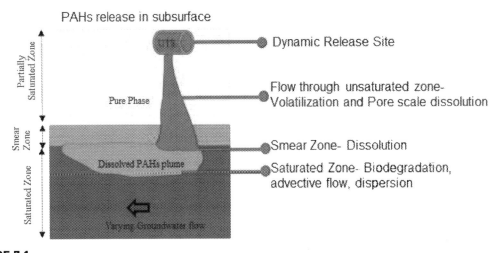

FIGURE 7.1

Fate and transport of PAHs in subsurface. *PAH*, Polycyclic aromatic hydrocarbon.

2019a, b, c). Therefore to develop the knowledge and understanding on the multiscale-associated problems are the most important challenge among the scientific communities(Amrit et al., 2019; Gupta, 2020a, 2020b; Gupta & Sharma, 2018; Gupta & Yadav, 2019, 2020, Himanshu et al., 2017, 2018, 2021; Kumari et al., 2019; Mauget et al., 2021; Basu et al., 2020; Gupta & Bhargava, 2020; Gupta & Yadav, 2019; Gupta & Yadav, 2017). Thus the focus of this chapter is to provide a better understanding on fate, transport, and bioremediation of PAHs in subsurface environment. Thus this manuscript has direct practical importance in different remediation and contamination management planning, research, and field application.

7.2 PAHs fate and transport mechanisms

Much of the problem is attributed to the facile migration of hydrocarbons to the subsurface or a deeper where PAHs slowly dissolve into soil—water and vapor mixing in soil—air (Kuiper et al., 2004). Processes include source zone mass transfer due to simultaneous volatilization and dissolution; advection, dispersion, and diffusion in air and water phases influenced by time-variant infiltration, geological heterogeneity, and preferential flows; volatility driven gas—liquid (air—water/PAHs) phase partitioning resulting in a dynamic inter-action of dissolved-phase and vapor-phase plumes; phase partitioning due to sorption to solid phases and the air—water interface; and biodegradation and chemical reaction. It is clear that with so many operative processes, an inte-grated understanding of them is required to allow the robust prediction of fate and transport in the unsaturated zone. These processes require collective understanding to enable risks to be assessed and remediation works appro-priately assigned (Rivett et al., 2011).

A shallow unconfined aquifer is mostly characterized by fluctuating water table (Lee & Chrysikopoulos, 1995). The water-table elevation and capillary fringe vary due to the changing water stage in unsaturated zones. Diurnal stage fluctuations induce a rapid and significant response in the surround-ing land mass over shorter distances and damp at further distances (Williams & Oostrom, 2000). Seasonal fluctuations may occur when water uptake by vegetation, evaporation, surface recharge, and groundwater with-drawal varies over the course of the year (Zhang et al., 1998). Variations in water table can also occur over longer time if average groundwater with-drawal/discharge rates exceed the average water recharge rates or vice versa (Lakshmi et al., 1998). The water-table dynamics can greatly affect the dis-tribution, particularly in the vertical direction and degradation process of LNAPLs (Mercer & Cohen, 1990) in addition to the soil moisture and tem-perature distribution. The LNAPLs floating near the water table are susceptible to smearing up and down due to the water-table fluctuations. The LNAPLs move downward as the water level falls, leaving behind a residual fraction in the unsaturated zone in the form of isolated ganglia. Conversely, a rise of water table leads to an upward migration of LNAPLs resulting in the entrapment of LNAPLs and air by snap-off or bypassing mechanism in smear zone below the groundwater table (Chatzis et al., 1983; Kechavarzi et al., 2005). On the other hand, the entrapped LNAPLs reduce the free-phase pool and, thus, mitigate the likelihood of the pure LNAPLs migration to down-gradient receptors (Fry et al., 1997). The entrapped air in saturated zone provides additional oxygen to LNAPL-degrading microbes. Also, the pulse of oxygen introduced by lowering the water table exposes the microbes to the air in the soil pore spaces and enhances biodegradation without the injection of oxygen. At the same time, the entrapment of LNAPLs increases their water interfacial area resulting in an enhanced dissolution of the hydrocarbons and hence, enlargement of plume size migrating in the direction of water flow (Muller et al., 1998). Nevertheless, the dissolution process of PAHs as a result of groundwater flow in saturated zone is by far considered the main limiting factor of the contaminant removal from the source zone (Abriola et al., 2004; Kamon et al., 2006; Powers et al., 1991). Furthermore, the wider spreading of PAHs in response to the falling and rising of water table exposes them to more microorganisms. The rise of water level also exposes the upper dryer regions of a polluted site to the moisture necessary to sustain microbial metabolism and growth. Thus fate and transport of the accumulated PAHs pool and its associated discontinuous ganglia/blobs in the smear zone are important considerations in the bioremediation of polluted sites due to additional chemical and hydraulic heterogeneity in space and time introduced by the fluctuating water table.

7.3   Studies investigated PAHs behaviors in laboratory domain

Laboratory studies that have examined fate and transport of PAHs plumes under unsaturated-zone conditions are quite rare. Powers et al. (1991) investigated the mass transfer and transport mechanisms for LNAPLs pool and associated residual blobs under constant water-table condition. These studies contributed to the understanding of the LNAPLs mass transfer near water table and, thus, provide some insight into estimating the time needed for the complete dissolution of LNAPL pools as well as residual blobs in the unsaturated zone. However, the oversimplified assumptions of steady-state flow and/or static water-table conditions rarely exist in the field, particularly in coastal regions. Furthermore, Williams and Oostrom (2000) performed an intermediate-scale flow cell experiment with a fluctuating water table to study the effect of entrapped air on dissolved oxygen transfer and transport by taking relative permeability influence into account. Later, Oostrom et al. (2006) conducted a 2D experiment to investigate the migration behavior of two LNAPLs having different viscosity under variable water-table conditions. The results of this experiment showed that more viscous mobile LNAPL, subject to variable water-table conditions, does not necessarily float on the water table and may not appear in an observation well.

Recently, Gupta et al. (2019) performed a series of 2D laboratory experiments to investigate the fate and transport of PAHs (toluene) under dynamically fluctuating groundwater table. A two-dimensional sand tank setup was fabricated and filled homogenous with Indian sand (fine) for this purpose. Two wells were installed in sand tank, one in upstream side and another in downstream side to maintain the flow of groundwater and three layers of sampling ports (in saturated, unsaturated, and headspace zone) which make this setup novel and more realistic to field conditions. An auxiliary column was attached with the inlet of upstream well and outlet of the downstream well to maintain the groundwater-table dynamics in sand tank setup. Three levels of groundwater fluctuation (slow, rapid, and general) were maintained and pure phase toluene was spilled from top surface. The analysis of the results shows that a large-size pure phase toluene pool gets developed under rapid groundwater-table fluctuation condition, which significantly contributes to high dissolution rate and large dissolved plume with high toluene concentrations. Similarly, another unpublished study performed by Gupta and Yadav (2019) adds a new dimension in laboratory scale experiment by investigating the flow of toluene in 3D sand tank setup under varying groundwater flow conditions. A series of 3D sand tank experiments were performed to investigate fate and transport of toluene in subsurface under varying groundwater flow velocities. Toluene was spilled in top of tank filled with Indian sand homogenously, and groundwater velocities were maintained by providing influx from upstream well using peristaltic pump. Periodically, toluene concentrations were observed by collecting samples from three layers sampling ports and analyzed using gas chromatography—mass spectrometry. Finally, numerical simulation runs were performed using HYDRUS 3D modeling. The results of this study established a liner relationship between groundwater flow velocity and (1) dissolution rate, (2) Peclet number, and (3) biodegradation rate. The findings of this study may help one to estimate the PAHs movement and to frame bioremediation plan for real field conditions having a wide range of dynamically groundwater conditions. The peer-review literature on PAHs fate and transport in the subsurface is listed in Table 7.1.

Table 7.1 Summary of laboratory experimental studies on polycyclic aromatic hydrocarbons fate and transport reported in the peer-reviewed literature.

References	Title of paper	Experimental design/setup	Methodology	Results/comments/gaps
Rühle et al. (2015)	Response of Transport Parameters and Sediment Microbiota to Water Table Fluctuations in Laboratory Columns	Plexiglas column (14-cm ID, 50-cm long)	The columns were continuously infiltrated from the top with a constant water flux (0.19 ± 0.01 cm h^{-1}) and operated in a downward flow mode using tracer	The 16 inflow points were distributed evenly, ensuring homogeneous groundwater distribution over the entire diameter of the column
Søvik et al. (2002)	Transport and Degradation of Toluene and o-Xylene in an Unsaturated Soil with Dipping Sedimentary Layers	Column (10-cm ID, 0.5-m long)	The toluene concentration was applied in the column having Gardermoen delta soil	A flow rate of 0.8 mL h^{-1} was maintained
Legout et al. (2009)	Experimental and Modeling Investigation of Unsaturated Solute Transport with Water-Table Fluctuation	Plexiglas column (0.3-m ID, 1-m long)	Experiments reproducing one-dimensional vertical flow and tracer transport were performed at two infiltration rates, 12 and 22 mm h^{-1}	The experimental apparatus with the locations of the six tensiometers (TS)
Henry et al. (2002)	Two-dimensional modeling of flow and transport in the vadose zone with surfactant-induced flow	2D sand tank 234×153 cm (equipped with seven injection ports at the inlet and seven extraction ports)	Water was supplied to the microporous tubing from a constant head reservoir. The point source where a flow rate of 210.5 cm day^{-1} into the region was applied over a length of 1.9 cm (parallel to the long direction of the flow cell) during contaminant application	Modeling integrated with HYDRUS 2D
Luo et al. (2015)	Effects of carrier bed heterogeneity on hydrocarbon migration.	2/3D heterogeneous domain modeling	The MigMOD mode was used to simulate the formation of hydrocarbon migration pathways in heterogeneous carrier beds	Reservoir heterogeneity is likely the main reason for the irregular distribution of hydrocarbons in the study area and other oilfields
Abreu et al. (2009)	Simulated soil vapor intrusion attenuation factors including biodegradation for petroleum hydrocarbon	Three-dimensional numerical model simulations	Homogeneous soil properties and steady-state conditions were simulated that incorporating the natural attenuation from source points	The vapor intrusion depends upon the source–surface distances, and biodegradation was significant for low concentration

Continued

Table 7.1 Summary of laboratory experimental studies on polycyclic aromatic hydrocarbons fate and transport reported in the peer-reviewed literature. *Continued*

References	Title of paper	Experimental design/setup	Methodology	Results/comments/ gaps
Ozgur et al. (2009)	Simulation of the vapor Intrusion Process for Nonhomogeneous Soils Using a Three-Dimensional Numerical Model	Three-dimensional numerical model simulations	A three-dimensional finite element model was used to investigate the importance of factors that could influence vapor intrusion when the site is characterized by nonhomogeneous soils	In layered geologies, a lower permeability and diffusivity soil layer between the source and building often limit vapor intrusion rates, even if a higher permeability layer near the foundation permits increased soil—gas flow rates into the building
Gupta et al. (2019)	Assessment of LNAPL in Subsurface under Fluctuating Groundwater Table Using 2D Sand Tank Experiments	2D sand tank experiment and numerical modeling	Three levels of groundwater fluctuation (slow, rapid, and general) were maintained and pure phase toluene was spilled from top surface	High dissolution and volatilization were observed under fluctuating water-table conditions

7.4 Polishing PAH-polluted site using subsurface-constructed wetlands

Conventional treatment of polluted natural resources (wastewater/polluted soil—water) by a treatment plant typically involves physical and chemical processes for removing major contaminants before disposal. Such treatment of PAH-polluted sites is too expensive, and the infrastructure for treatment prior to disposal is insufficient in small cities and may be nonexistent in the rural and other remote communities. Constructed wetlands (CWs) are cost-effective approach to polish polluted sites. Subsurface-constructed wetlands (SSCW) are shallow open-water bodies and, hence, are strongly affected by the prevailing climate and weather. SSCWs may enhance the existing geo-chemical processes by accepting PAHs as carbon sources to native microorganisms. A detail description of geochemical process acting at PAHs polluted sites are discussed next.

1. *Nitrification:* Nitrification is a biological process that converts the ammonia into nitrate and is generally followed by denitrification that involves the reduction of nitrates to nitrogen gas (Fig. 7.2). This sequence of the reactions is one of the most important processes in SSCW for N-removal followed by sulfate reduction from the soil—water system (Bastviken et al., 2003). Nitrification can be modeled in two-step process using CW2D model of HYDRUS by

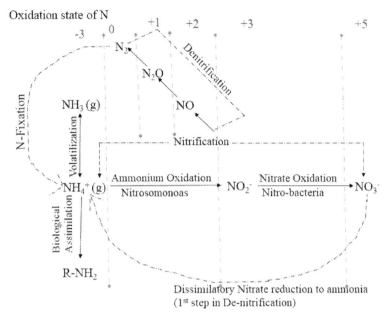

FIGURE 7.2
Geochemical pathways of N species in wetland system.

incorporating, (1) aerobic growth of *Nitrosomonas* on ammonium ion, and (2) aerobic growth of *Nitrobacter* on nitrite ion. In the first step, *Nitrosomonas* consumes ammonium ion (NH_4^+) and oxygen to produce NO_2^-. Inorganic phosphorus along with traces of NH_4^+ is also incorporated in the biomass.

$$rc_{step1} = \mu_{ANs} \frac{C_{O_2}}{K_{ANs,O_2} + C_{O_2}} \frac{C_{NH_4}}{K_{ANs,NH_4} + C_{NH_4}} \frac{C_{IP}}{K_{ANs,IP} + C_{IP}} C_{XANs}$$

Lysis of *Nitrosomonas* produces organic matter, NH_4^+ as follows:

$$rc_{step1A} = b_{XANs} C_{XANs}$$

In the second step, microbes consume nitrite and oxygen to produce NO_3^-. Ammonium are incorporated in the biomass as follows:

$$rc_{step2} = \mu_{ANb} \frac{C_{O_2}}{K_{ANb,O_2} + C_{O_2}} \frac{C_{NO_2}}{K_{ANb,NO_2} + C_{NO_2}} \bigg|_{N,ANb} C_{XANb}$$

Modeling of CW processes involved aerobic growth of Nitrobacter on Nitrate followed by the lysis of Nitrobacter. In the lysis of Nitrobacter, OM, NH_4^+, and IP are as follows:

$$rc_{step1A} = b_{XANb}C_{XANb}$$

2. *Denitrification:* In denitrification process, nitrate-based growth of heterotrophs can be modeled on radially degradable chemical oxygen demands (COD). This process consumes NO_3^- and OM and reduces dinitrogen (N_2).

$$rc_{DN-Nitrate} = \mu_{DN} \frac{K_{DN,O_2}}{K_{DN,O_2} + C_{O_2}} \frac{C_{NO_3}}{K_{DN,NO_3} + C_{NO_3}} \frac{K_{DN,NO_2}}{K_{DN,NO_2}} \frac{C_{CR}}{K_{DN,CR} + C_{CR}} \bigg|_{N,DN} C_{XH}$$

On the other hand, in nitrite-based growth of heterotrophs on readily biodegradable COD, the process consumes NO_2^-, OM CR, NH_4^+, and IP and produces N_2.

$$rc_{DN-Nitrite} = \mu_{DN} \frac{K_{DN,O2}}{K_{DN,O2} + C_{O2}} \frac{C_{NO2}}{K_{DN,NO2} + C_{NO2}} \frac{C_{CR}}{K_{DN,CR} + C_{CR}} \bigg|_{N,DN} C_{XH}$$

where lysis produces OM, NH_4^+, and IP and assumed to represent the sum of all decay and sink processes. Refer Tables 7.2 and 7.3 for values associated with different mechanisms.

$$rc_{DN} = b_H C_{XH}$$

3. *Ammonia volatilization:* It refers to the general physicochemical process for obtaining the ammonia either from ammonification reactions or from wastewater, which is transferred from water to the atmosphere. The pH of system plays an important role in

Table 7.2 Components defined in the reactive transport module.

I	Symbol	Unit	Description
1	O_2	$mg_{O2}\,L^{-1}$	DO
2	CR	$mg_{COD}\,L^{-1}$	Readily biodegradable COD
3	CS		Slowly biodegradable COD
4	CI		Inert chemical oxygen demand
5	XH		Heterotrophic microorganism
6	XAN_s		*Nitrosomonas* spp. (autotrophic bacteria 1)
7	XAN_b		*Nitrosomonas* spp. (autotrophic bacteria 2)
8	NH_4N	$mg_N\,L^{-1}$	Ammonium ion (NH_4^+)
9	NO_2N		Nitrite ion (NO_2^-)
10	NO_3N		Nitrate ion (NO_3^-)
11	N_2N		Dinitrogen gas (N_2)
12	IP		Phosphorus (inorganic)

COD, *Chemical oxygen demand;* DO, *dissolved oxygen.*
Adopted from Langergraber, G., & Šimůnek, J. (2005). Modeling variably saturated water flow and multicomponent reactive transport in constructed wetlands. Vadose Zone Journal, 4(4), 924–938 (Langergraber & Šimůnek, 2005).

Table 7.3 Kinetic parameters (at 20°C, values in brackets at 0°C).

Parameter		Value
Hydrolysis		
K_h	Hydrolysis rate constant (d^{-1})	3 (2)
K_x	Coefficient for hydrolysis (mg$_{COD, CS}$ mg$_{COD, BM}^{-1}$)	0.1
Heterotrophic microorganism (aerobic growth)		
μ_H	Maximum growth rate on CR in aerobic conditions (d^{-1})	6 (3)
b_H	Rate constant for lysis (d^{-1})	0.4 (0.2)
$K_{het, O2}$	Oxygen coefficient (mg$_{O2}$ L^{-1})	0.2
$K_{het, CR}$	Substrate coefficient (mg$_{COD, CR}$ L^{-1})	2
$K_{het, NH4N}$	Ammonium ion coefficient (mg$_{NH4N}$ L^{-1})	0.05
$K_{het, IP}$	Phosphorus coefficient (mg$_{IP}$ L^{-1})	0.01
Heterotrophic microorganism (denitrification)		
μ_{DN}	Maximum denitrification rate (d^{-1})	4.8 (2.4)
$K_{DN, O2}$	Oxygen coefficient (mg$_{O2}$ L^{-1})	0.2
$K_{DN, NO3N}$	NO$_3^{-1}$ coefficient (mg$_{NO3N}$ L^{-1})	0.5
$K_{DN, NO2N}$	NO$_2^{-1}$ coefficient (mg$_{NO2N}$ L^{-1})	0.5
$K_{DN, CR}$	Substrate coefficient (mg$_{COD, CR}$ L^{-1})	4
$K_{DN, NH4N}$	Ammonium ion coefficient (mg$_{NH4N}$ L^{-1})	0.05
$K_{DN, IP}$	Phosphorus coefficient (mg$_{IP}$ L^{-1})	0.01
Nitrification step 1 (Nitrosomonas)		
μ_{ANs}	Maximum growth rate on NH$_4^+$, aerobic condition (d^{-1})	0.9 (0.3)
b_{ANs}	Rate constant for lysis (d^{-1})	0.15 (0.05)
$K_{ANs, O2}$	Oxygen coefficient (mg$_{O2}$ L^{-1})	1
$K_{ANs, NH4N}$	Ammonium ion coefficient (mg$_{NH4N}$ L^{-1})	0.5
$K_{ANs, IP}$	Phosphorus coefficient (mg$_{IP}$ L^{-1})	0.01
Nitrification step 2 (Nitrosomonas)		
μ_{ANb}	Maximum aerobic growth rate on NO$_2^-$ (d^{-1})	1 (0.35)
b_{ANb}	Rate constant for lysis (d^{-1})	0.15 (0.05)
$K_{ANb, O2}$	Oxygen coefficient (mg$_{O2}$ L^{-1})	0.1
$K_{ANb, NO2N}$	NO$_2^-$ coefficient (mg$_{NO2N}$ L^{-1})	0.1
$K_{ANb, NH4N}$	Ammonium ion coefficient (mg$_{NH4N}$ L^{-1})	0.05
$K_{ANb, IP}$	Phosphorus coefficient (mg$_{IP}$ L^{-1})	0.01

The term "coefficient" in the earlier table signifies for saturation/inhibition coefficient of respective elements.
Adopted from Langergraber, G., & Šimůnek, J. (2005). Modeling variably saturated water flow and multicomponent reactive transport in constructed wetlands. *Vadose Zone Journal, 4(4)*, 924–938.

controlling this process. For significant mechanism for ammonia removal, pH must be above 10. Mostly, the pH of CWs ranges from 7 to 8.5; thus ammonia volatilization is not a significant parameter in the case of CWs.

4. *Hydrolysis:* In SSCW, hydrolysis refers to a process by means of which biomolecules degrade due to reaction with water. It is a first-order kinetics that helps the microorganism to gain energy by breaking the contaminant molecule into daughter ones, which ultimately enhances the overall efficiency the system under consideration. On the contrary, it slightly increases the dissolved COD concentration in the wetlands.

$$rc_{hydrolysis} = K_h \frac{C_{CS}/C_{XH}}{K_X + C_{CS}/C_{XH}} C_{XH}$$

5. *Fermentation:* Fermentation is an exothermic and anaerobic process that includes the formation of valuable products by reduction−oxidation process. It does not include any external electron acceptor. It is a substrate-level phosphorylation.

6. *Respiration:* Respiration can be aerobic and anaerobic both. Unlike fermentation, respiration necessarily needs an external electron acceptor. It is electron transport−linked phosphorylation.

7. *Methanogenesis:* Final degradation of organic compounds forming methane by the microbes is known as methanogenesis. It is carried by anaerobic microorganism called methanogens. It is a dominating pathway for the degradation of organic compounds in SSCW due to the lack of oxidant in usually water logged soil. In organic matter−rich sediments, the availability of oxygen is limited only to the top few millimeters near the ground surface and near plant roots; hence, most of the mineralization of organic matter is carried out anaerobically. Purification of wastewater in SSCW is done by the combination of various geochemical processes which, in turn, govern by the hydrologic design of the wetlands.

7.5   Conclusive remark and future prospects

The focus of this chapter is to enhance the understanding of fate, transport, and bioremediation of PAHs in subsurface. Pollution of soil−water resources due to release hydrocarbons such as PAHs is a major concern in shallow aquifers regions. Hydrocarbons are well-recognized, toxic, organic compounds, especially carcinogenic to human health. Once released in subsurface, they start getting partition in gas phase due to volatilization and dissolving in pore water. In a partially saturated zone, hydrocarbon liquids behave multiphase flow along with air and water. At water table, dissolved phase starts moving down-gradient locations with groundwater, which forms a large plume. Due to diffusion and dispersion, different concentration zone formed in subsurface. Groundwater-table fluctuation and flow velocity play significant roles in dissolution and subsequent spreading of dissolved plume. A growing literature is available on fate and transport of PAHs in experimental domain; however, only few studies have reported the role of varying subsurface conditions. Based on the literature review, following future directions have been highlighted:

1. There is urgent need to investigate fate and transport of PAHs in field scale and under varying environmental conditions.
2. Combined impact of dissolution and volatilization needs to be verified for multiscale domain, especially in heterogeneous study domain.
3. The impact of groundwater flow regimes and dynamically fluctuating water-table conditions must be considered in the remediation of PAH-polluted sites.
4. The role of native microbes in natural attenuation and source zone attenuation must be considered in future studies.
5. The performance of plant-assisted bioremediation and SSCWs will be helpful for the effective management of PAH-polluted sites.
6. It is important to investigate bioclogging and its impact of future removal effective of the CW.
7. The application of machine learning and artificial intelligence will be more effective tools for the estimation and forecasting of future PAHs loading in subsurface.

References

Abreu, L. D. V., Ettinger, R., & Todd, M. A. (2009). Simulated soil vapor intrusion attenuation factors including biodegradation for petroleum hydrocarbons. *Ground Water Monitoring and Remediation, 29*(1), 105−117.

Abriola, L. M., Scott, A. B., John, L., & Charles, L. G. (2004). Volatilization of binary non-aqueous phase liquid mixtures in unsaturated porous media. *Vadose Zone Journal, 3*(2), 645.

Amrit, K., Mishra, S. K., Pandey, R. P., Himanshu, S. K., & Singh, S. (2019). Standardized precipitation index-based approach to predict environmental flow condition. *Ecohydrology, 12*(7) e2127.

Bastviken, S. B., Eriksson, P. G., Martins, I., Neto, L., & Tonderski, K. (2003). Potential nitrification and denitrification on different surfaces in a constructed treatment wetland. *Wetlands and Aquatic Processes, 32*(6), 2414−2420.

Basu, S., Yadav, B.K., Mathur, S., Gupta, P.K. (2020). In situ bioremediation of LNAPL polluted vadose zone: Integrated column and wetland study. *CLEAN - Soil, Air, Water,* 2000118.

Chatzis, I., Morrow, N. R., & Lim, H. T. (1983). Magnitude and detailed structure of residual oil saturation. *Society of Petroleum Engineering Journal, 23,* 311−325.

Directorate General of Commercial Intelligence and Statistics, Ministry of Commerce and Industry, Government of India. (2015). *Annual report 2015*.

Fry, V. A., Selker, J. S., & Gorelick, S. M. (1997). Experimental investigations for trapping oxygen gas in saturated porous media for in situ bioremediation. *Water Resources Research, 33*(12), 2687–2696.

Gupta, P. K., & Yadav, B. K. (2017). Bioremediation of non-aqueous phase liquids (NAPLs) polluted soil and water resources. In R. N. Bhargava (Ed.), Environmental pollutants and their bioremediation approaches. Florida, USA: CRC Press, Taylor and Francis Group, ISBN 9781138628892.

Gupta, P. K., & Sharma, D. (2018). Assessments of hydrological and hydro-chemical vulnerability of groundwater in semi-arid regions of Rajasthan, India. *Sustainable Water Resources Management, 1*(15), 847–861.

Gupta, P. K., & Yadav, B. K. (2019). In R. P. Singh, A. K. Kolok, & L. S. Bartelt-Hunt (Eds.), *Subsurface processes controlling reuse potential of treated wastewater under climate change conditions". Water Conservation, Recycling and Reuse: Issues and Challenges*. Singapore: Springer, 9789811331787 (ISBN).

Gupta, P. K., & Yadav, B. K. (2019). 3-D laboratory experiments on fate and transport of light NAPL under varying groundwater flow conditions. *ASCE Journal of Environmental Engineering, 146*. Available from https://doi.org/10.1061/(ASCE)EE.1943-7870.0001672.

Gupta, P.K., Yadav, B.K., (2019). "Remediation and management of petrochemical polluted sites under climate change conditions". In Bhargava, R.N. (Ed.), Environmental contaminations: Ecological implications and management. Springer Nature Singapore Pte Ltd. 9789811379048

Gupta, P. K., & Yadav, B. K. (2020). Three-dimensional laboratory experiments on fate and transport of light NAPL under varying groundwater flow conditions. *ASCE Journal of Environmental Engineering, 146*(4)04020010.

Gupta, P. K., Yadav, B., & Yadav, B. K. (2019). Assessment of LNAPL in subsurface under fluctuating groundwater table using 2D sand tank experiments. *ASCE Journal of Environmental Engineering, 145*. Available from https://doi.org/10.1061/(ASCE)EE.1943-7870.0001560.

Gupta, P. K. (2020). Pollution load on Indian soil-water systems and associated health hazards: A review. *ASCE Journal of Environmental Engineering, 146*(5)03120004.

Gupta, P. K. (2020). Fate, transport and bioremediation of biodiesel and blended biodiesel in subsurface environment: A review. *ASCE Journal of Environmental Engineering, 146*(1)03119001.

Gupta, P. K., Kumari, B., Gupta, S. K., & Kumar, D. (2020). Nitrate-leaching and groundwater vulnerability mapping in North Bihar, India. *Sustainable Water Resources Management, 6*, 1–12.

Gupta, P.K., & Bhargava, R.N., (2020). Fate and transport of subsurface pollutants. Springer, eBook ISBN-978-981-15-6564-9. https://www.springer.com/gp/book/9789811565632.

Henry, E. J., Smith, J. E., & Warrick, A. W. (2002). Two-dimensional modelling of flow and transport in the vadose zone with surfactant-induced flow. *Water Resources Research, 38*(11), 33–1 – 33–16. Available from http://doi.wiley.com/10.1029/2001WR000674.

Himanshu, S. K., Pandey, A., & Shrestha, P. (2017). Application of SWAT in an Indian river basin for modeling runoff, sediment and water balance. *Environmental Earth Sciences, 76*(1), 1–18.

Himanshu, S. K., Pandey, A., & Patil, A. (2018). Hydrologic evaluation of the TMPA-3B42V7 precipitation data set over an agricultural watershed using the SWAT model. *Journal of Hydrologic Engineering, 23*(4)05018003.

Himanshu, S. K., Ale, S., Bordovsky, J. P., Kim, J., Samanta, S., Omani, N., & Barnes, E. M. (2021). Assessing the impacts of irrigation termination periods on cotton productivity under strategic deficit irrigation regimes. *Scientific Reports, 11*(1), 1–16.

Kamath, R., Rentz, J. A., Schnoor, J. L., & Alvarez, P. J. J. (2004). Phytoremediation of hydrocarbon-contaminated soils: Principles and applications. *Studies in Surface Science and Catalysis, 151*, 447–478.

Kamon, M., Li, Y., Flores, G., Inui, T., & Katsumi, T. (2006). Experimental and numerical study on migration of LNAPL under the influence of fluctuating water table in subsurface. *Annuals of Disaster Prevention Research Institute*, Kyoto Univ., No. 49 B.

Kechavarzi, C., Soga, K., & Illangasekare, T. H. (2005). Two-dimensional laboratory simulation of LNAPL infiltration and redistribution in the Vadose Zone. *Journal of Contaminant Hydrology, 76*(3–4), 211–233.

Kuiper, I., Lagendijk, E. L., Bloemberg, G. V., & Lugtenberg, B. J. J. (2004). Rhizoremediation: A beneficial plant-microbe interaction bioremediation: A natural method. *Molecular Plant-Microbe Interactions, 17*(1), 6–15.

Kumar, R., Sharma, P., Verma, A., Jha, P. K., Singh, P., Gupta, P. K., Chandra, R., & Prasad, P. V. V. (2021). Effect of physical characteristics and hydrodynamic conditions on transport and deposition of microplastics in riverine ecosystem. *Water, 13*, 2710.

Kumari, B., Gupta, P. K., & Kumar, D. (2019). In-situ observation and nitrate-N load assessment in Madhubani District, Bihar, India. *Journal of Geological Society of India, 93*(1), 113–118.

Lakshmi, N. R., Han, W., & Banks, M. K. (1998). Mass loss from LAPL pools under fluctuating water table conditions. *Journal of Environmental Engineering, 124*(12), 1171–1177.

Langergraber, G., & Šimůnek, J. (2005). Modeling variably saturated water flow and multicomponent reactive transport in constructed wetlands. *Vadose Zone Journal, 4*(4), 924–938.

Lee, K. Y., & Chrysikopoulos, C. V. (1995). Numerical modeling of three-dimensional contaminant migration from dissolution of multicomponent NAPL pools in saturated porous media. *Environmental Geology, 26*(3), 157–165.

Legout, C., Molenat, J., & Hamon, Y. (2009). Experimental and modeling investigation of unsaturated solute transport with water-table fluctuation. *Vadose Zone Journal, 8*(1), 21.

Luo, X., Hu, C., Xiao, Z., Zhao, J., Zhang, B., Yang, W., Zhao, H., Zhao, F., Lei, Y., & Zhang, L. (2015). Effects of carrier bed heterogeneity on hydrocarbon migration. *Marine and Petroleum Geology, 68*(3), 120–131. Available from http://linkinghub.elsevier.com/retrieve/pii/S0264817215300672.

Margesin, R., Zimmerbauer, A., & Schinner, F. (2000). Monitoring of bioremediation by soil biological activities. *Chemosphere, 40*(4), 339–346.

Mauget, S. A., Himanshu, S. K., Goebel, T. S., Ale, S., Lascano, R. J., & Gitz. (2021). Soil and soil organic carbon effects on simulated Southern High Plains dryland Cotton production. *Soil and Tillage Research, 212*, 105040.

Mercer, J. W., & Cohen, R. M. (1990). A review of immiscible fluids in the subsurface: Properties, models, characterization, and remediation. *Journal of Contaminant Hydrology, 6*, 107–163.

Muller, R., Antranikian, G., Maloney, S., & Sharp, R. (1998). Thermophilic degradation of environmental pollutants. In G. Antranikian (Ed.), *Biotechnology of Extremophiles* (Vol. 61, pp. 155–169). Advances in Biochemical Engineering/Bio-technology.

Oostrom, M., Hofstee, C., & Wietsma, T. W. (2006). Behaviour of a viscous LNAPL under fluctuating water table conditions. *Soil and Sediment Contamination, 15*, 543–564.

Ozgur, B., Kelly, G. P., & Eric, M. S. (2009). Simulation of the vapor intrusion process for nonhomogeneous soils using a three-dimensional numerical model. *Ground Water Monitoring and Remediation, 29*(1), 92–104. Available from http://www3.interscience.wiley.com/journal/122264246\abstract/npapers2://publication/uuid/562134E8-48E2-489F-9323-662C4BD7879E.

Powers, S. E., Loureiro, C. O., Abriola, L. M., & Weber, W. J. (1991). Theoretical study of the significance of non-equilibrium dissolution of nonaqueous-phase liquids in subsurface systems. *Water Resources Research*, *27*(4), 463–477.

Rivett, M. O., Gary, P. W., Rachel, A. D., & Todd, A. M. (2011). Review of unsaturated-zone transport and attenuation of volatile organic compound (VOC) plumes leached from shallow source zones. *Journal of Contaminant Hydrology*, *123*(3–4), 130–156. Available from http://www.ncbi.nlm.nih.gov/pubmed/21316792.

Rühle, F. A., Frederick von, N., Tillmann, L., & Christine, S. (2015). Response of transport parameters and sediment microbiota to water table fluctuations in laboratory columns. *Vadose Zone Journal*, *14*(5). Available from https://dl.sciencesocieties.org/publications/vzj/abstracts/14/5/vzj2014.09.0116.

Salleh, A. B., Ghazali, F. M., Rahman, R. N. Z. A., & Basri, M. (2003). Bioremediation of petroleum hydrocarbon pollution. *Indian Journal of Biotechnology*, *2*(3), 411–425.

Søvik, A. K., Alfnes, E., Breedveld, G. D., French, H. K., Pedersen, T. S., & Aagaard, P. (2002). Transport and degradation of toluene and O-xylene in an unsaturated soil with dipping sedimentary layers. *Journal of Environment Quality*, *31*, 1809–1823.

Williams, M. D., & Oostrom, M. (2000). Oxygenation of anoxic water in a fluctuating water table system: An experimental and numerical study. *Journal of Hydrology*, *230*(1–2), 70–85.

Zhang, Q., Davis, L. C., & Erickson, L. E. (1998). Effect of vegetation on transport of groundwater and non-aqueous phase liquid contaminants. *Journal of Hazardous Substance Research*, *1*, 1–20.

CHAPTER 8

Mathematical modeling of contaminant transport in the subsurface environment

Abhay Guleria and Sumedha Chakma

Department of Civil Engineering, Indian Institute of Technology Delhi, Delhi, India

8.1 Introduction

Contaminant transport modeling in the subsurface environment is a challenging area for the last few decades due to the involvement of complex physical, chemical, and biogeological processes taking place near the source and within the subsurface system (Essaid et al., 2015; Russo et al., 2010; Selim, 2014; Zheng & Bennett, 2002). The emanation of leachate from waste dumping sites, the release of toxic chemicals from mining operations and tailing ponds, leakage of toxic substances from chemical distribution pipes in factories, leaks from underground storage tanks, diffused pollution sources like an application of fertilizers and pesticides in agricultural farms are known as the major sources of soil and groundwater pollution (Blackmore et al., 2014; Gupta & Yadav, 2019; Moran et al., 2007; USEPA, 1997). Generally, it is not feasible to assess the fate and transport of a single contaminant or simultaneous multiple contaminants by direct in situ field sampling over longer periods. Therefore mathematical modeling proved to be a cost-effective and efficient approach to describe the fate and transport of contaminants in the complex porous system and to design remediation strategies as per in situ conditions (Guo & Brusseau, 2017a, 2017b; Guo et al., 2019a). In mathematical modeling, accuracy in the prediction becomes difficult when factors and/or mechanisms affecting contaminant transport are different at different spatial and temporal scales. For example, pore-scale heterogeneity caused by variations in soil texture and structure is the dominant factor and influences the model's transport behavior and simulation capabilities even at the laboratory column experiments (Raoof & Hassanizadeh, 2013; Rolle et al., 2012). The macroscale spatial variations in properties such as layering of soil strata, the presence of rock formation (fractures) affect transport behavior at the field scale (Gelhar et al., 1992). Further, scale-dependence of flow and transport parameters, for example, hydrodynamic dispersion coefficient (Gelhar et al., 1992; Pickens & Grisak, 1981) and mass

141

Advances in Remediation Techniques for Polluted Soils and Groundwater. DOI: https://doi.org/10.1016/B978-0-12-823830-1.00007-9

transfer coefficient (Gao et al., 2010; Logan, 1996; Swami et al., 2016, 2018) makes contaminant transport modeling a challenging task.

Mathematical models have been used to describe the physics of the processes primarily via partial and/or ordinary differential equations, and concentrations are computed by analytical/semianalytical approach or numerical method such as finite-difference method (FDM), finite-element method (FEM) (Zheng & Bennett, 2002). Several mathematical models emphasizing saturated (Brusseau et al., 1989; van Genuchten & Wierenga, 1976) and unsaturated porous media (Beegum et al., 2019; Morway et al., 2012; Raoof & Hassanizadeh, 2013; Sander & Braddock, 2005) have been developed in last decades. The most commonly used mathematical model is the conventional advection−dispersion equation (ADE).

However, the conventional ADE model failed to mimic the "anomalous" transport behavior in the complex porous system such as heterogeneous soil column, aquifer−aquitard system due to lumped parameter approach (Gao et al., 2009). Due to the limited simulation capabilities of ADE-based models, several higher complex models are developed, such as the mobile−immobile model (MIM), multiprocesses nonequilibrium (MPNE) model, to simulate the asymmetric transport behavior in the heterogeneous porous media (Brusseau et al., 1989; van Genuchten & Wierenga, 1976). It is observed that the higher modeling approaches such as the dual-porosity model with scale-dependent dispersion are found to be more precise in capturing anomalous behavior in the heterogeneous 1-D soil column (Gao et al., 2010; Sharma et al., 2016), two-dimensional (2-D) sand tank experiments (Gupta et al., 2019), 3-D lab-scale aquifer conditions (Gupta & Yadav, 2020), however, only limited to laboratory-scale conditions.

On the other hand, several mathematical models are developed and considered highly heterogeneous porous systems mimicking field scenarios. For example, anaerobic transformation reactions were incorporated to simulate the methyl tert-butyl ether/tert-butyl alcohol transport for Vandenberg Air Force Base California, USA site (Rasa et al., 2011). To measure diffusive mass transfer and biodegradation reactions, simulated results were compared with long-term monitored data (Rasa et al., 2011). Yang et al. (2017a) developed analytical solutions for contaminant diffusion between aquifer and single aquitard system and applied them to field-scale problems from sites of Connecticut (United States), Ontario (Canada), Essen (Belgium), Marcoule (France), Mont Terri (Switzerland), and laboratory-scale studies. The developed analytical approach accurately modeled the measured concentration profiles in aquitards and flux-averaged concentrations in the aquifers (Yang et al., 2017a). Based on collected 245 borehole geological data, a highly heterogeneous geostatistical model was developed using T-PROGS

(Guo et al., 2019a). Thereafter, a geostatistical data-based 3-D numerical flow and contaminant transport model was designed to evaluate the long-term performance of the pump and treat method for trichloroethylene (TCE)-contaminated site located in Tucson, Arizona, USA (Guo et al., 2019a). The results from the 3-D model closely matched with measured concentrations indicated that the site conditions and contaminant plume behavior were mimicked reasonably well (Guo et al., 2019a). However, it can be seen from literature that higher modeling approaches have not been implemented to heterogeneous field-scale conditions such as aquitard−aquifer system, high-permeability region (sandy aquifer) with low-permeability silt/clay lenses with reference to the Indian context.

Therefore there is a need to present the review of mathematical modeling studies that emphasize the transport dynamics of a contaminant in the heterogeneous porous media. Thus the focus of this chapter is to present the (1) framework of widely used contaminant transport models for the saturated porous systems, (2) category-wise mini-review of modeling studies with reference to the Indian context, and (3) investigating the transport behavior of conservative contaminant in the dissolved phase for various soil types via MIM model with time-dependent dispersion.

8.2 Contaminant transport models for saturated porous media

8.2.1 Conventional advection−dispersion model

In the conventional advection−dispersion model, porous media is represented by a single porosity system. For the conventional ADE model, the permeability field is assumed macroscopically homogeneous; furthermore, a single lumped value of dispersivity or dispersion coefficient is considered in this type of approach. The governing 1-D contaminant transport equation for saturated porous media considering unidirectional steady water flow is written as (Zheng & Wang, 1999):

$$\frac{\partial C}{\partial t} = D\left(\frac{\partial^2 C}{\partial x^2}\right) - v\left(\frac{\partial C}{\partial x}\right) - \left(\frac{\rho_b}{\theta} \cdot \frac{\partial S}{\partial t}\right) \tag{8.1}$$

where C is the contaminant concentration in the dissolved phase (M/L^3) at any time t, x is the spatial coordinate (L) considered along the fluid flow direction; D (L^2/T) represents hydrodynamic dispersion coefficient, v is the pore water velocity along x-direction, ρ_b is the bulk density of the porous system (M/L^3), S is the amount of contaminant sorbed per unit weight of the solid, θ is the porosity of the saturated porous system.

The equilibrium linear sorption isotherm is defined as:

$$S = K_d C \tag{8.2}$$

By substituting the value of S in the abovementioned equation, conventional ADE can be written as:

$$\frac{\partial C}{\partial t}\left(1 + \frac{K_d \rho_b}{\theta}\right) = D\left(\frac{\partial^2 C}{\partial x^2}\right) - v\left(\frac{\partial C}{\partial x}\right) \tag{8.3a}$$

$$R\frac{\partial C}{\partial t} = D\left(\frac{\partial^2 C}{\partial x^2}\right) - v\left(\frac{\partial C}{\partial x}\right) \tag{8.3b}$$

where retardation factor $R = (1 + \rho_b K_d/\theta)$, K_d is the sorption distribution coefficient of the linear sorption process (L^3/M). $R = 1$ is used for nonreactive contaminant.

8.2.2 Mobile–immobile model

The MIM conceptualization emphasized the division of heterogeneous porous media into mobile (flowing) or immobile (stagnant) regions (van Genuchten & Wierenga, 1976). The mobile region comprises a well-connected pore structure, whereas the immobile region represents the water content, which is irreducible. Contaminant transport in the mobile region is governed by advection and dispersion processes, whereas in the immobile region, it is majorly dominated by the diffusive gradient (Fig. 8.1). Due to hydraulic coupling and dynamic concentration behavior within the porous formation (in the time-dependent source), immobile regions start behaving as distributed sink/source components.

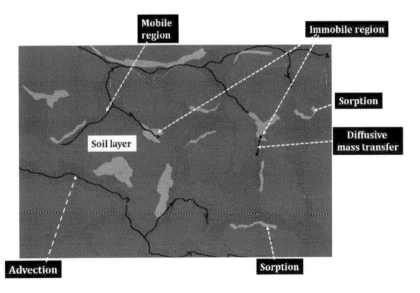

FIGURE 8.1

Schematic of transport processes assumed in the MIM model (*dissolved contaminant species through saturated porous media*). MIM, Mobile–immobile model.

Fig. 8.2 represents the conceptual diagram of a MIM, which divides contaminant concentration in the saturated porous media into four parts. C_m and S_m are dissolved- and sorbed-phase concentration in the mobile region, respectively. Under the local equilibrium assumption, sorption and desorption processes occur rapidly. Any change in the dissolved-phase mobile region concentration is thus assumed to be accompanied instantly by a corresponding change in sorbed-phase concentration of the mobile region. A similar phenomenon has been assumed for immobile region dissolved- (C_{im}) and sorbed-phase (S_{im}) concentrations. Due to the concentration gradient, there is a diffusive mass transfer between mobile and immobile regions.

The governing equations of the 1-D MIM model considering linear sorption isotherm are written as (van Genuchten & Wierenga, 1976):

$$\left(\theta_m + f\rho_b K_{d_m}\right)\frac{\partial C_m}{\partial t} = \theta_m D \frac{\partial^2 C_m}{\partial x^2} - v_m \theta_m \frac{\partial C_m}{\partial x} - \omega(C_m - C_{im}) - \left(\theta_m \mu_{lm} + f\rho_b K_{d_m}\mu_{sm}\right)C_m$$

(8.4a)

$$\left(\theta_{im} + (1-f)\rho_b K_{d_{im}}\right)\frac{\partial C_{im}}{\partial t} = \omega(C_m - C_{im}) - \left(\theta_{im}\mu_{\text{lim}} + (1-f)\rho_b K_{d_{im}}\mu_{sim}\right)C_{im} \qquad (8.4b)$$

where C_m and C_{im} are the concentrations in the mobile and immobile regions (M/L^3) at any time t; x is the spatial coordinate (L); D is the hydrodynamic dispersion coefficient (L^2/T); θ_m and θ_{im} are volumetric water contents of the mobile and immobile regions, respectively, and $\theta = \theta_m + \theta_{im}$; θ is the total volumetric water content of the porous media; v_m is the pore water

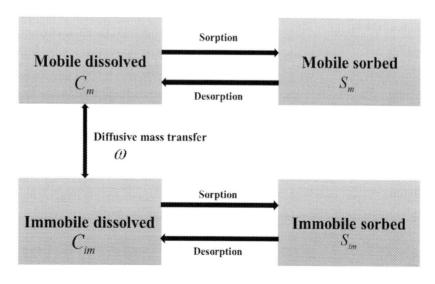

FIGURE 8.2
Conceptual diagram of a mobile—immobile model.

velocity (L/T); $v_m\theta_m$ is equal to q [flow rate (L/T)]; ω is the first-order mass transfer coefficient (T^{-1}); f and $(1-f)$ represent the fractions of sorption sites that equilibrate rapidly with the mobile and immobile regions, respectively; μ_{lm} and $\mu_{\lim}$ are the first-order dissolved-phase decay coefficients in the mobile and immobile regions, respectively; μ_{sm} and μ_{sim} are the first-order sorbed-phase decay coefficients in the mobile and immobile regions, respectively; K_{dm} = sorption distribution coefficient (L^3/M) in the mobile region; K_{dim} = sorption distribution coefficient (L^3/M) in the immobile region; ρ_b = bulk density of the porous medium (M/L^3).

8.2.3   Multiprocess nonequilibrium model

The MPNE model developed by Brusseau et al. (1989) incorporated both physical and chemical nonequilibrium processes.

"Physical Nonequilibrium" can be introduced into the system by combining the advective (mobile) and nonadvective (immobile) regions hydraulically. Due to the difference in hydraulic conductivity, a fraction of the contaminant mass driven by a diffusive gradient is moved to the immobile region. Due to reduced flow, immobile regions tend to act as distributed source/sink components. "Chemical Nonequilibrium" occurs in the system due to chemical interaction between the contaminant and soil grains. The processes that generate chemical nonequilibrium are intraparticle diffusion, rate-limited contact between the contaminant and particular sorption sites, and intrasorbent diffusion. The conceptual representation of the MPNE model is presented in Fig. 8.3. The advection is assumed only in the region of higher hydraulic conductivity (mobile region). The hydraulic conductivity of the immobile region

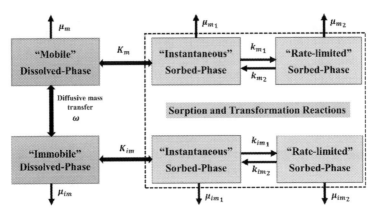

FIGURE 8.3

Conceptual diagram of MPNE model. *MPNE,* Multiprocesses nonequilibrium.

is low, and transport within the immobile region is assumed to be governed by diffusion only.

The one-dimensional contaminant transport equation for the mobile region of porous media can be written as:

$$\left(\theta_m + f\rho_m F_m K_m\right)\frac{\partial C_m}{\partial t} + f\rho_m\frac{\partial S_{m_2}}{\partial t} = \frac{\partial}{\partial x}\left(\theta_m D\frac{\partial C_m}{\partial x}\right) - q\frac{\partial C_m}{\partial x} - \omega(C_m - C_{im})$$

$$- \left(\mu_m\theta_m + \mu_{m_1}f\rho_m F_m K_m\right)C_m - \mu_{m_2}f\rho_m S_{m_2} \tag{8.5}$$

The transport equation for the immobile region can be written as:

$$\left(\theta_{im} + (1-f)\rho_{im}F_{im}K_{im}\right)\frac{\partial C_{im}}{\partial t} + (1-f)\rho_{im}\frac{\partial S_{im_2}}{\partial t} = \omega(C_m - C_{im})$$

$$- \left(\mu_{im}\theta_{im} + \mu_{im_1}(1-f)\rho_{im}F_{im}K_{im}\right)C_{im} - \mu_{im_2}(1-f)\rho_{im}S_{im_2} \tag{8.6}$$

Rate-limited sorption and transformation for the mobile and immobile regions are described by:

$$\frac{\partial S_{m_2}}{\partial t} = k_{m_2}[(1 - F_m)K_m C_m - S_{m_2}] - \mu_{m_2}S_{m_2} \tag{8.7}$$

$$\frac{\partial S_{im_2}}{\partial t} = k_{im_2}[(1 - F_{im})K_{im}C_{im} - S_{im_2}] - \mu_{im_2}S_{im_2} \tag{8.8}$$

where C_m and C_{im} are contaminant concentration in the dissolved phase for mobile and immobile regions (ML^{-3}), respectively; S_{m_2} and S_{im_2} are the rate-limited sorbed-phase concentration in the mobile and immobile regions (MM^{-1}), respectively, D is the dispersion coefficient (L^2T^{-1}), q is the flow rate (LT^{-1}), ω is the first-order mass transfer coefficient between mobile and immobile regions (T^{-1}), θ_m is the fractional volumetric water content of the mobile region, θ_{im} is the fractional volumetric water content of the immobile region, f is the fraction of sorption site related to the mobile region, ρ is the bulk density of the region (ML^{-3}), F_m and F_{im} are the mass fraction of sorbent for which sorption is instantaneous in the mobile and immobile regions, respectively, K_m and K_{im} are the equilibrium sorption coefficient for the mobile and immobile regions (L^3M^{-1}), respectively, μ_m and μ_{im} are the first-order dissolved-phase transformation coefficients in the mobile and immobile regions (T^{-1}), respectively, μ_{m_1} and μ_{im_1} are the instantaneous sorbed-phase transformation coefficients for the mobile and immobile regions, respectively (T^{-1}), μ_{m_2} and μ_{im_2} are the transformation coefficients for rate-limited sorbed-phase domains of mobile and immobile regions (T^{-1}), respectively, and k_{m_2} and k_{im_2} are the first-order reverse sorption rate coefficients (T^{-1}) for the mobile and immobile regions respectively, subscripts 1 and 2 represent instantaneous and rate-limited sorbed-phase domains, while subscripts m and im represent mobile and immobile regions, respectively.

8.2.4 Continuous-time random walk method

In the continuous-time random walk (CTRW) model the movement of solute particles is simulated by a joint probability density function. 1-D Laplace transformed concentration, $\check{C}(x,u)$ from CTRW model is computed as (Berkowitz et al., 2006; Cortis & Berkowitz, 2004):

$$u\tilde{C}(x,u) - C_i(x) = -\tilde{M}(u)\left[v_\psi \frac{\partial}{\partial x}\tilde{C}(x,u) - D_\psi \frac{\partial^2}{\partial x^2}\tilde{C}(x,u)\right] \tag{8.9}$$

where $\quad \tilde{M}(u) = t_{char}u\left(\dfrac{\tilde{\psi}(u)}{1-\tilde{\psi}(u)}\right)$

where u is the Laplace variable, C_i is the initial concentration, v_ψ and D_ψ are the transport velocity and generalized dispersion coefficient, respectively, and $\tilde{M}(u)$ is the memory function and represents anomalous transport behavior caused by unresolved heterogeneities of the porous media. Therefore the transport velocity and dispersion coefficient in the CTRW model are different from the conventional ADE model. t_{char} is characteristic time, $\tilde{\psi}(u)$ is the Laplace transform of $\tilde{\psi}(t)$ and known as "heart" of the CTRW formulation and characterizes the nature of contaminant movement (Cortis et al., 2017). The nonlocal behavior of the solute transport as caused by stagnation zones is also represented by $\tilde{M}(u)$. Large-scale heterogeneities are depicted by v_ψ, while local dispersion processes are represented by D_ψ (Dentz & Berkowitz, 2003). The dispersivity in the CTRW model can be defined as $\alpha_\psi = D_\psi/v_\psi$. Different formulations of CTRW as described in the literature are truncated power law model, modified exponential (eta) model, and classical ADE model (Cortis et al., 2017).

8.3 Categorization of mathematical modeling studies related to Indian groundwater and soil systems

In this section, a category-wise mini-review of modeling studies with reference to the Indian context is presented. Studies involving flow and contaminant transport models are divided broadly into three categories based on the approach used. These categories are analytical/semianalytical contaminant transport modeling, numerical transport modeling, and integrated analytical–numerical and simulation–optimization (S/O) approach-based studies.

8.3.1 Analytical contaminant transport modeling

Analytical or semianalytical solutions were obtained by applying differential and/or integral and/or transformation techniques to simplified contaminant transport equations. For example, Yadav et al. (2012) developed analytical solutions for isotropic and homogeneous porous media in 2-D spatial domain and found that the concentration profiles were significantly affected by periodic velocity fluctuations. However, analytical solutions were not applied to simulate the experimental data (Yadav et al., 2012). Similarly, contaminant dispersion in a forward and backward direction was analyzed for semiinfinite homogeneous porous media (Singh et al., 2020). The analytical solutions matched well with numerical solutions obtained using Crank–Nicolson FDM. Chaudhary et al. (2020) developed analytical solutions for a semiinfinite groundwater aquifer system assuming spatiotemporal-dependent seepage velocity and dispersion coefficient for homogeneous and heterogeneous. They applied for different geological formations, namely, sandstone, shale, and gravel. The derived analytical solutions were in good agreement with numerical solutions and performed well for data from existing literature (Chaudhary et al., 2020). However, in the study by Chaudhary et al. (2020) and Singh et al. (2020), the developed solutions were derived in 1-D spatial domain and were not tested for any laboratory-scale data/field-scale observations.

In contrast to the Indian context, complex scenarios were considered while deriving analytical or semianalytical solutions. Analytical solutions for aquitard–aquifer arrangement were developed to simulate the concentration profiles and diffusive flux into/out of low-permeability zones for flow chamber experiments (Yang et al., 2015). Yang et al. (2017b) used analytical solutions to assess the aquitard concentration profiles at field conditions. Three different scenarios, dissolution alone, dissolution plus back diffusion, and back diffusion alone, were considered. Aquifer concentration data of tetrachloroethene or perchloroethylene (PCE), TCE, and 1,2-dichlorethane at monitoring wells from Naval Air Station Jacksonville (Florida), Dover Air Force Base (Delaware), and Contaminated site (Connecticut) was well described by developed analytical solutions (Yang et al., 2017b). However, these models (Yang et al., 2015, 2017b) were limited to the saturated porous system and considered 1-D spatial domain in a vertical direction, ignoring the effect of transverse dispersion, which can cause mixing and spreading of contaminant even at the microscale level (Cirpka et al., 1999; Tartakovsky & Neuman, 2008). An integrated semianalytical and numerical transport model was developed to simulate matrix diffusion (Falta & Wang, 2017; Muskus & Falta, 2018). The developed method performed efficiently during the contaminant loading period with run time ranging from fractions of seconds to half a minute and suited for any numerical model because the matrix-diffusion flux term was added as a concentration-dependent source/sink term discretized equations (Muskus & Falta, 2018).

8.3.2 Numerical flow and contaminant transport modeling

Numerical modeling studies have addressed the transport behavior of heavy metals, chlorinated solvents in the groundwater aquifer system (Guo & Brusseau, 2017a; Guo et al., 2019a; Rasa et al., 2011). In several studies, conventional numerical models were modified as per the problem of interest despite available open-source and commercial simulators. For example, the effect of the interaction of sorption and biodegradation on the transport behavior of dissolved BTEX components along with multicomponent dissolution mass transfer was investigated (Valsala & Govindarajan, 2018). Rate-limited dissolution mass transfer dynamics of entrapped hydrocarbon (toluene) in 15-cm long saturated soil column was studied using mobile–immobile-based numerical modeling (Vasudevan et al., 2016). Unlike grid- or mesh-based methods, a meshless method such as the Radial point collocation method (RPCM) was developed to simulate multispecies transport problems in a confined aquifer system in which a linked first-order reaction network had been assumed (Anshuman & Eldho, 2020). Solutions obtained from RPCM were verified against a semianalytical solution for hypothetical 1-D and 2-D problems. Further RPCM simulated well concentrations for the field case study compared to FEM simulations (Anshuman & Eldho, 2020). The description of the transport model used, single/multispecies transport problem, and key observations of several mathematical modeling studies with reference to the Indian context are presented in Table 8.1.

On the other hand, higher mathematical models were applied to study the contaminant transport behavior at field-scale conditions. For example, a 2-D finite element numerical model (HydroGeoSphere) was employed to investigate the plume persistence of dense nonaqueous phase liquid (DNAPL) in the aquitard–aquifer arrangement (Chapman & Parker, 2005). Steady-state groundwater flow conditions were simulated using MODFLOW 2000, and solute transport code (SEAM3D) was coupled to NAPL dissolution terms to compute the aqueous and NAPL phase concentrations (Mobile et al., 2012). Further, different modified versions of MODFLOW and MT3DMS have been developed in recent years. For example, local 1-D modeling approach was coupled with conventional 3-D MT3DMS to simulate diffusion in low-permeability soils (Carey et al., 2015). The Markov chain–based stochastic method was used to generate the hydraulically heterogeneous domain, and a 3-D numerical model was used to evaluate the long-term performance of the pump and treat system for TCE contaminated site (Guo et al., 2019a). Guo et al. (2019a, 2019b) implemented a geostatistical approach coupled with a numerical contaminant transport model to simulate pump and treat operations. However, in these studies, transverse dispersion coefficients in horizontal and vertical directions were assumed as a lumped value without any inverse optimization procedure (Guo et al., 2019a, 2019b).

Table 8.1 Summary of mathematical modeling-based studies (Indian context).

Reference	Focus of study	Laboratory setup/ field-scale study 1-D/2-D/3-D	Contaminant (single/ multispecies)	Porous media	Flow and transport model and parameters	Experimental/ observed field data simulated?
Rao et al. (2011)	Assessed groundwater contamination due to the dumpsite in Ranipet, Tamil Nadu, India	Field-scale	TDS	Five hydrogeological units Distributed in the model; single lumped value of porosity, specific yield, specific storage used	MODFLOW and MT3DMS (Visual MODFLOW – software) Groundwater levels were predicted via calibrating groundwater flow model under steady-state conditions Observed and simulated TDS values were compared too	Yes
Sharma and Srivastava (2012)	Investigated the effects of heterogeneity on reactive transport through porous media	Hypothetical example (1-D), field case study (2-D)	Single species Dissolved contaminant	Homogenous and heterogeneous porous media	MPNE model (2-D) + distance- or time-dependent dispersion function (in-house code)	Partially yes
Sharma et al. (2012)	Reactive transport behavior in the coupled fracture-porous block porous system	Synthetic example 1-D fracture and 2-D porous block	Single species-reactive dissolved contaminant	Fracture–porous matrix block system	2-D MPNE model for porous matrix block + 1-D ADE model for fracture and coupled with equations of matrix block. (in-house code) The impact of fractured permeable and impermeable formation was studied	No
Yadav et al. (2012)	Developed an analytical solution for 2-D contaminant transport problem considering periodic flow conditions	2-D synthetic example	Single species contaminant (conservative and reactive)	Saturated porous media with periodic seepage velocity assumption	2-D advection–dispersion equation accounting equilibrium linear sorption process (in-house analytical solution) Derived solutions were not implemented for any observed lab- or field-scale data	No

Continued

Table 8.1 Summary of mathematical modeling-based studies (Indian context). *Continued*

Reference	Focus of study	Laboratory setup/ field-scale study 1-D/2-D/3-D	Contaminant (single/ multispecies)	Porous media	Flow and transport model and parameters	Experimental/ observed field data simulated?
Rao et al. (2013)	Applied groundwater flow and transport model in the basaltic terrain (Bagalkot district, Karnataka, India)	Field-scale study (3-D)	TDS	Five hydrogeological units were considered; however, single lumped value of porosity, specific yield, specific storage used	MODFLOW and MT3DMS (Visual MODFLOW 4.1 – software) Calibration was done using observed head data; however, TDS-simulated values were not compared with field data	Yes
Vasudevan et al. (2014)	Numerical simulation of dissolution and transport of toluene for soil column experiments	1-D soil column	Single species— NAPL toluene	Uniformly filled sand with an average grain size of 0.60 mm	ADE model + two-site sorption + dissolution (in-house code) Lumped model of dispersion function 4 different dissolution MTC were tested	Partially yes
Borah and Bhattacharjya (2015)	Development of GMS-ANN-based S/O method for identification of contamination source	Hypothetical confined aquifer	Single species contaminant	Homogeneous and isotropic condition	GMS (GMS 7.1, Aquaveo) software External simulator GMS linked with optimization model in MATLAB environment	Partially yes Results were compared with published literature
Swami et al. (2016)	Analyzed the behavior of mass transfer coefficient for conservative solute using mobile—immobile model	2-D tank setup, laboratory scale	Single species— chloride	Stratified porous media (natural soil— gravel; silt sand)	MIM Model + constant dispersion function (in-house code) Different values of transport parameters were estimated for different downgradient observation points	Yes
Vasudevan et al. (2016)	Investigated dissolution mass transfer of entrapped hydrocarbon in saturated subsurface system	Synthetic example (15 cm, 1-D soil column)	Single species— toluene	Uniformly filled fine sand	MIM model + two-site sorption + dissolution + biodegradation based on Monod kinetics (in-house code) Constant dispersivity function	No

Reference	Objective	Scale/Case study	Species/Contaminant	Media	Model/Software	Validated
Gandhi et al. (2017)	Development of virus source identification model considering variable number and location of contaminant source	3D unconfined aquifer hypothetical problems	Virus transport with equilibrium sorption process	Flow and transport parameters were taken assuming homogeneous sand soil type	MODFLOW and MT3DMS (GMS software) linked with optimization model using MATLAB	Partially Yes Results validated with published data for synthetic example
Jahangeer et al. (2017)	Evaluated the effects of drainage flux, recharge influx on nitrate movement in sand tank experiment.	2-D sand tank, laboratory scale (6 m × 2 m)	Single species — nitrate	Homogeneous sand	Nonlinear Richards' equation dependent upon moisture content and pressure head was used to simulate the water flow in the coupled unsaturated and saturated region Single porosity model of HYDRUS-2D (software) + constant drainage flux (bottom side) or constant recharge flux (top boundary)	Yes
Yadav et al. (2018)	Development of simulation optimization based approach for in situ bioremediation of BTEX-contaminated site	Hypothetical 2-D aquifer case study	BTEX + oxygen (electron acceptor)	Homogeneous porous media	2-D finite-difference model—BIOPLUME III	No
Gupta et al. (2019)	Numerical modeling of LNAPL was done using HYDRUS 2-D model	2-D sand tank setup, laboratory scale	Single species — LNAPL (toluene)	Homogeneous sand	HYDRUS (2-D) software Single lumped value of dispersivity (longitudinal, transverse) and mass transfer coefficient used	Yes
Guleria et al. (2019)	Investigated the nonreactive solute transport behavior by numerical modeling using MIM model with time-dependent dispersion	Laboratory scale 2-D tank setup, 1-D numerical model	Single species — chloride	Stratified porous media (natural soil—gravel; silt sand)	MIM model + time-dependent dispersion function (in-house code)	Yes
Leichombam and Bhattacharjya (2019)	Development of new hybrid optimization methodology when locations and flux of contaminant source were unknown	Synthetic case study based on literature (2-D)	Conservative contaminant	Homogeneous and isotropic aquifer conditions	Groundwater flow process—MODFLOW Transport equations (FDM)—MATLAB	No

Continued

Table 8.1 Summary of mathematical modeling-based studies (Indian context). *Continued*

Reference	Focus of study	Laboratory setup/ field-scale study 1-D/2-D/3-D	Contaminant (single/ multispecies)	Porous media	Flow and transport model and parameters	Experimental/ observed field data simulated?
Majumder and Eldho (2019)	Proposed a reactive transport model consisting of AEM, RWPT, and kernel density simulator	2-D Hypothetical aquifer cases	Single species Radium-228; TCE	Hydrological features like pumping wells, river, injection wells, and inhomogeneities were incorporated	Advection–dispersion reaction + linear adsorption model—Lagrangian approach (in-house code)	No
Sathe and Mahanta (2019)	Studied the arsenic contamination in the Bongaigaon and Darrang district, Assam, India using numerical flow and transport model	Field scale (3-D model)	Single species— arsenic	Heterogeneous subsurface system (clay, silt, fine sand, coarse sand, sandy gravel, coarse sand with granule, boulder with gravel)	MODFLOW-MT3DMS (3-D software) + 2D lithological and 3D stratigraphy model Calibration was done using observed groundwater well data; however, simulated GW level followed the trend of observed GW level and exact matching was not done	Yes
Valsala and Govindarajan (2019)	Numerically investigated colloid–microbe BTEX transport	Synthetic aquifer 2-D domain (10 × 5 m)	Multispecies— benzene, toluene, ethylbenzene, xylene, colloids, microbe	Saturated homogeneous aquifer	Two-phase (mobile–immobile) with 10 constituents' system—multicomponent cotransport model (colloid, dissolved BTEX, microbes, and dissolved oxygen) (in-house code)	No
Anshuman and Eldho (2020)	Developed RPCM to simulate multispecies transport problem in confined aquifer system	Hypothetical 1-D and 2-D problems	Linked first-order multispecies	Homogenous and heterogeneous aquifer with spatially varying transmissivity	Meshless method—RPCM—in-house code FEM-based model COMSOL (software)	No
Chaudhary et al. (2020)	Analyzed the contaminant transport in semiinfinite groundwater reservoir	1-D study + field-scale example	Single species contaminant	Homogeneous and heterogeneous structure site	Advection–dispersion equation with space- and time-dependent dispersion function	Partially yes Results from the developed model were compared with already published data

Reference	Objective	Scale	Contaminant	Media	Model/Method	Field validation
Gupta and Yadav (2020)	Numerical modeling of LNAPL was done using HYDRUS for saturated and unsaturated region	3-D tank setup, laboratory scale	Single species—LNAPL (toluene)	Homogeneous sand with grain size 0.5–1.0 mm	HYDRUS 3-D model (software) Single lumped value of dispersivity (longitudinal, transverse) and mass transfer coefficient used	Yes
Gupta et al. (2020)	Derived groundwater vulnerability map of Samastipur, Darbhanga, and Madhubani districts of Bihar state, India due to nitrate leaching	Field scale (1-D numerical modeling)	Single species—nitrate-N	Vertical depth of 15 m (alluvial sandy, clay, loam soil)	Richard's equation–based soil water flow and solute transport model (HYDRUS-1D software) Site-specific lithology is considered; however, limited to a vertical direction A load of nitrate concentrations and vulnerability maps was presented using GIS framework	Yes
Natarajan et al. (2020)	Investigated the transport behavior of dissolved contaminant using constant, distance- and time-dependent dispersion function	Synthetic example (1-D)	Single species	Heterogeneous porous media	ADE + Freundlich isotherm + (constant continuous, pulse-type, exponential decaying, sinusoidal varying source) + (constant, distance- time-dependent dispersion)	No
Pal and Chakrabarty (2020)	Assessed ANN models for the simulation of groundwater contaminant	2-D synthetic example	Conservative contaminant	Homogeneous porous media	US Geological Survey—SUTRA Model	No
Pathania and Eldho (2020)	Designed an optimal bioremediation system based on S/O model	2-D hypothetical aquifer condition	BTEX	Homogeneous porous media	Meshless element-free Galerkin method—BIOEFGM (in-house code)	Partially Yes BTEX plume from BIOEFGM and RT3D were compared
Pathania et al. (2020)	Developed an optimal in situ bioremediation approach (comprising meshless FEM and PSO) for contaminated aquifer site	Synthetic field-scale large aquifer with irregular boundaries	BTEX	Homogeneous soil conditions and lumped value of porosity, hydraulic conductivity used Field-scale conditions with irregular boundaries	Meshless element-free Galerkin method—BIOEFGM (2-D) coupled with particle swarm optimization—In-house code in MATLAB	Partially yes

Continued

Table 8.1 Summary of mathematical modeling-based studies (Indian context). *Continued*

Reference	Focus of study	Laboratory setup/ field-scale study 1-D/2-D/3-D	Contaminant (single/ multispecies)	Porous media	Flow and transport model and parameters	Experimental/ observed field data simulated?
Sharma et al. (2020)	Studied the conservative solute transport behavior in soil column	15-m long horizontal soil column 1-D model	Single species — chloride	15-m long homogeneous and heterogeneous soil column (fine and coarse sand, natural soil)	ADE model and spatial fractional ADE (in-house code) Parameter estimation was done using Bees algorithm–based procedure	Yes
Singh et al. (2020)	Investigated the impacts of forward–backward dispersion process on solute transport for semiinfinite system	1-D hypothetical example	Single species contaminant	Homogeneous porous media	ADE with zero-order decay, sorption, and first-order production (analytical and numerical solution) Sinusoidally and sigmoidally varying seepage velocity considered mimicking the realistic groundwater-level scenarios; however, model was not used in simulation of observed data	No

ADE, Advection–dispersion equation; AEM, analytical element method; ANN, artificial neural network; FDM, finite-difference method; FEM, finite-element method; GIS, geographic information system; GMS, groundwater modeling system; GW, groundwater; LNAPL, Light nonaqueous phase liquid; MIM, mobile–immobile model; MPNE, multiprocesses nonequilibrium; MTC, mass transfer correlations; NAPL, nonaqueous phase liquid; PSO, particle swarm optimization; RPCM, radial point collocation method; RWPT, random walk particle tracking; S/O, simulation–optimization; TCE, trichloroethylene; TDS, total dissolved solid.

8.3.3 Integrated analytical—numerical or simulation—optimization approach

It is observed that new methods are emerging with the advancement in computational facilities. The coupling of analytical models with numerical simulators is evolving to solve complex flow and contaminant transport problems. S/O-based methods are preferred over direct simulation models to solve contaminant transport problems for heterogeneous geological settings. For example, the optimization model (Levenberg—Marquardt and particle swarm optimization, PSO) was coupled with FDM-based 1-D transport model to estimate the mass transport parameters for single species (acetone, chloroform, fluoride) transport problem (Bharat et al., 2009). It is observed that the FDM—PSO model does not require an initial guess and gradient of the objective function and performs better in comparison to the gradient-based optimization method (Bharat et al., 2009). Similarly, a genetic algorithm (GA) was coupled with a 1-D analytical model of pathogen transport in saturated porous media and tested for a hypothetical example (Agrawal et al., 2013). Although varying first-order inactivation rate coefficient and hydrodynamic dispersion coefficient over second order of magnitude on either side of true value, the established GA-analytical model converged to true solutions. (Agrawal et al., 2013). Yadav et al. (2018) developed an extreme learning machine (ELM)-based model to replace the BIOPLUME III and coupled the simulation model with PSO for optimum in situ bioremediation design. The proposed ELM—PSO approach proved to be efficient in the remediation of a site contaminated by BTEX compound; however, the influence of linear and nonlinear sorption (Freundlich and Langmuir) process was ignored, keeping the retardation factor equal to unity for hypothetical 2-D aquifer case study (Yadav et al., 2018).

Artificial neural network (ANN)-based S/O methods are emerging for designing remediation measures for any contaminated site. For example, in the study by Pal and Chakrabarty (2020), 87 ANN models like cascade-forward backpropagation, feed-forward backpropagation, radial basis function neural network, and generalized regression neural network were assessed to estimate the concentration in the water supply wells for conservative contaminant transport problem in 2-D hypothetical aquifer. Results reported that the five ANN models based on cascade-forward backpropagation performed well among 87 ANN models; however, the study was limited to single species and was not tested for reactive contaminants for irregular aquifer domain (Pal & Chakrabarty, 2020). Pathania and Eldho (2020) integrated the meshless element-free Galerkin method (BIOEFGM) with PSO to design an optimal in situ bioremediation system. BIOEFGM was found to be efficient in comparison to FEM—PSO based on optimal bioremediation cost. The developed approach was tested for 2-D hypothetical aquifer conditions and neglected the impact of dissolution and transformation reactions (Pathania & Eldho, 2020). An S/O model consisting of an ANN and a Gray wolf optimizer (GWO) was developed for groundwater remediation, and found that the ANN model performed better than AEM–RWPT–KDE model (Majumder & Eldho, 2020).

Further, the stability and convergence behavior of the ANN-GWO model was higher than ANN-differential evolution and ANN–PSO models; however, the developed model was not tested and calibrated for long-term monitoring field-scale data (Majumder & Eldho, 2020). It is observed from Table 8.1 that most of the mathematical modeling studies with reference to the Indian context were limited to soil column experiments and tank experiments and not applied to real field-scale conditions with long-term observed data. None of the investigated studies have emphasized the impact of diffusion fluxes into or out of the low-permeability region (silt or clay lenses) on the contaminant transport behavior in the high-permeability region (sandy aquifer).

8.4   Contaminant transport modeling in the subsurface environment using mobile–immobile model

The contaminant transport behavior in the saturated subsurface system was studied via numerical simulations using MIM model. The nonreactive contaminant in the dissolved phase was considered for a hypothetical aquifer case study assuming steady-state flow conditions. Governing equations of the MIM model with asymptotic time-dependent dispersion function (Eqs. 8.4a, 8.4b and 8.10a, 8.10b) were used to calculate dissolved-phase concentrations at various downgradient locations from an input source. The 2-D saturated porous system of $100 \text{ m} \times 50 \text{ m}$ with a natural hydraulic gradient of 0.001 was considered for various soil types (gravel, sand, silt, and clay). Further, the influence of the fraction of immobile regions on the contaminant breakthrough curves (BTCs) was analyzed by conducting sensitivity analysis for various soil types. Governing equations were solved using the FDM adopting the Crank–Nicolson scheme. The in-house code implementing the iterative method (Gauss–Seidel method) was implemented to compute the contaminant concentration. The schematic of the spatial domain, including boundary conditions used, is shown in Fig. 8.4. Input parameters used are presented in Table 8.2. The hydraulic conductivity values for various soil layers are adapted from Guo et al. (2019c).

Asymptotic time-dependent dispersion:

$$D_L(t) = \alpha_L v \left(\frac{t}{t + K_{asymptotic}} \right) + D_m \tag{8.10a}$$

$$D_T(t) = \alpha_T v \left(\frac{t}{t + K_{asymptotic}} \right) + D_m \tag{8.10b}$$

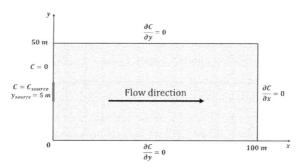

FIGURE 8.4

Schematic diagram of numerical domain.

Table 8.2 Input parameters used in numerical modeling.

Parameter	Value	Parameter	Value
L	100 m	Δt	1 day
B	50 m	ω	0.01 day^{-1}
Δx	1 m	C_0	1 mg L^{-1}
Δy	0.50 m	α_L	10 m
$t_{simulate}$	500 day	α_T	1.0 m
t_{pulse}	180 day	$K_{Asymptotic}$	100 day
D_m	1E − 05 m^2 day^{-1}		

where α_L and α_T are the longitudinal and transverse dispersivity of the porous system, respectively (L); D_m is the molecular diffusion coefficient of solute (L^2/T); $D_L(t)$ and $D_T(t)$ are asymptotic time-dependent dispersion function in longitudinal and transverse direction, respectively (L^2/T); $K_{asymptotic}$ (T) is the asymptotic time-dependent dispersion coefficient, which is equal to mean travel time.

First, numerical simulations were conducted assuming a fraction of 15% immobile regions for all the soil types, which lead to $\theta_m = 85\%$ of θ and $\theta_{im} = 15\%$ of θ. Results for a fraction of 15% immobile regions are shown in this subsection. The concentration distribution of nonreactive contaminants in the gravel, sand, silt, and clay soil type immediately after the source removal (at 181st day) is shown in Fig. 8.5. It is observed that contaminant advected to the largest longitudinal distance of ~60 m for gravel soil layer (Fig. 8.5A), while for sand soil layer, contaminant plume front goes up to ~40-m distance. Further, for silt soil type, contaminant plume front advected up to ~8 m, while advected to a smallest longitudinal distance of ~3 m for clay layer representing the dominance of diffusion over advection

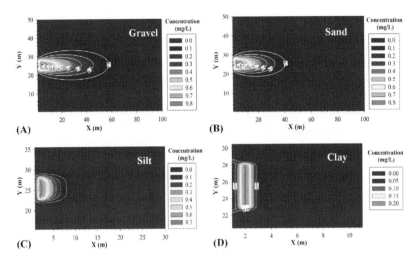

FIGURE 8.5

Contaminant distribution (mg L^{-1}) instantly after source removal at time $=$ 181st day for (A) gravel, (B) sand, (C) silt, and (D) clay soil types.

in low hydraulic conductivity soil type. The maximum value of 0.225 mg L^{-1} of mobile dissolved-phase concentration in the clay soil type is observed as compared to 0.848 mg L^{-1} for the gravel layer. Thus it can be observed that the impact of immobile regions on the contaminant transport dynamics is predominant in the low hydraulic conductivity soil layer (silt, clay) as compared to the high hydraulic conductivity soil layer (sand, gravel). Fig. 8.6 shows the spatial distribution of contaminant concentration at time $=$ 500 day. The peak value of 0.107 mg L^{-1} is observed at $\sim$85 m from a source location in the longitudinal direction for the gravel soil layer, while the peak value of 0.069 mg L^{-1} observed at $\sim$2 m for clayey soil shows that the center of mass of contaminant plume travels second-order larger distance in gravel in comparison to the clay soil layer. It is observed that the shape of a contaminant plume in the high hydraulic conductivity soil layer (gravel, sand) after 500 days varies significantly as compared to low hydraulic conductivity soil layer (silt, clay), depicting the dominance of advection overdispersion and mass transfer processes in the gravel and sand soil layers. The influence of diffusion and transverse dispersion on the plume is predominant in the low hydraulic conductivity soil layer (silt and clay) compared to the high hydraulic conductivity soil layer (sand, gravel), as the plume stretched in the transverse direction dominantly.

Fig. 8.7 shows the simulated BTC along the central line at $y = 25$ m for different observation points (OBS) along longitudinal direction for gravel and

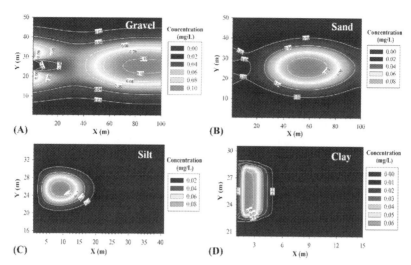

FIGURE 8.6

Contaminant distribution (mg L^{-1}) at time $= 500$ day for (A) gravel, (B) sand, (C) silt, and (D) clay soil types.

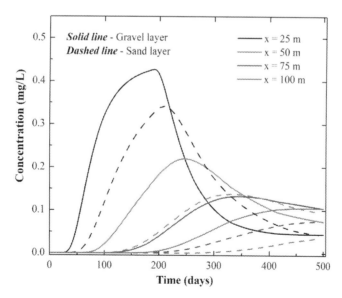

FIGURE 8.7

Simulated BTC at $y = 25$ m for different locations along longitudinal direction for gravel and sand soil types. *BTC*, Breakthrough curve.

sand soil types. The early breakthrough point is observed for OBS at $x = 25$ m (near to source) for gravel soil type as compared to sand layer; however, a sharp falling limb of BTC is seen for gravel soil layer, showing the sudden drop of concentration which is due to the removal of the source. It is observed that the peak value of contaminant concentration decreases with longitudinal distance depicting the dominance of dispersion at large distances far away from source for both the sand and gravel soil layers. The BTCs are dispersed more for observation points located far away from the source depicting the dilution of the plume by dispersion processes for large travel distant locations. The simulated BTC along the central line at $y = 25$ m for different OBS along longitudinal direction for the silt and clay soil layer is shown in Fig. 8.8. It is observed that the peak value of concentration drops by approximately second-order of magnitude for OBS located at $x = 25$ m, $y = 25$ m for the case of silt soil layer as compared to gravel and sand layers (Fig. 8.8A). To capture the very small values of concentration, BTC is plotted in a semilog scale for the silt and clay soil layers. The lowest value of concentration from 10^{-100} to 10^{-20} mg L^{-1} observed for clayey soil shows the diffusion-dominated contaminant transport behavior (Fig. 8.8B). Further, the effect of fraction of immobile region (in the porous media) on contaminant transport behavior is studied by conducting sensitivity analysis for high (sand) and low (clay) hydraulic conductivity soil layer (Fig. 8.9). The fraction of immobile region is varied from 5% to 25%, which lead to different values of θ_m and θ_{im} of an immobile fraction, as shown in Table 8.3. It is observed that the rising limb of BTC is more sensitive to an immobile fraction as compared to the falling limb as seen from BTC at $x = 25$ m, $y = 25$ m for sand soil type (Fig. 8.9A). Similar behavior is observed for the BTC at $x = 50$ m, $y = 25$ m for the sand layer. On the other hand, for low hydraulic conductivity soil layer (clay), equal sensitiveness of concentration toward immobile fraction is observed (Fig. 8.9B); however, for both the soil types, a large variation in the concentration of BTC is not observed with the change in immobile fraction. In the future the sensitivity of transport parameters can be conducted based on spatial or temporal moments of contaminant concentrations.

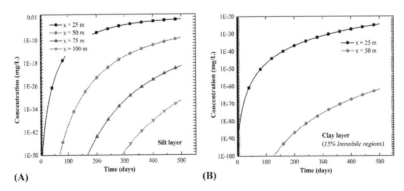

(A)   **(B)**

FIGURE 8.8

Simulated BTC at $y = 25$ m for different locations along longitudinal direction for (A) silt and (B) clay soil types. *BTC*, Breakthrough curve.

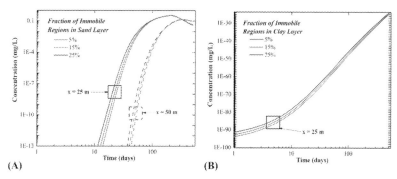

FIGURE 8.9

Temporal variation of concentration with different values of fraction of immobile regions for (A) sand and (B) clay soil types.

Table 8.3 Physical parameters of different soil types.

Soil type	Hydraulic conductivity[a] (m day^{-1})	Bulk density, ρ_b (g cm^{-3})	Porosity	Darcy's velocity (m day^{-1})	Fraction of Immobile regions (%)	θ_m	θ_{im}
Gravel	124	2.39[b]	0.50[b]	0.124	5	0.475	0.025
					15	0.425	0.075
					25	0.375	0.125
Sand	45	1.82[c]	0.32[c]	0.045	5	0.304	0.016
					15	0.272	0.048
					25	0.240	0.080
Silt	3	1.63[c]	0.40[c]	3E − 03	5	0.38	0.02
					15	0.34	0.06
					25	0.30	0.10
Clay	0.13	2.19[d]	0.44[d]	1.3E − 04	5	0.418	0.022
					15	0.374	0.066
					25	0.33	0.11

[a]Guo et al. (2019c).
[b]Chaudhary et al. (2020).
[c]Swami et al. (2016).
[d]Singh et al. (2020).

8.5 Conclusion and future directions

The emphasis of this chapter was to present the framework of trending contaminant transport models along with a mini-review on modeling studies with reference to the Indian context. It is hoped that the key observations deduced from this chapter will enhance the knowledge of readers regarding the current trending modeling approaches. Therefore the following recommendations are listed next:

1. It is observed that the impact of immobile regions on the contaminant transport dynamics is predominant in the low hydraulic conductivity soil layer (silt, clay) as compared to the high hydraulic conductivity soil layer. Thus it is stated that multiple-porosity approach such as the MIM model can be implemented for field-scale conditions to check the applicability of such methods with reference to the Indian context.
2. The effects of molecular diffusion and transverse dispersion on the plume evolution are observed to be predominant in the low hydraulic conductivity soil layer compared to the high hydraulic conductivity layer, as revealed from higher plume spreading in the transverse direction.
3. The influence of distance- and/or time-dependent flow and transport parameters need to be considered in the field-scale studies. Special attention can be given while modeling contaminant transport behavior in the contrasting permeability regions (low- and high-permeability zones, aquifer—aquitard, fracture—matrix system).
4. The numerical simulations presented in this study highlighted the applicability of the 2-D MIM model with time-dependent dispersion for various soil types. This type of modeling approach can be used for scenarios such as rapid rise and fall of water table levels during premonsoon and monsoon seasons and where variability in dispersion and dilution processes has been observed.

This chapter is limited to mathematical models focusing on dissolved-phase contaminant transport behavior in the saturated subsurface system. A review of transport modeling studies emphasizing unsaturated porous media, non-aqueous phase liquids (LNAPLs and DNAPLs), agricultural soil, and water contamination due to pesticide can be carried out in future studies. It gives a broader aspect of the mathematical modeling practices in the Indian context.

Acknowledgment

The first author (AG) would like to thank the Indian Institute of Technology Delhi, India, for supporting this study. In addition, we wish to thank the editors and anonymous reviewers for their suggestions and critical comments that helped in improving the chapter.

References

Agrawal, R., Nandwani, N., & Hari Prasad, K. S. (2013). Pathogen transport in groundwater—Estimation of transport parameters. *ISH Journal of Hydraulic Engineering, 19*(3), 250—256.

Anshuman, A., & Eldho, T. I. (2020). Meshfree radial point collocation-based coupled flow and transport model for simulation of multi-species linked first order reactions. *Journal of Contaminant Hydrology, 229*(November 2019), 103582, Elsevier.

Beegum, S., Šimůnek, J., Szymkiewicz, A., Sudheer, K. P., & Nambi, I. M. (2019). Implementation of Solute Transport in the Vadose Zone into the 'HYDRUS Package for MODFLOW'. *Groundwater, 57*(3), 392−408.

Berkowitz, B., Cortis, A., Dentz, M., & Scher, H. (2006). Modeling non-Fickian transport in geological formations as a continuous time random walk. *Reviews of Geophysics, 44*(2), 1−49, Wiley Online Library.

Bharat, T. V., Sivapullaiah, P. V., & Allam, M. M. (2009). Swarm intelligence-based solver for parameter estimation of laboratory through-diffusion transport of contaminants. *Computers and Geotechnics, 36*(6), 984−992.

Blackmore, S., Smith, L., Ulrich Mayer, K., & Beckie, R. D. (2014). Comparison of unsaturated flow and solute transport through waste rock at two experimental scales using temporal moments and numerical modeling. *Journal of Contaminant Hydrology, 171*(2014), 49−65, Elsevier BV.

Borah, T., & Bhattacharjya, R. K. (2015). Development of unknown pollution source identification models using GMS ANN−based simulation optimization methodology. *Journal of Hazardous, Toxic, and Radioactive Waste, 19*(3), 04014034.

Brusseau, M. L., Jessup, R. E., & Rao, P. S. C. (1989). Modeling the transport of solutes influenced by multiprocess nonequilibrium. *Water Resources Research, 25*(9), 1971−1988.

Carey, G. R., Chapman, S. W., Parker, B. L., & McGregor, R. (2015). Application of an adapted version of MT3DMS for modeling back-diffusion remediation timeframes. *Remediation Journal, 25*(4), 55−79.

Chapman, S. W., & Parker, B. L. (2005). Plume persistence due to aquitard back diffusion following dense nonaqueous phase liquid source removal or isolation. *Water Resources Research, 41*(12), 1−16.

Chaudhary, M., Thakur, C. K., & Singh, M. K. (2020). Analysis of 1-D pollutant transport in semi-infinite groundwater reservoir. *Environmental Earth Sciences, 79*(1), 24.

Cirpka, O. A., Frind, E. O., & Helmig, R. (1999). Numerical simulation of biodegradation controlled by transverse mixing. *Journal of Contaminant Hydrology, 40*(2), 159−182.

Cortis, A., & Berkowitz, B. (2004). Anomalous transport in 'Classical' soil and sand columns. *Soil Science Society of America Journal, 68*(5), 1539−1548.

Cortis, A., Emmanuel, S., Rubin, S., Willbrand, K., & Nissan, A. (2017). *The CTRW Matlab toolbox v4.0: A practical user's guide.*

Dentz, M., & Berkowitz, B. (2003). Transport behavior of a passive solute in continuous time random walks and multirate mass transfer. *Water Resources Research, 39*(5), 1−20.

Essaid, H. I., Bekins, B. A., & Cozzarelli, I. M. (2015). Organic contaminant transport and fate in the subsurface: Evolution of knowledge and understanding. *Water Resources Research, 51*(7), 4861−4902.

Falta, R. W., & Wang, W. (2017). A semi-analytical method for simulating matrix diffusion in numerical transport models. *Journal of Contaminant Hydrology, 197*, 39−49, Elsevier BV.

Gandhi, B. G. R., Bhattacharjya, R. K., & Satish, M. G. (2017). Simulation−optimization-based virus source identification model for 3D unconfined aquifer considering source locations and number as variable. *Journal of Hazardous, Toxic, and Radioactive Waste, 21*(2), 04016019.

Gao, G., Feng, S., Zhan, H., Huang, G., & Mao, X. (2009). Evaluation of anomalous solute transport in a large heterogeneous soil column with mobile-immobile model. *Journal of Hydrologic Engineering, 14*(9), 966−974.

Gao, G., Zhan, H., Feng, S., Fu, B., Ma, Y., & Huang, G. (2010). A new mobile-immobile model for reactive solute transport with scale-dependent dispersion. *Water Resources Research, 46*(8), 1−16.

Gelhar, L. W., Welty, C., & Rehfeldt, K. R. (1992). A critical review of data on field-scale dispersion in aquifers. *Water Resources Research, 28*(7), 1955−1974, Wiley Online Library.

Guleria, A., Swami, D., Sharma, A., & Sharma, S. (2019). Non-reactive solute transport modelling with time-dependent dispersion through stratified porous media. *Sādhanā, 44*(4), 81, Springer India.

Guo, Z., & Brusseau, M. L. (2017a). The impact of well-field configuration and permeability heterogeneity on contaminant mass removal and plume persistence. *Journal of Hazardous Materials, 333*(October), 109−115, Elsevier BV.

Guo, Z., & Brusseau, M. L. (2017b). The impact of well-field configuration on contaminant mass removal and plume persistence for homogeneous vs layered systems. *Hydrological Processes, 31*(26), 4748−4756.

Guo, Z., Brusseau, M. L., & Fogg, G. E. (2019a). Determining the long-term operational performance of pump and treat and the possibility of closure for a large TCE plume. *Journal of Hazardous Materials, 365*(June 2018), 796−803, Elsevier.

Guo, Z., Fogg, G. E., Brusseau, M. L., LaBolle, E. M., & Lopez, J. (2019b). Modeling groundwater contaminant transport in the presence of large heterogeneity: A case study comparing MT3D and RWhet. *Hydrogeology Journal, 27*(4), 1363−1371.

Guo, Z., Fogg, G. E., & Henri, C. V. (2019c). Upscaling of regional scale transport under transient conditions: Evaluation of the multirate mass transfer model. *Water Resources Research, 55*(7), 5301−5320.

Gupta, P. K., & Yadav, B. K. (2019). Remediation and management of petrochemical-polluted sites under climate change conditions. *Environmental Contaminants: Ecological Implications and Management* (pp. 25−47).

Gupta, P. K., & Yadav, B. K. (2020). Three-dimensional laboratory experiments on fate and transport of LNAPL under varying groundwater flow conditions. *Journal of Environmental Engineering, 146*(4), 04020010.

Gupta, P. K., Kumari, B., Gupta, S. K., & Kumar, D. (2020). Nitrate-leaching and groundwater vulnerability mapping in North Bihar, India. *Sustainable Water Resources Management, 6*(3), 48.

Gupta, P. K., Yadav, B., & Yadav, B. K. (2019). Assessment of LNAPL in subsurface under fluctuating groundwater table using 2D sand tank experiments. *Journal of Environmental Engineering, 145*(9), 04019048.

Jahangeer., Gupta, P. K., & Yadav, B. K. (2017). Transient water flow and nitrate movement simulation in partially saturated zone. *Journal of Irrigation and Drainage Engineering, 143*(12), 04017048.

Leichombam, S., & Bhattacharjya, R. K. (2019). New hybrid optimization methodology to identify pollution sources considering the source locations and source flux as unknown. *Journal of Hazardous, Toxic, and Radioactive Waste, 23*(1), 04018037.

Logan, J. D. (1996). Solute transport in porous media with scale-dependent dispersion and periodic boundary conditions. *Journal of Hydrology, 184*(3−4), 261−276.

Majumder, P., & Eldho, T. I. (2019). Reactive contaminant transport simulation using the analytic element method, random walk particle tracking and kernel density estimator. *Journal of Contaminant Hydrology, 222*(December 2018), 76−88, Elsevier.

Majumder, P., & Eldho, T. I. (2020). Artificial neural network and grey wolf optimizer based surrogate simulation-optimization model for groundwater remediation. *Water Resources Management, 34*(2), 763−783.

Mobile, M. A., Widdowson, M. A., & Gallagher, D. L. (2012). Multicomponent NAPL source dissolution: evaluation of mass-transfer coefficients. *Environmental Science & Technology, 46*(18), 10047−10054.

Moran, M. J., Zogorski, J. S., & Squillace, P. J. (2007). Chlorinated solvents in groundwater of the United States. *Environmental Science & Technology, 41*(1), 74−81.

Morway, E. D., Niswonger, R. G., Langevin, C. D., Bailey, R. T., & Healy, R. W. (2012). Modeling variably saturated subsurface solute transport with MODFLOW-UZF and MT3DMS. *Ground Water, 51*(2), no-no.

Muskus, N., & Falta, R. W. (2018). Semi-analytical method for matrix diffusion in heterogeneous and fractured systems with parent-daughter reactions. *Journal of Contaminant Hydrology, 218* (October), 94−109, Elsevier.

Natarajan, N., Vasudevan, M., & Kumar, G. S. (2020). Simulating scale dependencies on dispersive mass transfer in porous media under various boundary conditions. *Iranian Journal of Science and Technology, Transactions of Civil Engineering, 44*, 375−393.

Pal, J., & Chakrabarty, D. (2020). Assessment of artificial neural network models based on the simulation of groundwater contaminant transport. *Hydrogeology Journal, 28*(1−2), 1−17.

Pathania, T., & Eldho, T. I. (2020). Optimal in situ bioremediation system design for contaminated groundwater using meshless EFGM simulation and PSO. *World Environmental and Water Resources Congress 2020* (pp. 115−124). Reston, VA: American Society of Civil Engineers.

Pathania, T., Eldho, T. I., & Bottacin-Busolin, A. (2020). Optimal design of in-situ bioremediation system using the meshless element-free Galerkin method and particle swarm optimization. *Advances in Water Resources, 144*(July), 103707.

Pickens, J. F., & Grisak, G. E. (1981). Scale-dependent dispersion in a stratified granular aquifer. *Water Resources Research, 17*(4), 1191−1211, Wiley Online Library.

Rao, G. T., Rao, V. V. S. G., Ranganathan, K., Surinaidu, L., Mahesh, J., & Ramesh, G. (2011). Assessment of groundwater contamination from a hazardous dump site in Ranipet, Tamil Nadu, India. *Hydrogeology Journal, 19*(8), 1587−1598.

Rao, G. T., Rao, V. V. S. G., Surinaidu, L., Mahesh, J., & Padalu, G. (2013). Application of numerical modeling for groundwater flow and contaminant transport analysis in the basaltic terrain, Bagalkot, India. *Arabian Journal of Geosciences, 6*(6), 1819−1833.

Raoof, A., & Hassanizadeh, S. M. (2013). Saturation-dependent solute dispersivity in porous media: Pore-scale processes. *Water Resources Research, 49*(4), 1943−1951.

Rasa, E., Chapman, S. W., Bekins, B. A., Fogg, G. E., Scow, K. M., & Mackay, D. M. (2011). Role of back diffusion and biodegradation reactions in sustaining an MTBE/TBA plume in alluvial media. *Journal of Contaminant Hydrology, 126*(3−4), 235−247.

Rolle, M., Hochstetler, D., Chiogna, G., Kitanidis, P. K., & Grathwohl, P. (2012). Experimental investigation and pore-scale modeling interpretation of compound-specific transverse dispersion in porous media. *Transport in Porous Media, 93*(3), 347−362.

Russo, A., Johnson, G. R., Schnaar, G., & Brusseau, M. L. (2010). Nonideal transport of contaminants in heterogeneous porous media: 8. Characterizing and modeling asymptotic contaminant-elution tailing for several soils and aquifer sediments. *Chemosphere, 81*(3), 366−371.

Sander, G. C., & Braddock, R. D. (2005). Analytical solutions to the transient, unsaturated transport of water and contaminants through horizontal porous media. *Advances in Water Resources, 28*(10), 1102−1111.

Sathe, S. S., & Mahanta, C. (2019). Groundwater flow and arsenic contamination transport modeling for a multi aquifer terrain: Assessment and mitigation strategies. *Journal of Environmental Management, 231*(April 2018), 166−181, Elsevier.

Selim, H. (2014). Transport & fate of chemicals in soils: Principles & applications. Boca Raton: CRC Press.

Sharma, P. K., Agarwal, P., & Mehdinejadiani, B. (2020). Study on non-Fickian behavior for solute transport through porous media. *ISH Journal of Hydraulic Engineering, 00*(00), 1−9.

Sharma, P. K., & Srivastava, R. (2012). Concentration profiles and spatial moments for reactive transport through porous media. *Journal of Hazardous, Toxic, and Radioactive Waste, 16*(2), 125−133.

Sharma, P. K., Sekhar, M., Srivastava, R., & Ojha, C. S. P. (2012). Temporal moments for reactive transport through fractured impermeable/permeable formations. *Journal of Hydrologic Engineering, 17*(December), 1302−1314.

Sharma, P. K., Shukla, S. K., Choudhary, R., & Swami, D. (2016). Modeling for solute transport in mobile−immobile soil column experiment. *ISH Journal of Hydraulic Engineering, 22*(2), 204−211, Taylor & Francis.

Singh, M. K., Singh, R. K., & Pasupuleti, S. (2020). Study of forward−backward solute dispersion profiles in a semi-infinite groundwater system. *Hydrological Sciences Journal, 65*(8), 1416−1429, Taylor & Francis.

Swami, D., Sharma, P. K., & Ojha, C. S. P. (2016). Behavioral study of the mass transfer coefficient of nonreactive solute with velocity, distance, and dispersion. *Journal of Environmental Engineering, 143*(1), 1−10.

Swami, D., Sharma, P. K., Ojha, C. S. P., Guleria, A., & Sharma, A. (2018). Asymptotic behavior of mass transfer for solute transport through stratified porous medium. *Transport in Porous Media, 124*(3), 699−721, Springer Netherlands.

Tartakovsky, A. M., & Neuman, S. P. (2008). Effects of Peclet number on pore-scale mixing and channeling of a tracer and on directional advective porosity. *Geophysical Research Letters, 35* (21), L21401.

USEPA. (1997). State source water assessment and protection programs guidance, final guidance. *Report number 816-R-97-009.* Washington, DC, USA.

Valsala, R., & Govindarajan, S. K. (2018). Interaction of dissolution, sorption and biodegradation on transport of BTEX in a saturated groundwater system: Numerical modeling and spatial moment analysis. *Journal of Earth System Science, 127*(4), 53.

Valsala, R., & Govindarajan, S. K. (2019). Co-colloidal BTEX and microbial transport in a saturated porous system: numerical modeling and sensitivity analysis. *Transport in Porous Media, 127*(2), 269−294.

van Genuchten, M. T., & Wierenga, P. J. (1976). Mass transfer studies in sorbing porous media I. Analytical solutions. *Soil Science Society of America Journal, 40*(4), 473−480.

Vasudevan, M., Suresh Kumar, G., & Nambi, I. M. (2014). Numerical study on kinetic/equilibrium behaviour of dissolution of toluene under variable subsurface conditions. *European Journal of Environmental and Civil Engineering, 18*(9), 1070−1093, Taylor & Francis.

Vasudevan, M., Suresh Kumar, G., & Nambi, I. M. (2016). Numerical modelling on rate-limited dissolution mass transfer of entrapped petroleum hydrocarbons in a saturated sub-surface system. *ISH Journal of Hydraulic Engineering, 22*(1), 3−15.

Yadav, B., Mathur, S., Ch, S., & Yadav, B. K. (2018). Simulation-optimization approach for the consideration of well clogging during cost estimation of in situ bioremediation system. *Journal of Hydrologic Engineering, 23*(3), 04018001.

Yadav, R. R., Jaiswal, D. K., & Gulrana. (2012). Two-dimensional solute transport for periodic flow in isotropic porous media: An analytical solution. *Hydrological Processes, 26*(22), 3425−3433.

Yang, M., Annable, M. D., & Jawitz, J. W. (2015). Back diffusion from thin low permeability zones. *Environmental Science & Technology, 49*(1), 415−422.

Yang, M., Annable, M. D., & Jawitz, J. W. (2017a). Forward and back diffusion through argillaceous formations. *Water Resources Research, 53*(5), 4514−4523.

Yang, M., Annable, M. D., & Jawitz, J. W. (2017b). Field-scale forward and back diffusion through low-permeability zones. *Journal of Contaminant Hydrology, 202*(February), 47−58, Elsevier.

Zheng, C., & Bennett, G. D. (2002). *Applied Contaminant Transport Modeling*. New York: Wiley-Interscience.

Zheng, C., & Wang, P.P. (1999). *MT3DMS: A modular three-dimensional multi-species transport model for simulation of advection, dispersion, and chemical reactions of contaminants in groundwater systems; documentation and user's guide*. Alabama Univ University.

CHAPTER 9

Impacts of climatic variability on subsurface water resources

Pankaj Kumar Gupta[1], Brijesh Kumar Yadav[2] and Devesh Sharma[3]

[1]Wetland Hydrology Research Laboratory, Faculty of Environment, University of Waterloo, Waterloo, ON, Canada, [2]Department of Hydrology, Indian Institute of Technology Roorkee, Roorkee, India, [3]Department of Atmospheric Sciences, Central University of Rajasthan, Ajmer, India

9.1 Introduction

World's water resources are under severe stress by experiencing the climate changes during the last few decades. Observational records and climate projections provide abundant evidence that like other ecosystems, subsurface water resources are vulnerable and have the potential to be strongly affected by climate change with wide-ranging consequences (Green et al., 2011). Impact of climate change on the subsurface water resources is further compounded due to accelerating demands arising from increasing population, urbanization, deforestation, and intensification of agriculture. Subsurface water resources stored in confined and unconfined aquifers, including vadose zone moisture content, have both physical and functional interaction with ground surface and its associated environmental conditions (Holman, 2006). Thus the subsurface water resource is highly vulnerable to extreme climate events, sea-level rise, and increasing pollution loads (Bovolo et al., 2009; Gupta, 2020). With the growing threat of more frequent and intense climatic disasters, including, but not limited to, droughts, floods, and pollution, there is a strong need to assess the impact of climate change on management of subsurface water resources.

Any alternation in climatic conditions directly affects the soil−water−atmospheric interaction resulted in quantitative and qualitative changes in subsurface water resources. The climatic variabilities affect the subsurface physical interaction by changing the land use/cover pattern and soil moisture regimes (Bates et al., 2008). The functional interaction is mainly caused by increasing evapotranspiration (ET) rates resulting from rising temperature (Woldeamlak et al., 2007). All these changes are responsible for reduced infiltration rates

Advances in Remediation Techniques for Polluted Soils and Groundwater. DOI: https://doi.org/10.1016/B978-0-12-823830-1.00003-1

arising from precipitation ultimately decreasing the groundwater recharge. Furthermore, shifting of seasons and rainfall intensity significantly influence the timing and magnitude of subsurface recharge consequently changing the subsurface water storage and flow conditions (Van Dijck et al., 2006). Sherif and Singh (1999) found that higher variability in precipitation and in temperature reduces subsurface recharge in general. The rising temperature due to climate change also causes high evaporation losses from streams/rivers resulting in increased hydraulic gradient between river and groundwater systems (Burnett et al., 2006). This leads to enhanced discharge rate of groundwater toward the streams/rivers system (Wang et al., 2009). These variations not only change the subsurface yield or discharge but also modify the groundwater flow regime, for example, gaining streams may suddenly become losing streams, and movement of groundwater divides. The reduced subsurface recharge and increased groundwater discharge alter the moisture flow pattern affecting the biogeochemical characteristics of the subsurface system (Green et al., 2011).

Studies have also reported considerable changes in subsurface thermal regimes with ambient temperature in long term (Taniguchi et al., 2007). In subsurface environment, heat is transported by the plant—soil—atmospheric interactions in form of convective fluxes. Thus the heat transport in subsurface is a function of the ambient atmospheric temperature, land cover and land use patterns, subsurface thermal properties, and the soil moisture flow patterns (Miyakoshi et al., 2005). The increasing ambient temperature due to climate change and land use disturbances causes increased heat gradient between the ground surface and subsurface, and between aquifer and stream/rivers (Dragoni & Sukhija, 2008). This enhances the heat flow toward the groundwater resources either from ground surface or stream/rivers, causing subsurface water warming. Shallow groundwater resources are highly vulnerable to subsurface water warming due to enhanced evaporation of pore water from partially saturated zone (Taniguchi et al., 2007). This can change soil—water storage, groundwater flow patterns, and geochemical reactions of subsurface. Changes in biogeochemical interactions may take place due to subsurface water warming that can affect the pollution load on soil and groundwater resources. The heat propagated to sufficient depth can also alter metabolic actions of soil microbiota responsible for degrading subsurface pollutants.

Impacts of climate change on subsurface water quality are not investigated thoroughly, and very few studies were found relevant. Green et al., (2011) showed that the indirect impacts of climate change are more likely to affect subsurface water quality issues. The geochemical makeup of subsurface environment is under direct influences of the surface or atmospheric conditions, and thus the variations in ambient temperatures, precipitations, streamflow, and other dominating variables affect the subsurface water quality

(Klein & Nicholls, 1999; Sherif & Singh, 1999; Pierson et al., 2001; IPCC, 2007a, 2007b; Kundzewicz et al., 2007; Ranjan et al., 2006). At the same time the pollutions due to release of several pollutants are a major concern to subsurface water quality. Natural attenuation of these pollutants is likely to be affected by environmental variability as microorganisms present in subsurface work differently under various environmental conditions. Further, rapid groundwater table fluctuations along with high pore-water velocities are expected in shallow aquifers under climate change conditions (Dobson et al., 2007) which ultimately affect subsurface water resources. Changing groundwater flow velocity and groundwater table dynamics causes more strong advective transport and enhances the dissolution rate of immiscible pollutants.

Climate-driven changes in subsurface water resources play more prominent role in (1) semiarid and arid regions, where climate variability and changes accelerate high ET rates and cause the significant subsurface water losses that ultimately lead to the high salinity problems; (2) coastal regions, where the sea-level rises cause the reversal of hydraulic gradient between the sea and groundwater. This reversal of hydraulic gradient causes the sea water intrusion toward aquifer system, which may cause the high salinity and deplete the quality of fresh groundwater (Bear & Cheng, 1999); (3) temperate and polar areas, melting of snowpack/glaciers plays a crucial role in groundwater recharge.

The main focus of this chapter is to present a general overview on impacts of climate variability on subsurface water resources. The impact of climatic variables on subsurface water resources storage volume is discussed first. Thereafter, the role of varying climatic conditions on groundwater flow, geo-hydrological characteristics, and fate and transport of pollutants is discussed thoroughly. Different aspects of modeling and practical approaches to quantify the climate change impacts are also reviewed comprehensively. Finally, a methodological framework is proposed considering the technical and socioeconomic aspects to mitigate the impact of climate change conditions on subsurface water resources. The information provided in this chapter is of direct use in planning for management of subsurface soil—water resources under site prevailing dynamic environmental conditions.

9.2 Impacts on atmospheric boundary

At global scale, variations in precipitation caused by El Nino Southern Oscillation (ENSO) and Pacific Decadal Oscillation (PDO) generally result in variations in streamflow and associated subsurface recharge (Redmond &

Koch, 1991; Simpson et al., 1993). Elevated ENSO precipitation could also increase subsurface recharge. Hanson et al. (2004) and Venencio (2002) found that the increased precipitation and streamflow related to ENSO increase the subsurface recharge in California and Argentina. Similarly, the monsoon in Indian subcontinent significantly amplified the subsurface recharge by increased precipitation.

Climatic variability at regional scale can change water table levels in aquifers due to fluctuating subsurface recharge flux (Zektser & Loaiciga, 1993). Scanlon et al. (2006) reviewed subsurface recharge from precipitation/irrigation and found that an average of 3%−15% of surface water is able to reach the groundwater resources for its recharge. They also highlighted that the subsurface recharge is a function of evaporation and other variables, which may lead in reversal of hydraulic gradients toward gaining or losing streams. Changnon et al. (1988) established the relationship between the monthly precipitation and shallow groundwater table across Illinois for the duration of 1960−84. They found that the monthly precipitation significantly affects the recharge rate and consecutively groundwater level in the study area. Other than precipitation, the land use and land cover pattern, which are directly affected by the climatic variabilities, can alter the subsurface recharge. Shifts in natural vegetation due to climate variability may also influence subsurface recharge. Bromley et al. (1997) suggested that recharge may contribute to increased soil moisture content below the root zone of millet fields in Africa.

A drastic change in subsurface water recharge occurs during periods of snowfall and its melting. Precipitation in form of snow in cold regions creates impervious layer of frozen soil that acts as a barrier to subsurface recharge. In other hands, melting water from snow-covered topography commonly leads to accelerate subsurface recharge by providing a steady source of water to the downstream areas (Grasby & Chen, 2005). Earman et al. (2006) showed that the snowmelt is the major contributor of recharge in the western mountain regions of the United States. Dettinger et al. (2004) predicted that snow water amount will decline by 3%−79%, resulting in significant changes to spatial distribution and amount of subsurface recharge by the end of the 21st century in Sierra Nevada. Heavy recharge rates from the Himalayan snowmelt regions are also reported in Indo-Gangetic aquifer systems. As snowmelt-based runoff increases, contribution to subsurface recharge increases from the streamflow in the Indo-Gangetic basin as reported by Singh and Arora (2007). Thus the climatic variability affects subsurface recharge significantly due to alternation in precipitation, evaporation, land use/cover, glaciation, snowfall/melts, etc. The surface recharge also affects the water storage and soil moisture distribution in subsurface as elaborated next.

9.3 Impacts on water storage and flow pattern

Subsurface water storage represents the portion of water that penetrates into the variably saturated zones and is stored in pore space. Based on the availability of soil moisture, subsurface is divided into three main zones, that is, unsaturated or partially saturated zone, capillary zone, and saturated zones as shown in Fig. 9.1.

Generally, subsurface water storage and moisture flow depend on the hydrogeological conditions of the area. However, climatic variability in form of precipitation, ET, and surface water conditions substantially controls water storage and soil moisture content (Jasper et al., 2006). Alley (2001) reported critical impacts of climate change and variability on subsurface water storage. This is noted that the increased precipitation enriched the subsurface water storage by amplified recharge rate, but at the same time, vegetation growth and intense pumping of water may deplete the subsurface water storage. The increased ET under climate change conditions magnifies the water losses from subsurface via soil—plant—atmospheric interactions. Especially in semi-arid and arid regions, ET is a dominating climatic variable for excess loss of subsurface water. This condition may also accelerate due to intensive withdrawal of subsurface water that can significantly affect the water balance of the area. Few studies highlighted that subsurface water storage is strongly influenced by the intensive withdrawal than the climatic variability in most regions. A comparative account of the role of ET, root water uptakes, and direct withdrawal on subsurface water storage is represented in Fig. 9.2 considering the water budget approach.

Soil moisture flow through partially saturated zone depends on the soil matric potential, which is less than atmospheric pressure and known as zone

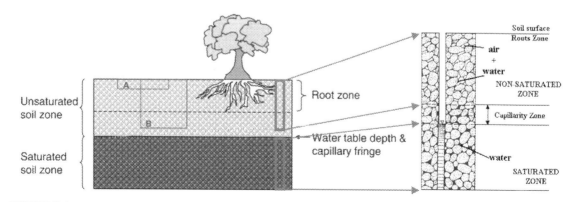

FIGURE 9.1

Distinction between unsaturated and saturated zones. Labels (A) and (B) denote two distinct soil moisture volumes.

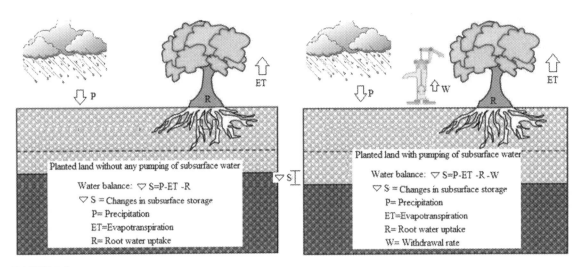

FIGURE 9.2
Representation of subsurface water storage scenarios in: (1) planted land without any pumping and (2) with pumping of subsurface water. The term R is a sink term representing root water uptake from vadose zone.

of suction head that binds water to soil solids in this zone. Therefore the climatic boundary conditions strongly affect the soil moisture flow pattern in subsurface system. Seneviratne et al. (2010) provided an extensive review of interactions between soil moisture content and climatic boundary conditions. Subsurface moisture flow and water storage responses to direct climate-driven changes in precipitation and recharge rates are likely to be slower than surface water responses. Soil–water system shows comparatively greater temporal stability to changes in climate conditions. As a result, short-term, higher frequency variations in recharge can be effectively filtered out at aquifer scale (Bredehoeft, 2011). These factors coupled with the relatively modest magnitude of the long-term changes may make it difficult to distinguish storage changes due to local climate variability, anthropogenic impacts, and/or the global climate change events (Bredehoeft, 2011; Kuss & Gurdak, 2014; Loaiciga, 2009). Direct climate-related storage responses in soil–water system that serve as sources of water supply are likely to be sensitive to (1) volumetric scale and hydraulic properties of the (un)saturated zones, (2) depth of the aquifer, and (3) length of the flow path between the point of recharge and supply well. In very large and deep zones, more time may require for recharging flux to reach groundwater storage particularly in downgradient sides from the recharging zone. Therefore storage conditions in shallow aquifers with shorter flow paths, often present in higher elevation settings, are likely to be the most sensitive to direct changes in recharge.

Subsurface through which soil moisture flow takes place represents the complex vadose zone of soil—plant—water—atmospheric continuum that directly affects the underlying groundwater quantity and quality. Thus root water uptake and return flow from root zone significantly affect the subsurface water storage and soil moisture flow regimes. Spatial and temporal dynamic process of vadose zone is influenced deeply by plant, soil, and prevailing climatic conditions. When root systems span through soil layers of different moisture contents, water is excreted by roots biomass according to root density and soil moisture availability (Burgess et al., 2001; Yadav & Mathur, 2008).

9.4 Impacts on ground—surface water interactions

Changes in patterns of flow in subsurface are the earliest and most noticeable direct consequences of climate change on ground—surface water interactions (Brouyere et al., 2004). The indirect impacts of climate change, like increase in groundwater pumping, could lead to large reductions in natural groundwater discharge to surface water resources. Thus climate change causes the reversal of the hydraulic head between groundwater and surface water streams (Dams et al., 2007). In general, perennial streams gain water from base flow through their sides when they flow through groundwater system. Accelerated ET and reduced precipitations due to climatic variation can cause the reversal of hydraulic head (Burnett et al., 2006). On contrary, snowmelt contributes additional water in steams that raise the steam water level inducing intrusion of stream water in aquifers (Walvoord & Striegl, 2007). Fig. 9.3 shows the behavior of saturated zones by reversal of the hydraulic head between groundwater and surface streams. Damming of streams and pumping of aquifers have dramatically altered natural stream/groundwater interactions in areas of heavy water demands for irrigation, industry, and domestic.

Further, damming raises groundwater table in upstream causing some losing streams to become gaining streams and lowers the water table downstream, causing gaining streams to become losing type. Heavy pumping has caused

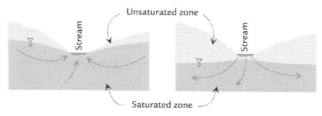

FIGURE 9.3
Cross sections of a gaining stream (left) and a losing stream (right).

some small gaining streams to become losing type because of lowering of groundwater table. Furthermore, hydrogeological settings play a key role in determining the streamflow recession and base flow characteristics of a watershed, which, in turn, significantly influence streamflow sensitivity to climate change.

9.5 Impacts on subsurface water quality

Impact of climate change on groundwater quality is not concluded directly due to high degree of uncertainty involved in the correlation. Inferences from relevant studies are listed next.

- Recharge rate, which keeps changing due to direct or indirect climate impacts, determines the degree of groundwater quality responses to climate change (Alley, 2001).
- Extreme rainstorm intensity related to climate change may increase downward flux of solutes (e.g., nitrate, chloride) present in vadose zone (Taylor et al., 2013). Alternatively, intense storms could produce precipitation rates that quickly exceed soil infiltration capacities. The related reductions in recharge could, in turn, reduce leaching rates.
- In arid and semiarid areas, where larger reservoirs of soluble nitrate have accumulated over time in vadose zone due to intensive long-term agricultural production and fertilization practices, increased storm intensity related to climate change may flush additional nitrate mass to underlying groundwater table (Dragoni & Sukhija, 2008; Gurdak et al., 2007; Kløve et al., 2014).
- With increase in temperature due to climate change, evaporation, atmospheric CO_2 concentration, plant growth, and crop water demand are expected to increase. Under these circumstances, profitable·growing season may lengthen; in some settings, farmers may even decide to increase the total number of planting/harvest cycles per year (Himanshu et al., 2021). These changes are likely to lead to increase in demand for irrigation water, fertilizer, and pesticides applied to crops that ultimately increase potential for the chemical leaching to underlying groundwater. Since different crops have different fertilization requirements and leaching potentials, changes in crop types in response to climate change could lead to an increase in subsurface water quality impacts (Earman & Dettinger, 2011).
- Reductions in depth of water table and shortening of residence times of solutes in the soil column and vadose zone due to an increase in irrigation rate could also increase the vulnerability of shallow aquifers to contamination (Dobson et al., 2007).

- Increase in rainfall intensity and surface flooding expected to accompany under climate change may drive to expansion of surface water control infrastructure for routing larger volume of runoff to subsurface (Amrit et al., 2019; Himanshu et al., 2018). This could potentially increase the amount of dissolved toxic chemicals and nutrients in runoff ultimately increasing the vulnerability of shallow aquifers to contamination (Clifton et al., 2010; Green et al., 2011).
- Ficklin et al., (2010) suggested that climate-related reductions in subsurface recharge rates in semiarid irrigated settings like the San Joaquin Valley, California could lead to a reduction in the leaching of agricultural chemicals to the underlying groundwater system.
- Temperature is a key factor in reaction kinetics and dissolved oxygen concentrations, and hence, small change in subsurface water temperatures could have a significant impact on subsurface water quality (Gunawardhana & Kazama, 2012). Increased subsurface temperature may alter the geochemical processes (particularly redox reactions) that can exert control on the dissolved concentration and mobility of a wide variety of chemical contaminants (Basu et al., 2020; Gupta & Yadav, 2017, 2019a, 2019b, 2020; Gupta & Bhargava, 2020) (e.g., nutrients, trace metals, iron, and manganese) (Destouni & Darracq, 2009). This may be of particular concern for supply wells that derive their water from riverbank infiltration. As stream temperatures rise and dissolved oxygen concentrations decrease, the reducing conditions created in near-stream groundwater could in turn mobilize solid-phase iron and manganese (Kurylyk et al., 2014). Rising temperatures could potentially increase soil mineralization rates of organic N to nitrate, leading to an increase in potential for nitrate leaching to the water table in agricultural areas that receive large surface loads of N-bearing fertilizers (Stuart et al., 2011).
- Unexpected changes in base flow to streams are likely to play a major role in altering subsurface water quality. Groundwater base flow can help to dilute contaminant concentrations in streams. If base flow declines due to changes in climate, the beneficial effects of dilution in rivers and streams may be reduced.
- Deeper subsurface water has typically experienced a longer residence time in subsurface. Longer the subsurface water in contact with the geologic matrix, more mineral dissolution is expected. As a result, older and deeper subsurface water typically has a higher dissolved mineral content (Freeze & Cherry, 1979).
- Deep subsurface water may also have a lower dissolved oxygen concentration than shallow water, potentially leading to reducing conditions that can dissolve and mobilize metals such as iron or manganese.

A large number of chemical substances often find their way into the environment either released from agricultural practices or leaked from industrial and municipal waste disposal sites. Modern agriculture uses an unprecedented number of chemicals, both in plant and animal production. A broad range of fertilizers, pesticides, and fumigants are now routinely applied to agricultural lands, making agriculture one of the most significant sources for nonpoint source pollution. Likewise, dairy farms are responsible for groundwater nitrate pollution in form of nonpoint sources. The industrial waste like petroleum hydrocarbons released in (sub)surface represents point source pollution cases.

When released in sufficient amount at the (sub)surface, the pollutants move downward through the unsaturated zone by soil moisture flow, gravity, and diffusion. The concentration gradient causes the molecular diffusion of pollutant in soil–water system. The pollutants like nonaqueous phase liquids (NAPLs) in partially saturated zone (Yadav & Hassanizadeh, 2011) cause multiphase partitioning (i.e., air, aqueous, solid, and pure phase). Thus pore air is also contaminated due to volatilization of these hydrocarbon pollutants in partially saturated zones. Likewise, the mass of the pollutants in solid phase causes adsorption on soil solids and serve as long-term pollutants in subsurface. The pure phase light NAPL pollutants are retained in capillary zone and then start dissolving in aqueous phase in moving groundwater resources. Hydrocarbon pollutants like dense NAPL penetrate the groundwater table and keep moving in downward direction till they are retained by impermeable layer or rock strata (USEPA US Environmental Protection Agency, 1995) and the dissolved and pure phase NAPLs move to surrounding locations due to advection, diffusion, and dispersion mechanisms of mass transport (Dobson et al., 2007; Powers et al., 1991). The following text describes the effects of climate change on fate and transport of these pollutants in subsurface.

- Environmental variability directly affects the subsurface water resources in form of frequent fluctuations in groundwater table and its flow velocities particularly in shallow unconfined aquifers. Rapid groundwater table fluctuations along with high pore-water velocities can enhance the mobilization of NAPL pollutants considerably (Dobson et al., 2007).
- Dynamics of groundwater table accelerate (up)downward movement of the NAPL pool causing their entrapment in pore space, which ultimately increases their coverage area. Pollutants entrapped in the form of isolated blobs or ganglia in pore spaces increase water interfacial area responsible for their enhanced dissolution rates in aqueous phase (Soga et al., 2004). Thus pollutants trapped in the porous media act as long-lasting sources of groundwater pollution (Yadav & Hassanizadeh, 2011).

- Large subsurface zones are polluted by advective and dispersive flux under high groundwater flow velocities caused by climatic variations. The fast groundwater velocity also enhances dissolution of pollutants, which may increase the pollution load in downgradient locations (Gupta & Yadav, 2017).

- Ratio of soil moisture and air in unsaturated zone has direct effect on pollutant\s fate and transport under varying climatic conditions (Yadav & Hassanizadeh, 2011). The low soil moisture content results in greater air-filled porosity, which should improve oxygen mass transfer to pollutants-degrading microbial assemblage. However, there is likely to be a trade-off between improved oxygen availability and soil moisture content (Alvarez & Illman, 2006; Arora et al., 1982).

- Likewise, the mass transfer of pollutants through soil is dependent on soil—water content (English & Loehr, 1991). At high soil moisture contents, pollutants mass transfer is impeded due to reduced air-filled porosity and partitioning of pollutants into soil—water phase (Papendick & Campbell, 1981).

- However, pollutant movement is get retarded by adsorption on organic and/or mineral components of soil solids when air-filled porosity increases at low water contents (Petersen et al., 1994).

- Likewise, temperature plays a significant role in controlling the nature and extent of microbial metabolisms that are responsible for degradation of several pollutants (Yadav & Hassanizadeh, 2011). Bioavailability and solubility of pollutants are also temperature dependent. Low-temperature conditions usually result in increased viscosity, reduced volatilization, and decreased water solubility of pollutants, and thus delayed onset of biodegradation process (Margesin & Schinner, 2001).

- Temperature changes at Earth's surface propagate slowly downward into the porous media beneath the surface and modify the ambient thermal regime in subsurface environment. Thus subsurface temperatures provide direct relation to temperature changes that have occurred at the surface due to climate change (Davis et al., 2010). (Smerdon et al., 2009). This increased heat in subsurface creates geothermal anomalies and enhances water losses by evaporation from subsurface (Chapman et al., 1992). Subsurface warming also causes the alternation in nutrient cycling, especially nitrogen and carbon cycle (Yadav & Hassanizadeh, 2011).

- To sum up the climatic variation response to dynamic soil moisture flow, nature of underlying groundwater flow, ambient temperature profile, and pollutant regimes affect not only the microbial degradation rate but also the transport of soil air and substrate throughout the variably saturated zone.

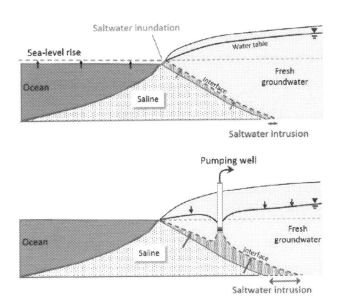

FIGURE 9.4
Schematic diagram representing the saltwater interface in coastal settings.

Climate change is responsible for raising the sea level annually by 1 cm, which directly accelerates the sea water intrusion in coastal aquifers. Furthermore, groundwater overdraft causes the sea–water interface to move toward inland primarily during high demand of freshwater (Fig. 9.4). Following impacts of sea-level rise on saline water intrusion in coastal zones are observed.

- The increased hydraulic head due to sea-level rises not only accelerates the saline water intrusion by moving the interface but also inundates low elevation portions of a coastline with seawater (Mauger, 2015).
- The coastal storm surges related to climate change can infiltrate downward, salinizing the subsurface water resources.
- It is expected that climate-driven droughts could lower the average recharge of freshwater. Reduction in recharge subsequently can increase saltwater intrusion in areas already susceptible to such problems.

9.6 Methodological framework for evaluating climate change impacts on subsurface

Climate change impacts on regional water resources are particularly critical for developing countries like India and are expected to intensify the situation in already water-stressed regions of the country. Predictions of climate response using coarse resolution models may be useful at large scales; however, for better management of regional water resources, impacts need to be assessed at finer scales especially for the groundwater resources. Thus methodologies are required to develop reliable regional climate change impact assessments.

Global atmospheric circulation models (GCMs) have been used directly to simulate hydrological components under prevailing climatic conditions and to predict the impact of climatic change on water resources at macroscale level. The use of direct GCM-derived hydrological output is helpful in evaluating different mechanisms involved in hydrological cycle, including subsurface water resources (Randall et al., 2007). Thus the numerical parameterization of the local hydrologic cycle using a GCM-derived data coupled with local hydrological models may help in regional-level climate change assessment. Dynamic downscaling approach like variable-resolution global models can support in better representation of subsurface water regimes (Hewitson & Crane, 2006; Wilby & Wigley, 1997). Further, regional-scale climatic variables such as precipitation and temperature are related to station-scale meteorological series that are evaluated by statistical downscaling approaches. Forecasting of climate change impacts needed to be made at multidecadal to century-long timescales, and therefore changes in subsurface conditions that are directly related to human activity or natural climatic cycles like the ENSO and the PDO are a prerequisite (Kuss & Gurdak, 2014).

Similarly, soil–water–plant–atmospheric relation is crucial to estimate the quantity and quality of subsurface water under varying climatic conditions. Thus the consideration of climatic variability and boundary conditions is required to estimate the various subsurface mass fluxes for climate change impact assessments. Role of land use/cover changes shall be incorporated at regional/local-level hydrological model. The qualitative assessments incorporating the fate and transport of subsurface pollutant are needed under climatic variation and boundary conditions using various mechanistic approaches. Fig. 9.5 shows the methodological framework for investigating the impact of climate change on subsurface water resources using pore- to global-scale models.

9.7   Conclusion and recommendations

Meteorological observations and related projections provide strong evidence that human society and ecosystem are vulnerable to climate change with wide-ranging consequences. Across the world, from the tropics to the poles, climate change is altering not only the average magnitude of precipitation

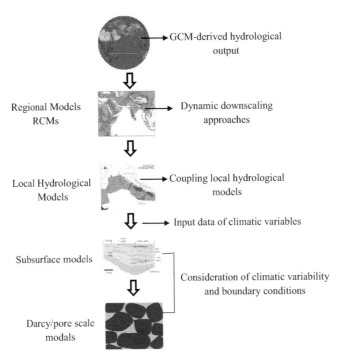

FIGURE 9.5
Suggested methodological framework for assessing the impact of climate changes on subsurface water resources.

and evaporation but also affecting seasonal pattern adversely that has reduced total water recharge to subsurface. Stress on subsurface water resources has also been increasing day by day due to additional demands from agriculture, domestic, and industrial needs. In this chapter, quantitative assessment of subsurface water storages and flow under climate change conditions are elaborated and exemplified first. The role of varying climatic conditions on geochemical makeup and pollutant transport is discussed next thoroughly. Different approaches to evaluate the climate change impacts on subsurface water resources are also reviewed comprehensively. A methodological framework is here suggested to evaluate global- to regional/local-level impact of climate change on subsurface water resources. Further, predicting future subsurface water availability and developing effective adaptation strategies for its sustainable management are required urgently under changing climate conditions.

Acknowledgment

Authors are extremely grateful Remwasol Remediation Technologies Private Limited.

References

Alley, W. M. (2001). Ground water and climate. *Ground Water, 39*(2), 161.

Alvarez, P. J. J., & Illman, W. A. (2006). *Bioremediation and natural attenuation, process fundamentals and mathematical models.* Wiley-Interscience, ISBN-10 0-471-65043-9.

Amrit, K., Mishra, S. K., Pandey, R. P., Himanshu, S. K., & Singh, S. (2019). Standardized precipitation index-based approach to predict environmental flow condition. *Ecohydrology, 12*(7) e2127.

Arora, H. S., Cantor, R. R., & Nemeth, J. C. (1982). Land treatment: A viable and successful method of treating petroleum industry wastes. *Environmental International, 7*, 285−291.

Basu, S., Yadav, B.K., Mathur, S., Gupta, P.K. (2020). In situ bioremediation of LNAPL polluted vadose zone: Integrated column and wetland study. *CLEAN - Soil, Air, Water*, 2000118.

Bates, B., Kundzewicz, Z. W., Wu, S., & Palutikof, J. P. (2008). *Climate change and water. Technical paper VI of the Intergovernmental Panel on Climate Change* (p. 210) Geneva: Intergovernmental Panel on Climate Change Secretariat.

Bear, J., & Cheng, H. D. (1999). *Seawater intrusion in coastal aquifers − Concepts, methods and practices* (p. 625) Dordrecht, Boston, London: Kluwer Academic Publisher.

Bovolo, C. I., Parkin, G., & Sophocleous, M. (2009). Groundwater resources, climate and vulnerability. *Environmental Research Letters, 4*(3), 035001.

Bredehoeft, J. D. (2011). Monitoring regional groundwater extraction: The problem. *Ground Water, 49*(6), 808−814.

Bromley, J., Edmunds, W. M., Fellman, E., Brouwer, J., Gaze, S. R., Sudlow, J., & Taupin, J. D. (1997). Estimation of rainfall inputs and direct recharge to the deep unsaturated zone of southern Niger using the chloride profile method. *Journal of Hydrology, 189*, 139−154.

Brouyere, S., Carabin, G., & Dassargues, A. (2004). Climate change impacts on groundwater resources: Modelled deficits in a chalky aquifer, Geer Basin, Belgium. *Hydrogeology Journal, 12* (2), 123−134.

Burgess, S. S. O., Adams, M. A., Turner, N. C., White, D. A., & Ong, C. K. (2001). Tree roots: Conduits for deep recharge of soil water. *Oecologia, 126*, 158−165.

Burnett, W. C., Aggarwal, P. K., Aureli, A., Bokuniewicz, H., Cable, J. E., Charette, M. A., Kontar, E., Krupa, S., Kulkarni, K. M., Loveless, A., Moore, W. S., Oberdorfer, J. A., Oliveira, J., Ozyurt, N., Povinec, P., Privitera, A. M. G., Rajar, R., Ramessur, R. T., Scholten, J., ... Turner, J. V. (2006).). Quantifying submarine groundwater discharge in the coastal zone via multiple methods. *The Science of the Total Environment, 367*(2−3), 498−543.

Changnon, sA., Huff, F. A., & Hsu, C.-F. (1988). Relations between precipitation and shallow groundwater in Illinois. *Journal of Climate, 1*, 1239−1250.

Chapman, D. S., Chisholm, T. J., & Harris, R. N. (1992). Combining borehole temperature and meteorologic data to constrain past climate change. *Global and Planetary Change, 6*(2−4), 269−281.

Clifton, C., Evans, R., Hayes, S., Hirji, R., Puz, G., & Carolina, P. (2010). *Water and climate change: Impacts on groundwater resources and adaptation options.* World Bank.

Dams, J., Woldeamlak, S. T., & Batelaan, O. (2007). Forecasting land-use change and its impact on the groundwater system of the Kleine Nete catchment, Belgium. *Hydrology and Earth System Sciences Discussions, 4*(6), 4265−4295.

Davis, M. G., Harris, R. N., & Chapman, D. S. (2010). Repeat temperature measurements in boreholes from northwestern Utah link ground and air temperature changes at the decadal time scale. *Journal of Geophysical Research, 115*(5), 22.

Destouni, G., & Darracq, A. (2009). Nutrient cycling and N$_2$O emissions in a changing climate: The subsurface water system role. *Environmental Research Letters*, 4(3), 035008.

Dettinger, M. D., Cayan, D. R., Meyer, M. K., & Jeton, A. E. (2004). Simulated hydrologic responses to climate variations and change in the Merced, Carson, and American River Basins, Sierra Nevada, California, 1900−2099. *Climatic Change*, 62, 283−317.

Dobson, R., Schroth, M. H., & Zeyer, J. (2007). Effect of water-table fluctuation on dissolution and biodegradation of a multi-component, light nonaqueous-phase liquid. *Journal of Contaminant Hydrology*, 94, 235−248.

Dragoni, W., & Sukhija, B. S. (2008). *Climate change and groundwater* (p. 186) Geological Society of London, Special Publications No. 288.

Earman, S., & Dettinger, M. (2011). Potential impacts of climate change on groundwater resources − A global review. *Journal of Water and Climate Change*, 02(4), 213−229. Available from http://www.iwaponline.com/jwc/002/jwc0020213.htm.

Earman, S., Campbell, A. R., Newman, B. D., & Phillips, F. M. (2006). Isotopic exchange between snow and atmospheric water vapour: Estimation of the snowmelt component of groundwater recharge in the southwestern United States. *Journal of Geophysical Research*, 111(D9), D09302. Available from https://doi.org/10.1029/2005JD006470.

English, C. W., & Loehr, R. C. (1991). Degradation of organic vapors in unsaturated soil. *Journal of Hazardous Material*, 28, 55−64.

Ficklin, D. L., Luedeling, E., & Zhang, M. (2010). Sensitivity of groundwater recharge under irrigated agriculture to changes in climate, CO2 concentrations and canopy structure. *Agricultural Water Management*, 97(7), 1039−1050.

Freeze, R. A., & Cherry, J. A. (1979). *Groundwater* (1st Ed., p. 604)Englewood Cliffs, NJ: Prentice-Hall, Inc.

Grasby, S., & Chen, Z. (2005). Subglacial recharge into the Western Canada sedimentary basin − Impact of Pleistocene glaciation on basin hydrodynamics. *Geological Society of America Bulletin*, 117(3−4), 500−514.

Green, T. R., Taniguchi, M., Kooi, H., Gurdak, J. J., Allen, D. M., Hiscock, K. M., Treidel, H., & Aureli, A. (2011). Beneath the surface of global change: Impacts of climate change on groundwater. *Journal of Hydrology*, 405(3-4), 532−560.

Gunawardhana, L. N., & Kazama, S. (2012). Statistical and numerical analysis of the influence of climate variability on aquifer water levels and groundwater temperatures: The impacts of climate change on aquifer thermal regimes. *Global and Planetary Change*, 86−87, 66−78.

Gupta, P. K., & Yadav, B. K. (2017). Bioremediation of non-aqueous phase liquids (NAPL$_S$) polluted soil and water resources. In R. N. Bhargava (Ed.), Environmental pollutants and their bioremediation approaches. Florida, USA: CRC Press, Taylor and Francis Group, ISBN 9781138628892.

Gupta, P. K., & Yadav, B. K. (2019a). Subsurface processes controlling reuse potential of treated wastewater under climate change conditions. In R. P. Singh, A. K. Kolok, & L. S. Bartelt-Hunt (Eds.), Water conservation, recycling and reuse: Issues and challenges. Singapore: Springer, 9789811331787 (ISBN).

Gupta, P. K., & Yadav, B. K. (2019b). Remediation and management of petrochemical polluted sites under climate change conditions. In R. N. Bhargava (Ed.), Environmental Contaminations: Ecological implications and management. Springer Nature Singapore Pte Ltd., ISBN 9789811379048.

Gupta, P. K., & Yadav, B. K. (2020). Three-dimensional laboratory experiments on fate and transport of light NAPL under varying groundwater flow conditions. *ASCE Journal of Environmental Engineering*, 146(4)04020010.

Gupta, P. K. (2020). Pollution load on Indian soil-water systems and associated health hazards: A review. *ASCE Journal of Environmental Engineering, 146*(5)03120004.

Gupta, P.K., Bhargava, R.N., (2020). *Fate and transport of subsurface pollutants*, Springer, eBook ISBN-978-981-15-6564-9. Available from https://www.springer.com/gp/book/9789811565632.

Gurdak, J. J., Hanson, R. T., McMahon, P. B., Bruce, B. W., McCray, J. E., Thyne, G. D., & Reedy, R. C. (2007). Climate variability controls on unsaturated water and chemical movement, High Plains aquifer, USA. *Vadose Zone Journal, 6*(3), 533–547.

Hanson, R. T., Newhouse, M. W., & Dettinger, M. D. (2004). A methodology to assess relations between climatic variability and variations in hydrologic time series in the southwestern United States. *Journal of Hydrology, 287*, 252–269.

Hewitson, B. C., & Crane, R. G. (2006). Consensus between GCM climate change projections with empirical downscaling: Precipitation downscaling over South Africa. *International Journal of Climatology, 26*(10), 1315–1337.

Himanshu, S. K., Pandey, A., & Patil, A. (2018). Hydrologic evaluation of the TMPA-3B42V7 precipitation data set over an agricultural watershed using the SWAT model. *Journal of Hydrologic Engineering, 23*(4)05018003.

Himanshu, S. K., Ale, S., Bordovsky, J. P., Kim, J., Samanta, S., Omani, N., & Barnes, E. M. (2021). Assessing the impacts of irrigation termination periods on cotton productivity under strategic deficit irrigation regimes. *Scientific Reports, 11*(1), 1–16.

Holman, I. P. (2006). Climate change impacts on groundwater recharge-uncertainty, shortcomings, and the way forward? *Hydrogeology Journal, 14*(5), 637–647.

IPCC. (2007a). Climate change (2007). the physical science basis. In S. Solomon, et al. (Eds.), *Contribution of working group I to the fourth assessment report of the intergovernmental panel on climate change* (p. 996). Cambridge, and New York: Cambridge University Press.

IPCC. (2007b). Climate change (2007). impacts, adaptation and vulnerability. In M. L. Parry, O. F. Canziani, J. P. Palutikof, P. Jvd Linden, & C. E. Hanson (Eds.), *Contribution of working group II to the fourth assessment report of the intergovernmental panel on climate change*. Cambridge, and New York: Cambridge University Press.

Jasper, K., Calanca, P., & Fuhrer, J. (2006). Changes in summertime soil water patterns in complex terrain due to climatic change. *Journal of Hydrology, 327*(3–4), 550–563.

Klein, R. J. T., & Nicholls, R. J. (1999). Assessment of coastal vulnerability to climate change. *Ambio, 28*(2), 182–187.

Kløve, B., Ala-Aho, P., Bertrand, G., Gurdak, J. J., Kupfersberger, H., Kværner, J., Muotka, T., et al. (2014). Climate change impacts on groundwater and dependent ecosystems. *Journal of Hydrology, 518*, 250–266.

Kundzewicz, Z. W., Mata, L. J., Arnell, N. W., Doll, P., Kabat, P., Jimenez, B., Miller, K. A., Oki, T., Sen, Z., & Shiklomanov, I. A. (2007). Freshwater resources and their management. In M. L. Parry, O. F. Canziani, J. P. Palutikof, P. J. van der Linden, & C. E. Hanson (Eds.), *Climate change 2007: Impacts, adaptation and vulnerability* (pp. 173–210). Cambridge: Cambridge University Press.

Kurylyk, B. L., MacQuarrie, K. T., & Voss, C. I. (2014). Climate change impacts on the temperature and magnitude of groundwater discharge from shallow, unconfined aquifers. *Water Resources Research, 50*(4), 3253–3274.

Kuss, A. J. M., & Gurdak, J. J. (2014). Groundwater level response in U.S. principle aquifers to ENSO, NAO, PDO, and AMO. *Journal of Hydrology, 519*, 1939–1952. Available from http://www.sciencedirect.com/science/article/pii/S0022169414007616.

Loaiciga, H. A. (2009). Long-term climatic change and sustainable ground water resources management. *Environmental Research Letters, 4*, 11.

Margesin, R., & Schinner, F. (2001). Biodegradation and bioremediation of hydrocarbons in extreme environments. *Applied Microbiology and Biotechnology, 56*(5), 650–663.

Mauger, G. S. (2015). *State of knowledge: Climate change in puget sound. Report prepared for the puget sound partnership and the National Oceanic and Atmospheric Administration* (pp. 281–p). Seattle: Climate Impacts Group, University of Washington.

Miyakoshi, A., Taniguchi, M., Okubo, Y., & Uemura, T. (2005). Evaluations of subsurface flow for reconstructions of climate change using borehole temperature and isotope data in Kamchatka. *Physics of the Earth and Planetary Interiors, 152*(4), 335–342.

Papendick, R. I., & Campbell, G. S. (1981). Theory and measurement of water potential. In J. F. Parr, W. R. Gardner, & L. F. Elliott (Eds.), *Water potential relations in soil microbiology* (pp. 1–22). Madison, WI: Soil Science Society of America.

Petersen, L. W., Rolston, D. E., Moldrup, P., & Yamaguchi, T. (1994). Volatile organic vapor diffusion and adsorption in soils. *Journal of Environmental Quality, 23*, 799–805.

Pierson, W. L., Nittim, R., Chadwick, M. J., Bishop, K. A., & Horton, P. R. (2001). Assessment of changes to saltwater/freshwater habitat from reductions in flow to the Richmond river estuary, Australia. *Water Science and Technology: A Journal of the International Association on Water Pollution Research, 43*(9), 89–97.

Powers, S. E., Loureiro, C. O., Abriola, L. M., & Weber, W. J. (1991). Theoretical study of the significance of non-equilibrium dissolution of nonaqueous-phase liquids in subsurface systems. *Water Resources Research, 27*(4), 463–477.

Randall, D. A., Wood, R. A., Bony, S., Colman, R., Fichefet, T., Fyfe, J., Kattsov, V., Pitman, A., Shukla, J., Srinivasan, J., Stouffer, R. J., Sumi, A., & Taylor, K. E. (2007). Climate models and their evaluation. In S. Solomon, D. Qin, M. Manning, Z. Chen, M. Marquis, K. B. Averyt, M. Tignor, & H. L. Miller (Eds.), *Climate change 2007: The physical science basis. contribution of working group I to the fourth assessment report of the intergovernmental panel on climate change* (pp. 589–662). Cambridge, and New York: Cambridge University Press.

Ranjan, S. P., Kazama, S., & Sawamoto, M. (2006). Effects of climate and land use changes on groundwater resources in coastal aquifers. *Journal of Environmental Management, 80*(1), 25–35.

Redmond, K. T., & Koch, R. W. (1991). Surface climate and streamflow variability in the western United States and their relationship to large-scale circulation indices. *Water Resources Research, 27*, 2381–2399.

Scanlon, B. R., Keese, K. E., Flint, A. L., Flint, L. E., Gaye, C. B., Edmunds, M., & Simmers, I. (2006). Global synthesis of groundwater recharge in semiarid and arid regions. *Hydrological Processes, 20*, 3335–3370.

Seneviratne, S. I., Corti, T., Davin, E. L., Hirschi, M., Jaeger, E. B., Lehner, I., Orlowsky, B., & Teuling, A. J. (2010). Investigating soil moisture–climate interactions in a changing climate: A review. *Earth-Science Reviews, 99*(3–4), 125–161.

Sherif, M. M., & Singh, V. P. (1999). Effect of climate change on sea water intrusion in coastal aquifers. *Hydrological Processes, 13*(8), 1277–1287.

Simpson, H. J., Cane, M. A., Herczeg, A. L., Zebiak, S. E., & Simpson, J. H. (1993). Annual river discharge in southeastern Australia related to El-Nino Southern oscillation forecasts of sea-surface temperatures. *Water Resources Research, 29*, 3671–3680.

Singh, P., & Arora, M. (2007). Water resources potential of Himalayas and possible impact of climate. *Jalvigyan Sameeksha, 22*, 109–132.

Smerdon, J. E., Beltrami, H., Creelman, C., & Stevens, M. B. (2009). Characterizing land surface processes: A quantitative analysis using air-ground thermal orbits. *Journal of Geophysical Research, 114*(D15), D15102.

Soga, K., Page, J. W. E., & Illangasekare, T. H. (2004). A review of NAPL source zone remediation efficiency and the mass flux approach. *Journal of Hazardous Materials, 110*(1−3), 13−27. Available from https://doi.org/10.1016/j.jhazmat.2004.02.034.

Stuart, M. E., Gooddy, D. C., Bloomfield, J. P., & Williams, A. T. (2011). A review of the impact of climate change on future nitrate concentrations in groundwater of the UK. *Science of the Total Environment, 409*(15), 2859−2873.

Taniguchi, M., Uemura, T. J. K., & Jago-on, K. (2007). Combined effects of heat island and global warming on subsurface temperature. *Vadose Zone Journal, 6*(3), 591−596.

Taylor, R. G., Scanlon, B., Döll, P., Rodell, M., Van Beek, R., Wada, Y., Longuevergne, L., et al. (2013). Ground water and climate change. *Nature Climate Change, 3*(4), 322−329.

USEPA (U.S. Environmental Protection Agency). (1995). *Light nonaqueous phase liquids*. Office of solid waste and emergency response, Washington, DC. EPA/540/S-95/500.

Van Dijck, S. J. E., Laouina, A., Carvalho, A. V., Loos, S., Schipper, A. M., Van der Kwast, H., Nafaa, R., Antari, M., Rocha, A., Borrego, C., & Ritsema, C. J. (2006). Desertification in northern Morocco due to effects of climate change on groundwater recharge. In W. G. Kepner, J. L. Rubio, D. A. Mouat, & F. Pedrazzini (Eds.), *Desertification in the Mediterranean region: A security issue* (pp. 549−577). Dordrecht: Springer.

Venencio M. V. (2002). Climate variability and ground water resources. In *Sixth international conference on Southern Hemisphere meteorology and oceanology*, Long Beach, CA, Feb. 2002, 142−144.

Walvoord, M. A., & Striegl, R. G. (2007). Increased groundwater to stream discharge from permafrost thawing in the yukon river basin: Potential impacts on lateral export of carbon and nitrogen. *Geophysical Research Letters, 34*, L12402.

Wang, T., Istanbulluoglu, E., Lenters, J., & Scott, D. (2009). On the role of groundwater and soil texture in the regional water balance: An investigation of the Nebraska Sand Hills, USA. *Water Resources Research, 45*(10), W10413.

Wilby, R. L., & Wigley, T. M. L. (1997). Downscaling general circulation model output: A review of methods and limitations. *Progress in Physical Geography, 21*, 530−548.

Woldeamlak, S. T., Batelaan, O., & De Smedt, F. (2007). Effects of climate change on the groundwater system in the Grote-Nete catchment, Belgium. *Hydrogeology Journal, 15*(5), 891−901.

Yadav, B. K., & Hassanizadeh, S. M. (2011). An overview of biodegradation of LNAPLs in coastal (semi)-arid environment. *Water, Air, and Soil Pollution, 220*(1−4), 225−239. Available from https://doi.org/10.1007/s11270-011-0749-1.

Yadav, B. K., & Mathur, S. (2008). Modeling soil water extraction by plants using non-linear dynamic root density distribution function. *Journal of Irrigation and Drainage Engineering, 134*(4), 430−436.

Zektser, I. S., & Loaiciga, H. A. (1993). Groundwater fluxes in the global hydrologic cycle: Past, present, and future. *Journal of Hydrology, 144*, 405−427.

Microplastic in the subsurface system: Extraction and characterization from sediments of River Ganga near Patna, Bihar

Rashmi Singh, Rakesh Kumar and Prabhakar Sharma
School of Ecology & Environment Studies, Nalanda University, Rajgir, India

10.1 Introduction

10.1.1 Background

Patna is the sixth most polluted city in India (WHO, 2018b). Out of 900 megatons of daily solid waste generated in Patna city, 90 megatons are plastic wastes (Li et al., 2015), which are nonbiodegradable generally and everlasting (Thompson et al., 2004). Ganga being the second most polluted river in India is also the primary dumping site in North India (Wright & Kelly, 2017). Patna is one of the significant sources in contributing plastic wastes to rivers; therefore downstream of the Ganga River in Patna city may have more plastics as compared to its upstream and midstream. In India, very few studies have been done on plastic particles (Sruthy & Ramasamy, 2017; Kumar & Sharma, 2021a, 2021b; Kumar et al., 2021a, 2021b). Globally, plastic has been kept under the category of emerging pollutants, but this issue has not been taken seriously in India.

The word plastic is derivative of the Greek word "plastikos," which means moldable and capable of being shaped (Kamboj, 2016; PlasticEurope, 2018). The term "plastic" applies to synthetic organic carbon-based compounds made by polymerization and is composed of a large number of monomers that form the polymer structure (Hammer et al., 2012). Most of the plastic goods that are used today are derivative of fossil feedstocks, like natural gas, oil, or coal (PlasticEurope, 2017). With a combination of different monomers and their distinct polymer structure, various kind of plastic polymers is formed with different physical and chemical properties (Nicholson, 1997). At the time these plastics are manufactured, some additives are added (chemicals in nature) to them. These additives help the polymers to resist oxidative damage, heat and microbial degradation, as well as stabilizes the polymer structure and in this way, plastic performance enhances (Cole et al.,

Advances in Remediation Techniques for Polluted Soils and Groundwater. DOI: https://doi.org/10.1016/B978-0-12-823830-1.00013-4

2011; Teuten et al., 2009). Due to additive chemical compounds, plastics are long-lasting, lightweight, and display exceptional thermal and electrical insulation properties. Apart from these characteristics, plastics are very cheap. These plastic features are appropriate for the manufacturing of a wide variety of goods and use in various applications from food packing to the generation of renewable energies (Dris et al., 2015; Kunwar et al., 2016).

Plastic materials have become a dominant feature of modern life with disposal packing as the dominant form of practice (Galloway et al., 2017). The global demand for plastic products has amplified the productiveness of plastic. About 1.5 million tonnes (MT) of plastic production in 1950 have increased to 348 MT in 2017 (PlasticEurope, 2018). It has been found that till 2017, 8300 million metric tonnes plastics have been produced, counting from the first time when it was formed (Geyer et al., 2017). From the data, it has been found that the plastic industry has become a global production platform, with different countries producing large quantities of this so-called wonder material. Documented reports suggest that China, European Union, and North America generated an estimated 161 MT of plastics in 2013 (Geyer et al., 2017; PlasticEurope, 2016). According to recent data, it has been found that China alone produces 29.4% of total world plastics (PlasticEurope, 2018). There is an indication to suggest that plastic production is likely to increase with population growth (Geyer et al., 2017; Ryan, 2015). Therefore with the current dependence on plastic for everyday use, it is apparent that consumers will use more plastic products in the future to meet daily needs because this material provides a massive benefit to society (Wright & Kelly, 2017). Polyethylene (PE), polypropylene (PP), polyvinylchloride, polystyrene (PS), and PE terephthalate (PET) are the most commonly produced plastic polymers globally (Klein et al., 2015). It is estimated that plastic waste comprises 10% of the total globally produced solid wastes, containing plastic leftover (which are in large quantity) ending up in aquatic environments via direct disposal or discharge from mismanaged disposal sites and landfills. In 2010 approximately 5% of the generated plastic waste from 192 coastal countries was estimated to enter the aquatic environments from the terrestrial environment (Jambeck et al., 2015). Eriksen et al. (2014) expected that 5 trillion plastic debris were present in the marine ecosystem worldwide. Plastic particles entering the aquatic environments have the capacity to persist for centuries due to their low degradation speed and nature to undergo fragmentation (Eriksen et al., 2014).

Exposure to UV radiation, biological degradation and different mechanical factors can break larger plastic items into small plastics pieces (Galloway et al., 2017). These tiny fragments of plastics were first called "microplastic" (Thompson et al., 2004). To date, no such fixed magnitude for microplastic has been given based upon their size, but the term is typically applied to fragments having a diameter smaller than 5 mm. The size range differs in every study (Arthur et al., 2009; Cole et al., 2011; Wright et al., 2013; Kumar et al., 2021a, 2021b). Therefore to date, in most studies, plastics debris, size of which is less than 5 mm, has been defined as microplastic (Andrady, 2011; Arthur et al., 2009; Moore, 2008).

10.1.2 Occurrence, fate, types, and source of microplastic in the freshwater environment

A variety of size ranges of plastics enter the aquatic environment, most of these plastics never entirely "decompose" but breaks into smaller forms like filaments, fragments, pellets, etc. under ultraviolet light at relatively low temperatures (Nel & Froneman, 2015). Based on their origin, microplastics are divided into two groups, which are primary microplastic and secondary microplastic. Those microplastics which are manufactured intentionally for various industrial or domestic applications, like use in personal care products, toothpaste for the formation of preproduction pellets, or resin pellets that are being used in plastic industries, come in the category of primary microplastic. While those plastic formed due to the breaking of large plastic particles as a result of UV radiation or mechanical force are said to be secondary microplastics (Auta et al., 2017; Dris et al., 2015). Since last ten years, the amount of microplastics has increased exponentially in the aquatic environment (Anderson et al., 2016). These are ubiquitously present in the sediment and surface water (coastal area and offshore) of both marines as well as freshwater environments across the globe (Rocha-Santos & Duarte, 2015). The fact has been reported about the accumulation of microplastic in the shore sediment of eighteen countries from six continents, which indicates that contamination by microplastics has become a global problem (Browne et al., 2011). Microplastics can be transported to areas geographically distant from their source due to the tidal force and current of the ocean (Eriksen et al., 2014), which consequently spreads microplastics throughout the world's aquatic ecosystems. The existence of microplastic in the polar environment indicates the transportation of microplastics to remote ecosystems (Lusher et al., 2015). So far, studies have predominantly reported that microplastics exist in different forms such as fibers, fragments, pellets, and filaments in the aquatic environment, which can originate from diverse sources (Jiang et al., 2018). Microplastics in the freshwater environment enter into it from two primary sources, for example, aquatic-based and land-based sources (Ziajahromi et al., 2016).

Aquatic-based sources include those microplastics that are generated through direct dumping of plastic materials to the aquatic environment such as littering (e.g., discarded plastics materials), fishing (e.g., abandoned fishing lines), and shipping (e.g., accidentally spillage of plastic pellets) (Hammer et al., 2012). On the other hand, the sources through which microplastics make their way from the terrestrial environment to the waterways, land run-off, and Waste Water Treatment Plant effluent are Land-based (Eerkes-Medrano et al., 2015). It is assumed that fragments (a type of microplastic) are the result of the degradation of large plastic material in the aquatic environment. In the marine environment, various sizes and numbers with the different origins of plastics have affected numerous aquatic species (Gregory, 2009). Microplastics are inherently toxic for aquatic animals and can act as carriers for a massive range of pollutants (Scheurer & Bigalke, 2018).

There is quite a lot of literature that described contamination of marine water due to microplastic (Thompson et al., 2004), but only a few studies have been undertaken to describe the issue of microplastic contamination of freshwater (rivers and lakes) (Dris et al., 2015; Kumar et al., 2021a, 2021b, 2021c, 2021d). These studies not only address the freshwater contamination due to microplastic but also show that microplastics pollution is as grievous as in marine water (Dris et al., 2015; Klein et al., 2015; Scheurer & Bigalke, 2018). There are many pathways by which microplastics, either from primary or secondary sources, can enter an aquatic environment. These particles pass in either with rainwater coming from land or through wastewater treatment plant channels in the freshwater system. Apart from the skincare products (like cleansers with granulated polymers) through which microplastics can enter in the wastewater and finally to freshwater, laundry washing machine discharge is also a primary source. According to a study, it has been found that a single wash of laundry washing machine can produce 1900 fiber microplastic (Dris et al., 2015). Microplastic presence has also been observed in the vicinity of different industries, especially the paper industry. These plastic particles with water runoff or other sources can enter an aquatic environment and affect life from adversity. Some of the recent studies show that all the adverse effect which is faced by the marine organism is same for freshwater water organisms. In research done on fishes of freshwater of eleven French Stream, Imhof et al. (2013) observed that 12% of fishes had microplastics in their digestive system. It means that freshwater organisms also ingest microplastic. In the studies done on freshwater organisms, it was found that the vast spectrum of aquatic taxa is susceptible to intake of microplastics, for example, annelids, crustaceans, gastropods, and ostracods (Imhof et al., 2013).

India comes in the category of a few first countries of the world, which consumes the highest number of plastics as well as generates around 5.6 MT of plastic waste each year (Toxicslink, 2014). However, to date, none of the work has been done on freshwater sediments, rivers, or estuaries in India, addressing the presence of microplastics in them except a study conducted on Vembanad Lake of Kerala. This is a Ramsar site where water is brackish, and the presence of microplastic was ubiquitous in lake sediments (Sruthy & Ramasamy, 2017).

10.1.3 A potential concern for pollution and damage to aquatic biota caused by microplastics

Microplastics pollution in the aquatic environments is becoming a growing environmental concern not only due to its persistence and accumulation in the environment but also because of its capacity to adversely affect aquatic biota (Kumar et al., 2021b; Sruthy & Ramasamy, 2017). These particles may be flawed as food particles by aquatic animals with narrow choosiness between microplastics particles and their usual food (or prey) in a similar size range (Wright et al., 2013). Plastic debris has the capacity to adsorb organic pollutants. These pollutants along with plastic particles get ingested by organisms that impact aquatic fauna from higher trophic level to lower trophic level animals, and also the predators (Eriksen et al., 2014). The intake of microplastics by both low trophic and high trophic organisms, including species belonging to subphylum Vertebrata, like fishes, sea birds, turtles, and aquatic mammals, and invertebrates (e.g., zooplankton, lugworms, and mussels), have been observed. The intake of microplastics has often been documented with adverse health effects. The existence of microplastics in fish from marine as well as freshwater environments has also been reported (Kumar et al., 2021b; Lusher et al., 2013). Jabeen et al. (2017) found microplastics (mostly fibers) with concentration ranges of 1.1−7.2 items per individual in 36 species of benthopelagic fish from China. Moreover, it is recently reported that the microplastic fibers and fragments with the average concentration of 1.8 microplastic particles per fish in the stomach of 75%−100% of the studied fish were collected from the Northwest Atlantic (Jabeen et al., 2017).

The ingested microplastics can cause physical hazards such as gut blockage, reduced ability to avoid predators, changed eating habits, and decrease in energy levels, which affects the survival, growth, and reproduction of the exposed organisms (Wright et al., 2013). The Convention on Biological Diversity (CBD) noted that there are 663 species of marine organisms influenced by plastic particles (CBD, 2012). Further, entanglement can be considered another physical impact of microplastics on aquatic organisms that may cause physical distress and even death (Ziajahromi et al., 2017). It has also been suggested that all species of sea turtles, 45% of mammals living in the marine environment, and 21% of species belonging to seabirds may get damaged through physical injury from plastics. Apart from potential effect, physical effects of microplastics like toxicity may also arise because of discharged constituent toxins such as sorbed chemicals and plastic additives, which has the capability of causing carcinogenesis and endocrine effects in organisms (Wright et al., 2013). Trace organic contaminants (TrOCs) that occur in the water at shallow concentration can be sorbed to microplastics via partitioning (WHO, 2018a). It has been suggested that microplastics can act as a route to transfer sorbed TrOCs from the surrounding environment to the organism's body (Pandey, 2013). The extent to which microplastics and their associated contaminants harm organisms may vary depending on the susceptibility of species, the volume of ingested microplastics, type of pollutants, the kinetics of repartition of pollutants between plastic and tissues for which toxic microplastic is in the body of organisms (Andrady, 2011). There has also been some controversy regarding the role of microplastics to transfer sorbed TrOCs to aquatic organisms with some previous studies, suggesting an optimistic relationship between the ingestion of contaminated plastic debris (microplastic) and an increase in bioaccumulation of contaminants, such as PCBs (polychlorinated biphenyls) (Dubaish & Liebezeit, 2013), and polybrominated diphenyl ethers (Sruthy & Ramasamy, 2017). However, the modeling approach has predicted that microplastics may be capable of reducing the bioaccumulation of TrOCs depending on the contaminant gradient between the organism's tissues and the microplastic (Eriksen et al., 2014). Despite the physical and chemical impact of microplastics on organisms individually, microplastics can alter an organism's habitat and affect populations (Wright et al., 2013). Sruthy and Ramasamy (2017) investigated Kamilo beach in Hawaii, which was heavily polluted with microplastics, which can be seen as an excellent example of the population-level impacts of microplastics. It was found that higher amounts of microplastics increased the absorbency of the sediment and changed its maximum temperature. This caused sediment to become warmer and affect beach-dwelling organisms especially the sex determinant organisms, such as sea turtle eggs that are vulnerable to such changes (Sruthy & Ramasamy, 2017).

10.1.4 Microplastic and human health

In recent studies, microplastics have also been reported in the human body (Wright & Kelly, 2017). Food that is prepared for consumption by humans, like seafood, salt (Li et al., 2015), processed food, sugar and beverages, and beer (Liebezeit & Liebezeit, 2014), has been found with the presence of plastic particles. Microplastic reaches the human body either through diet or inhalation (Wright & Kelly, 2017). Microplastics are hydrophobic due to which they absorb and accumulate contaminants like polycyclic aromatic hydrocarbons, PCBs, or organochlorine pesticides (which are organic contaminants) to a higher degree (Ogata et al., 2009). Microplastics are added to processed food as additives. Through these sources, microplastics reach directly to the human body. This increased the ingestion of plastic debris (or microplastic) with an average amount of $40\,\mathrm{mg\,person^{-1}\,day^{-1}}$ (Powell et al., 2010). The exact harm caused by microplastic is still unknown well, but few studies have been done in this direction. All these studies suggest that microplastic can pass through cells and reach the circulatory or lymphatic system and accumulate in secondary organs (Atassi, 2013; Rieux et al., 2005). It can also affect the immune system of human beings or may harm the health of the body cell. Concerning the physical effects, biopersistence due to microplastics can cause a range of biological responses such as inflammation, genotoxicity, oxidative stress, apoptosis, and necrosis. If this condition continues, then tissue damage, carcinogenesis, and fibrosis can occur. Chemical effects can also be faced because of the different structures of the polymer itself. Leaching of the unbound chemical as well as unreacted residual monomer may lead to the accumulation of related hydrophobic organic contaminants. More studies are going on globally to know the effect of microplastic on the human body (Khan et al., 2015).

10.1.5 Sample collection practices and the estimation of microplastics

This section elucidates different techniques for collecting samples, separation of microplastics from sediments, and their identification, which were used for researching freshwater microplastic. Only a few studies have been conducted in North America, South America, Europe, and Asia, on microplastic pollution in riverbank and lakeshore sediments so far. The lakeshore and riverbank sediments were collected and analyzed on microplastics present in the sediment of various lakes such as Lake Huron (Canada, USA), Lake Geneva (Switzerland, France), Lake Garda (Italy) (Cauwenberghe & Janssen, 2014; Hodges et al., 1995; Liebezeit & Liebezeit, 2013), and in rivers such as Rivers Elique, Maipo (Chile), and Saint Lawrence River (Canada) (Powell et al., 2010; Rieux et al., 2005). In most of the cases, manual collection of samples was done except in Saint Lawrence River, where the grabbed sampler of $225\,\mathrm{cm^2}$ area was used. In Lake Garda, sampling of sediment was done using the random grid sample collection method (Imhof et al., 2012). Density separation was followed for the separation of microplastics, and zinc chloride solution was used for this density separation. In some cases, like Lake Huron, the visual inspection was used for identification at three different sites within the river bed (Zbyszewski & Corcoran, 2011).

It has been reported that microplastics collected from the surface water of Wuhan, China have a density higher than that of flowing water. For example, PET, nylon, and PS having a density of 1.37, 1.15, and $1.05\,\mathrm{g\,cm^{-3}}$ (Wang et al., 2017). There are several plastics, having a density in the range of

$1.02-1.03\,\mathrm{g\,cm^{-3}}$ (Filella, 2015). However, some of the studies also reported that the density of microplastics in the low range like PE $(0.89\,\mathrm{g\,cm^{-3}})$ to polyvinyl chloride $(1.58\,\mathrm{g\,cm^{-3}})$ exists in the marine ecosystem (Hidalgo-Ruz et al., 2012; Law, 2017; Stolte et al., 2015). Most commonly, PP, PE, PS, and polyamide (generally known as nylon) are the type of polymer, which are present in freshwater sediments (Quinn et al., 2017) and their densities are less than $1.2\,\mathrm{g\,cm^{-3}}$. Different salt is used for this purpose like sodium polytungstate $(1.4\,\mathrm{g\,cm^{-3}})$, zinc chloride, $ZnCl_2$ $(1.5-1.7\,\mathrm{g\,cm^{-3}})$, calcium chloride $(1.30-1.35\,\mathrm{g\,cm^{-3}})$, and sodium iodide (NaI) but sodium chloride (NaCl) is used for low-density plastics $(1.2\,\mathrm{g\,cm^{-3}})$ separation, whereas NaI and $ZnCl_2$ are used for a high-density particle (Claessens et al., 2011; Kumar et al., 2021a, 2021b). The efficiency for the separation of polyvinyl chloride particles is reported 100% with NaI solution. As of now, the application of NaI solution for particle separation is expensive than NaCl. The cost for 1 kg of NaI is reported as $388 (Fisherscientific, 2019a), while $60 costs for 1 kg of NaCl (Fisherscientific, 2019b). Several studies are conducted for the extraction of microplastics using $ZnCl_2$ as a density separator. However, it is reported that $ZnCl_2$ is a more hazardous density solution as compared to NaCl or NaI (Ivleva et al., 2017; Maes et al., 2017; Stolte et al., 2015). Choosing NaCl over other density separators is preferable because NaCl is not expensive, is widely available, and is environment friendly. It separates the microplastics that have a density of $\leq 1.2\,\mathrm{g\,cm^{-3}}$, common polymers present in water. So, considering all features of different density separators, NaCl is the most kosher salt used for separating the prominent microplastic through density separation.

10.1.6   Objectives of the study

Looking at different studies conducted in another part of India and the world, it can be assumed that microplastics are also present in the Indian freshwater ecosystem because of the effluents received from cities. So, the specific hypothesis of this study states that microplastic contents in soil sediments may be spatially distributed in the freshwater ecosystems which can be contributed from urbanization. The main goal of the current study is to develop an efficient as well as reliable method to study the accumulation of microplastics and classify the diverse type of microplastics based on shape, mass, color, and size from the soil sediment of the River Ganga. Therefore two objectives have been designed to perform the experiment on soil sediments which will be collected from the River Ganga near Patna. These two objectives are to develop a method for the extraction of microplastic from soil sediments in the River Ganga and characterize the microplastic extracted from different locations, for example, upstream, midstream, and downstream of the River Ganga near Patna.

10.2   Materials and method

10.2.1   Study area

Patna is the capital city of Bihar, situated at the bank of River Ganga. It has a geographical area of 517.3 km^2 with an estimated population of 20.5 lakhs. Three different locations that are upstream, Digha (25°39′10.82″N 85°5′26.01″E); midstream, Mainpura Diara (25°38′1.81″N 85°8′54.81″E); and downstream, Masjid Ghat (25°37′2.71″N 85°11′35.26″E) were chosen as sites for collecting samples in this study. On these samples, experiments were conducted for testing the hypothesis. It states that the microplastic contents may be present in the soil sediments of a freshwater ecosystem. Patna city may be one of the significant sources in contributing microplastic to the River Ganga. The downstream of River Ganga may have a higher concentration of microplastic content as compared to its upstream and midstream. In Fig. 10.1, three yellow dots (from right hand side, i.e., upstream to left hand side, i.e., downstream) represent the three respective locations at the bank of

FIGURE 10.1

Location of three different sampling sites as upstream (Digha), midstream (Mainpura Diara), and downstream (Masjid Ghat) at the bank of the River Ganga near Patna, Bihar. Source: *From Google Earth.*

River Ganga from where the sampling had been conducted in this study. These locations were upstream, midstream, and downstream of River Ganga in Patna. Google earth was used for taking the image and locating the sampling sites.

10.2.2 Sampling of soil sediments

The River Ganga in Patna flows through regions of high and low population density, agricultural field, industrial parts, etc. Therefore sewage pipe outlet from hospitals, industries, residential sectors, colleges, etc. discharges into the river. Hence, different types of wastewater, which may contain microplastics, get discharged into the River Ganga every day. The sampling of river sediments was performed at three different locations at the bank of the River Ganga. Deposits were collected from the depth of 2−3 cm using a stainless spoon from the shoreline of the river. Microplastics might likely have accumulated at the topsoil surface around the bank of the River Ganga. Around 2 kg of soil sediments were collected from all three sites in different zipper sealed bags. Soil samples were chosen over water samples for conducting this research because the sampling of soil sediments is far more comfortable than that of water. Another principal reason is that buoyant microplastic particles predominate over the nonbuoyant microplastic particle in the water samples, and microplastics might escape during water sample collection. However, this nonbuoyant microplastic particle is present in large volume along with buoyant microplastics in river shore sediments which can be collected by taking the soil samples for the experiment (Klein et al., 2015).

10.2.3 Sediments treatment and extraction of microplastics

The whole experiment was performed based on the density separation method of soil sediments in an aqueous solution of sodium chloride (NaCl). The soil sediments are denser than microplastics. Hence, when the sediment solution is left undisturbed for a particular period, the particles of high density settle down at the bottom of the vessel (cylinder), and lighter density particles float at the surface of the solution. Density separation is the utmost common and reliable technique for separating microplastics from soil sediments of rivers. In this method, samples having different densities are kept in a solution of intermediate density and particles having lighter density than solution float on the water surface, while other particles that settle low at the bottom are of higher density (Quinn et al., 2017). Sodium chloride (NaCl) helps the soil particles in the accumulation of sand/clay/soil particles, which leads the plastic particles of lighter density to float more freely on the surface of the water. oMre the amount of NaCl in the solution, the more the accumulation of clay (dense particles) takes place. Moreover, in this way, the two fractions are getting separated. Other salts, like zinc chloride and sodium iodide are also used in density separators, but NaCl is readily available, even cheaper than NaI and $ZnCl_2$ and nonhazardous to the environment (Quinn et al., 2017). So, NaCl was used in the density separation method in the current study. The density of moist soil sediment is 2.65 g cm^{-3} (in general), while the density of the polymer varies from 0.92 to 1.20 g cm^{-3} (Klein et al., 2015). Sample collection from a different location (i.e., up-, mid-, and downstream) was done at different times. After the complete processing of samples from one location in the stream, the collection of samples from another location was performed. For each location the experiment was repeated four times and their mean value was presented.

In each repetition, 25 g of samples were taken. So, for each stream location, 100 g of samples were used in the experiment. 25 g of the sample was put into a 1000 mL measuring cylinder. One liter of distilled water was mixed with the sample. Two grams of NaCl were further added to the solution. The sample was then plunged continuously for 2 min and left for 24 h undisturbed so that clay particles would settle down and lighter particles (e.g., microplastics) would float on the water surface. After an interval of 24 h, water samples were collected from the top 20 cm. In every measuring cylinder the water layer was almost transparent until the depth of 20−25 cm. The collection of the sample was done with a micropipette and was put in 1000 mL of a glass beaker. After the completion of density separation, filtration was done. A setup was created for conducting filtration of the collected samples. In this setup a conical flask (1000 mL) was attached with a vacuum pump (Tarsons Rockyvac, Model No- MV8000).

On the top of the flask an inverted glass tube was fixed with the help of a clamp. In between the tube and flask, the filter paper was kept for conducting the filtration. Two filters of different pore sizes (nylon membrane filters, pore size—0.45 μm; Whatman filter papers—pore size 11 μm) were taken for different and various rounds of filtration. Postdensity separation, water samples collected in the beaker were poured in a glass tube (by using the glass funnel), which was attached to the conical flask of the filtration unit. These water samples were filtered by filter paper that was placed in between flask and glass tube. In this set of filtration, first, the Whatman filter paper of pore size 11 μm was used for filtration. It was done to remove very fine clay particles from the sample solution, which did not get settle during density separation due to less density and mass of clay particles. Along with few clay particles, microplastics were also accumulated on the filter papers during the filtration process. The filter paper was replaced with a new filter paper of the same pore size (11 μm), after filtering 25−30 mL of the sample solution.

In the next round of filtration, all the filter papers of the previous set (on which filtered particles constituted of few clay particles and microplastics were attached) were taken and washed very carefully with the help of distilled water. The cleaned water was collected directly into the glass tube of the setup. This process was done very carefully and in such a way that all plastic particles present on the filter paper pass along with the distilled water directly to the glass tube of setup, while clay particles stuck at the filter paper. In general, the tendency of clay particles is sticky. It gets attached to the filter paper during the filtration process and does not go off quickly along with distilled water. Hence, in various rounds of rinsing of filter papers, plastic particles (microplastics) were collected on the new filter paper, while clay particles remained attached to the previous filter papers. This filtration process was repeated till the time there was no or negligible trace of clay particles left on filter paper. The sole aim of using a large pore size filter was to remove all clay particles through filtration or by accumulating them on filter paper (for those particles that did not get filtered because of the size larger than 11 μm). These clay particles did not have to reach the second round of filtration as in the next set of filtration, the weight of microplastic was measured which should be accurate and deprived of clay or any other particles than microplastics. In the next set of filtration, the nylon membrane filters of pore size 0.45 μm were used. Before using these filter membranes for filtration, their weight was measured on a microbalance (Mettler Toledo, MS 0STU). Post filtration, the nylon membrane filters were left in a desiccator (Tarsons) for drying. After 24 h when they were completely dried, the particles of filter paper were transferred to a bowl with the help of a needle. For destroying all kinds of natural debris/organic matter, a mixture of hydrogen peroxide solution (30%) and concentrated sulfuric acid (1:3 V:V) was added to the glass bowl. This reaction was conducted overnight. After the completion of 12 h, 300 mL of distilled water was poured into the bowl, and vacuum filtration was performed using nylon filter membrane. It was rinsed several times with distilled water until the whitish color of the solution (hydrogen peroxide and sulfuric acid) disappeared. Then, the filter paper was dried in a desiccator for 3 days. Each time filter paper was weighed before conducting the vacuum filtration. After 3 days the weight of the filter paper was measured on the microbalance, and the difference was the weight of the microplastic. This filter paper contains different kinds of microplastics.

All the residues left on filter paper, after the filtration process and drying for 3 days, were put on different glass slides. This was done with the use of a needle and forceps. While transferring the microplastics from filter paper to slides, it was essential to wear a mask and gloves. Coverslips were put on slides, and they were placed under the microscope (ZEISS) in turn. Then at the focus of 4 $\times$ and 10 $\times$, images were taken for all the microplastic present on the glass slide using the inbuilt Canon Camera with the Microscope. Later on, with the use of image software (ImageJ), microplastic's maximum and minimum sizes were measured. The standard reference of the scale was uploaded on ImageJ for measuring the accurate size of each microplastic particle. With the help of scale and this standard reference, the maximum and minimum sizes of microplastic were calculated. Microplastics were also categorized into different groups such as filaments, fragments, pallets, fibers, and films based on their structure and shape for analysis in this study.

10.2.4   Statistical analysis

Density separation was conducted in such a way that the primary experiment was performed along with three replications (i.e., four samples from each stream = primary experiment + triplicate) of different streams. Results were expressed in mean with respective standard deviation. The t-test and P-value test were performed with raw data to analyze the significant difference between different streams. Descriptive statistics one-way ANOVA test was performed to understand the significant difference at each stream in between the mass and abundance of microplastics. Significant levels were considered < 0.05 to be statistically significant.

10.3   Results and discussion

10.3.1   Method developed for the extraction of microplastic

A significant amount of plastic debris was obtained in the sediment samples. The average mass of microplastic obtained in three locations of River Ganga in 25 g of sediments ranged from 1.43 ± 0.26 mg to 2.45 ± 0.17 mg, which was much more than other studies conducted globally. Thompson et al. (2004), in one of their studies on the estuary of the United States, found an average number of 31 particles kg^{-1} of sediment (Thompson et al., 2004). In another study conducted on soil sediments of Swiss flood plain by Scheurer and Bigalke (2018), they found about 55.5 mg of microplastics kg^{-1} of sediment, that is, 593 particles kg^{-1} in terms of their number. In a different study conducted on the river Rhine-Main of Germany by Klein et al. (2015), the quantity of microplastic was varied from 21.8 to 932 mg kg^{-1}, that is, $228-3763$ particles kg^{-1} in terms of their number in the sediment. The quantity and number of microplastics found in all those sediments were far less than that of River Ganga in the present study.

Several rounds of filtration were performed in the study which removed most of the clay particles from the filtrate. As similar to Klein et al. (2015), the use of sulfuric acid and hydrogen peroxide could remove the organic materials entirely from the filtrate containing microplastic, and the exact

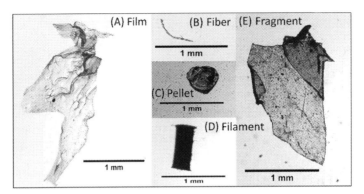

FIGURE 10.2

Different shapes of microplastics (A) Film, (B) Fiber, (C) Pellet, (D) Filament, and (E) Fragment found in soil sediments of River Ganga, identified using optical zoom microscope.

mass of microplastic was possible to measure (Klein et al., 2015). According to Jiang et al. (2018), the identification of microplastic was performed based on their shape, the result from this study came across a similar shape of microplastics which are shown in Fig. 10.2.

A total of six types of microplastic was found in the sediments in which one was not identified. These were fibers, filaments, films, pellets, and fragments. Those particles that did not come under any of these categories were kept in the class of others. Microplastic particles that were very thin or straight or fibrous are called fibers. Hard but cylindrical shape plastic particles are called filaments. Pellets are round plastic particles. Fragments are hard, jagged plastic particles. A film is a thin plane of flimsy plastic. Sruthy and Ramasamy (2017) stated one more type of plastic called foam, which is lightweight and sponge-like but absent in the current study. As a whole, this method of density separation (NaCl as extractor) on a self-designed setup, followed by several rounds of filtration and overnight reaction with sulfuric acid and hydrogen peroxide worked positively in extracting microplastic from soil sediments of the River Ganga.

10.3.2　Mass of microplastics

The mass of the obtained microplastics was different at all three locations of the River Ganga. In upstream the average value of microplastic mass was 1.43 ± 0.26 mg in 25 g of the sediment sample. In the midstream, the mean value of the mass of microplastic was 1.53 ± 0.33 mg, whereas, in the downstream, it has reached 2.45 ± 0.17 mg (Fig. 10.3). From the data, it is clear that microplastic was present in the river at all three locations. Their mass was continuously increasing from upstream to downstream. However, no

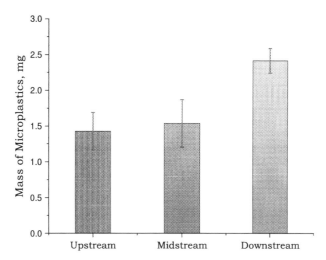

FIGURE 10.3

Average mass and standard deviation of microplastics (in 25 g of sediment sample) found at different locations of the River Ganga.

Table 10.1 Statistical parameters observed for the relationship between mass and quantity of microplastics at different streams.

Location	t-test[a]	P-value[a]	R-square[b]	Coeff variance[b]
Upstream	0.00257	.000088	0.93498	0.29856
Midstream	0.00398	.0002	0.91513	0.34693
Downstream	0.02084	.00425	0.76883	0.62108

[a]Paired sample t-test performed on four samples (experiment + triplicate) of different streams. At the 0.05 level, the difference in the population means is significantly different from the test differences (0).
[b]Descriptive statistics One-way ANOVA test performed on four samples (experiment + triplicate) of different streams. At the 0.05 level, the population means are significantly different.

consistent pattern was observed behind the increase in the mass of microplastic as discussed later in Table 10.1. The standard deviation of the average mean was different in each stream. The highest standard deviation was seen in the midstream (SD $= \pm 0.33$) which indicated that the microplastic mass found in this stream varied in the high range as compared to the other two streams. This deviation was followed by upstream and downstream, respectively. The sample collected for the experiment was taken from one place in each stream. The collected samples may be heterogeneous in their constituent. The collected sample may have a variable amount of plastic debris than the area just beside it because of the heterogeneity of soil, although 2 kg of sediment sample was collected for the study the experiment was performed only with 25 g of samples in each case due to the experimental limitation.

10.3.3   Number of microplastics

As similar to the mass, microplastic also varied the stream based on their number. The least number of microplastic was obtained upstream with an average of 144.75 ± 26.72, while 250.75 ± 96.28 number of microplastics were present in downstream. The average value of this plastic debris was 228.25 ± 48.81 in midstream in 25 g of sediment (Fig. 10.4). In short, $144-250$ particles per 25 mg of soil sediments (equivalent to $5,76,000-1,000,000$ particle kg^{-1}) in the sediments of River Ganga were observed. Standard deviation was maximum in downstream (± 96.28) followed by midstream (± 48.81) and upstream (± 26.72). However, the variation in the size of microplastic particles was maximum in downstream and least in upstream. The change in the microplastic particle number was directly proportional to population density, that is, urbanization. Where population density was thin, fewer microplastics were found but with the increase in the density of population, an increase in microplastic number was seen. The heterogeneity of the soil sample collected for research may be the reason for high standard deviation values. The mean value of mass (Fig. 10.3) and abundance (Fig. 10.4) were analyzed using paired sample t-test and P-value test with one-way ANOVA test to investigate statistical significance, such as R-square and coefficient of variance (Table 10.1). The statistical analysis rejected the null hypothesis over the alternative hypothesis as there is a difference in mean values of abundance and mass at each stream. As the number of microplastics was high in downstream, the reported mass of microplastics has a lower R-square value, high P-value, and coefficient of

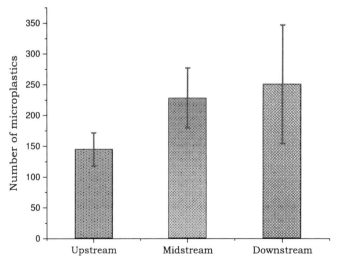

FIGURE 10.4

Mean value and standard deviation of the number of microplastics (in 25 g of sediment sample) present at different locations of the River Ganga.

variance in downstream. This might have happened due to the ratio of a number of microplastics found at downstream to midstream tends to decrease as compared with the ratio of a number of microplastics at midstream to upstream (Table 10.1). Sarkar et al. (2019) also reported that $99.27-409.86$ particle kg^{-1} of microplastic was abundant in the River Ganga. On the other hand, Singh et al. (2021) reported less count $(17-36$ particle $kg^{-1})$ of microplastics in sediments where high abundance $(380-684$ particle $L^{-1})$ in surface water.

10.3.4 Different types of microplastics in different streams

Based on appearance and shape, the observed microplastics were categorized into five different classes, fibers, filaments, films, pellets, and fragments. Those particles that did not fall under any of these categories were kept in the class of others. Based on different studies conducted in other countries, it was found that fibers are the most common microplastic found in soil sediments, followed by fragments (Wright & Kelly, 2017). In another study, film and foam were the dominant microplastics, but in the present study, pellets were prevailing in all the streams (Sruthy & Ramasamy, 2017). Pellets were the most dominant microplastics across the sample of upstream (66.75 ± 14.37), followed by fibers (24 ± 7.21), fragments (17.5 ± 8.14), filaments (16 ± 8.12), films (15 ± 7.10), and others (5.5 ± 4.71) (Fig. 10.5). Similarly, pellets were the most prevalent microplastics here as well, across

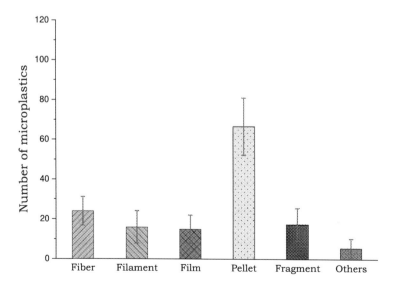

FIGURE 10.5

Mean value and standard deviation of different types of microplastics (in 25 g of sediment sample) found in upstream sediments of the River Ganga.

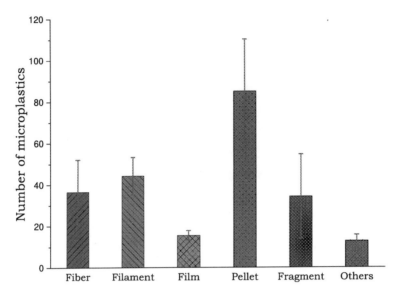

FIGURE 10.6

Mean value and standard deviation of different types of microplastics (in 25 g of sediment sample) found in midstream sediments of the River Ganga.

the samples of midstream (85 ± 25.23) followed by filaments (44.25 ± 8.9), fibers (36.5 ± 15.54), fragments (34.25 ± 20.47), films (15.5 ± 2.29), and others (12.75 ± 3.03) (Fig. 10.6). In addition, pellets were found as maximum in number about (71.25 ± 31.09) in the downstream, followed by fragments (55 ± 22.85), filaments (46.75 ± 22.06), films (45.25 ± 22.06), fibers (27.25 ± 7.79), and others (5.25 ± 1.29) (Fig. 10.7). The abundance of pellets was found the most dominant microplastic across upstream (46.11%) followed by midstream (37.23%) and downstream (28.41%). However, the abundance of films (18.04%) and fragments (21.93%) increased in downstream, whereas filaments (19.38%) and fragments (15.005%) were reported maximum in midstream, it may be because of films, filaments, and fragments were contributed by Patna city into the River Ganga.

10.3.5   Color of microplastics at different locations

Primarily four different colors of microplastics were present in the streams of Ganga (Fig. 10.8). These colors were black, transparent, yellow, and pink. Fig. 10.8A depicted the mean abundance of black microplastics across all three streams. It was found that the microplastics of black color were dominant over other microplastics in upstream (83 ± 12.67) followed by transparent (60.25 ± 17.17) and yellow (1.5 ± 2.38). The downstream also followed the

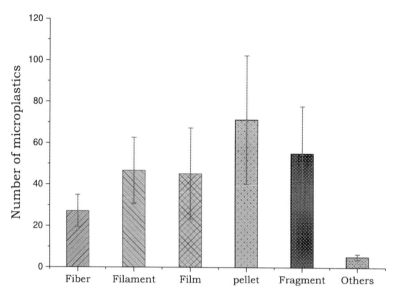

FIGURE 10.7

Mean value and standard deviation of different types of microplastics (in 25 g of sediment sample) found in downstream sediments of the River Ganga.

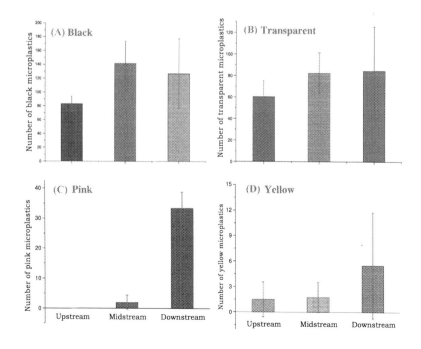

FIGURE 10.8

Mean value and standard deviation for (A) black, (B) transparent, (C) pink, and (D) yellow color of microplastics (in 25 g of sediment sample) at different locations of the River Ganga.

same trend, as black microplastic leading to their mean numbers (127 ± 58.34) over other plastics. But, among the streams, the black microplastic was highest in midstream (142 ± 36.95) (Fig. 10.8A). Pink microplastic was not observed in upstream. Similarly, midstream had followed the same pattern for an abundance of different colors of microplastic. Only a few numbers of pink microplastics were observed in midstream, which was suddenly increased at downstream (Fig. 10.8C). It was observed among three streams that the number of pink microplastic was maximum in downstream (33.5 ± 6.13). Transparent microplastic was the second highest in number (84.5 ± 47.64). The number of yellow particles was the least in downstream (5.5 ± 7.18) (Fig. 10.8D).

Black plastic is consumed in the highest number among other plastics. It may be because their manufacturing cost is lower than other plastics. In each shop, whether vegetable or ration or slaughterhouse, black plastics are used. The rate of degradation of black plastic is the slowest. So, these may be the reason that their number was maximum in downstream. Therein, no significant trends for the presence of different colors of microplastic were found in three streams. Overall, black plastic was the most dominant color, but its number decreased downstream from the average of 142 in midstream to 127.25 in downstream. Transparent and yellow microplastics showed very less increase from upstream to midstream to downstream. In the case of the pink microplastic particle, it was not present in upstream. It was seen first in midstream, and its number increased in downstream. These data indicate that city Patna is a contributor of microplastic in the freshwater ecosystem of the River Ganga.

10.3.6   Size of microplastics

All these particles detected in this study had unique shapes and sizes. Table 10.2 describes the size range of different kinds of microplastic observed in the River Ganga. In soil sediments at upstream of River Ganga, the size of microplastic was comparatively lower than that of midstream and downstream. But no statistical difference or pattern was found in the size

Table 10.2 Particles size (in mm) of different kinds of microplastics observed in soil sediment samples collected from different locations of the River Ganga.

Particulars	Upstream (mm)	Midstream (mm)	Downstream (mm)
Pellet	0.005–0.24	0.037–0.28	0.026–0.457
Fiber	0.081–0.917	0.239–3.429	0.152–2.141
Filament	0.316–1.932	0.233–1.872	0.1–2.141
Film	0.136–2.211	0.845–2.583	0.175–2.226
Fragment	0.049–1.469	0.475–1.767	0.325–1.946
Others	0.028–0.958	0.142–0.769	1.009–1.947

difference of microplastic of midstream and downstream. In general, the size of microplastics (specifically fibers and film types) present in the midstream were larger as compared to upstream and downstream (Table 10.2). Apart from these two types of microplastic, the larger size range was witnessed in downstream. Results state that microplastics are present in massive quantities in River Ganga, like other European rivers (Klein et al., 2015). Mass of plastic debris had also varied in different stream locations. In upstream where the population was low as compared to midstream and downstream, the mass of microplastic was also less than the other two stream locations. An increase in the mass of the microplastic took place in ascending order depending upon the increase in population density.

Microplastic mass found upstream was only 1.43 ± 0.26 mg in 25 g of the soil sediment sample. However, midstream had a dense population, and the mean value of the mass of microplastic was 1.53 ± 0.33 mg, while downstream had the densest population, and in this stream, the mean of microplastic's mass increased to 2.45 ± 0.17 mg. A direct link can be seen in population thickness with an increase in the mass of microplastic in soil sediment through this study. In Ganga sediments the size range of microplastics varied from 0.005 mm of pellets (smallest size of microplastic) to 3.43 mm of fibers (largest size of microplastic). In all three streams, pellets were a dominating category over other microplastic types. Moreover, its range was 0.005−0.28 mm. It can be said that Patna has a smaller size of plastics in greater quantity, while the large size of particles is less in quantity.

10.3.7   Possible source of microplastics

Somewhere in the middle of midstream and upstream, Bihar's most significant and busiest hospital, Patna Medical College and Hospital, is situated. It releases a huge amount of wastewater directly into the River Ganga without proper treatment. This water reaches downstream from where samples were collected. The population is also dense in downstream. Several institutions, like Patna University, National Institute of Technology, etc. are situated at the bank of the river. Downstream from where samples were collected in this study was located only at a distance of few kilometers from these places. Some of the small-scale paper mills were also located around these institutions. Paper industries contribute to the formation of microplastic particles as well (Dubaish & Liebezeit, 2013). Market areas with many big showrooms as well as small shops were found in these localities as well. Untreated wastewater from hospitals, laboratories of universities, paper factories, market areas, and beauty parlors (facial cleanser, cosmetics, etc.) can also be a significant contributor of microplastic in downstream. Therefore urbanization is the possible contributor of microplastics to the soil sediments of the River Ganga.

In the present study, pellets were maximum in number in all the streams. After pellets, no strict pattern was observed in the number or mass of microplastic particles. In the case of upstream, fiber was the second highest in number. In midstream, filament (44.25 ± 8.95) was less than pellets (85 ± 25.23). Fragments (55 ± 22.85) were the second highest microplastic in number after pellets (71.25 ± 31.09) in downstream. So, the reason behind this may depend upon the type of waste which was being introduced to the River Ganga by the various source, which may be different from that of waste of other countries. Along with an increase in number and mass, different colors of microplastic were also observed. There was no pink microplastic in upstream, but it was first witnessed in midstream (2 ± 2.83) and its number increased suddenly in the downstream (33.5 ± 6.14). Black, transparent, and yellow microplastic numbers also increased from midstream to downstream in ascending order. The result, generated after conducting this study, indicates that the number of microplastic increased in each stream based on their mass and number. Downstream had the highest number of microplastic, while upstream had the least number of plastic debris. Pellets were the dominating microplastic in all the streams. Among the size of all kinds of microplastic found during the study, pallets were the smallest in size, while fibers were the largest. As explained earlier, three different colors of microplastics (black, transparent, and yellow) were observed in all the streams, but pink was absent in upstream. It was first seen in midstream, and its number increased in downstream. So, it can be concluded that Patna city is the main contributor of microplastic in the River Ganga. Hence, the hypothesis of this study cannot be rejected by the researcher that microplastic is present in all three locations of River Ganga with the difference in their mass, number, size, and type. Ultimately, its number and mass increased from upstream to downstream so the River Ganga has microplastic and Patna city is a significant contributor to it. Moreover, the method developed for working on this hypothesis also proved to be correct.

10.4   Conclusion

Plastic materials, due to their substantial societal benefits, have become a dominant feature of modern life. Its production and consumption have increased tremendously from the time it was first introduced. But constituents of plastic are very harmful to entire biota (both aquatic and terrestrial) as well as human beings. In India, microplastic has not been kept under the category of pollutant. No attention on microplastic in freshwater has been drawn till now, unlike in another part of the world. There is no significant data present on the contamination of freshwater by microplastic in India.

Few studies have been conducted to find the impact of microplastic on the life of aquatic organisms. To date, no standard method has been developed for sampling, identifying, and separating microplastic from the soil or water sample. Thus, standard protocol should be developed in priority for finding the exact amount of microplastic in freshwater, and an improved comparative study can be done. The current study describes microplastic in freshwater sediments of the River Ganga and the best possible developed methodology with the provided resources (density separation, using NaCl, and filtration). After experimenting with three different locations of the River Ganga near Patna, by density separation, followed by filtration, the highest concentration of microplastic was found in downstream, which indicates that human activities might be a crucial step in polluting the River Ganga with microplastic. Although it is not directly visible by the naked eye, the shapes of microplastics found in the sediments were pellets, fibers, fragments, films, and filaments using the optical zoom microscope. The primary colors were black and transparent. Few particles of pink and yellow color were also present. Overall, the result highlights that the River Ganga is severely polluted with microplastic because of the significantly urbanized city, Patna; and the method used for extracting microplastic worked positively. This study provides essential data for any future research which will be conducted on freshwater sediments of the River Ganga intensively. People and government will come to know about freshwater contamination with microplastic. Some policies and management systems may come into the practice for reducing the consumption and manufacture of plastic products.

References

Anderson, J. C., Park, B. J., & Palace, V. P. (2016). Microplastics in aquatic environments: Implications for Canadian ecosystems. *Environmental Pollution (Barking, Essex: 1987), 218*, 269–280.

Andrady, A. L. (2011). Microplastics in the marine environment. *Marine Pollution Bulletin, 62*(8), 1596–1605.

Arthur C., Baker J. E., & Bamford H.A. (2009) In *Proceedings of the international research workshop on the occurrence, effects, and fate of microplastic marine debris*, September 9–11, 2008, University of Washington Tacoma, Tacoma, WA, USA.

Atassi, M. (2013). *Immunobiology of proteins and peptides V: Vaccines mechanisms, design, and applications.* Springer Science and Business Media.

Auta, H. S., Emenike, C., & Fauziah, S. (2017). Distribution and importance of microplastics in the marine environment: A review of the sources, fate, effects, and potential solutions. *Environment International, 102*, 165–176.

Browne, M. A., Crump, P., Niven, S. J., Teuten, E., Tonkin, A., Galloway, T. S., & Thompson, R. (2011). Accumulation of microplastic on shorelines woldwide: Sources and sinks. *Environmental Science & Technology, 45*(21), 9175–9179.

Cauwenberghe, L. V., & Janssen, C. R. (2014). Microplastics in bivalves cultured for human consumption. *Environmental Pollution (Barking, Essex: 1987), 193*, 65–70.

Claessens, M., De Meester, S., Van Landuyt, L., De Clerck, K., & Janssen, C. R. (2011). Occurrence and distribution of microplastics in marine sediments along the Belgian coast. *Marine Pollution Bulletin, 62*(10), 2199−2204.

Cole, M., Lindeque, P., Halsband, C., & Galloway, T. S. (2011). Microplastics as contaminants in the marine environment: A review. *Marine Pollution Bulletin, 62*(12), 2588−2597.

Dris, R., Imhof, H., Sanchez, W., Gasperi, J., Galgani, F., Tassin, B., & Laforsch, C. (2015). Beyond the ocean: Contamination of freshwater ecosystems with (micro-) plastic particles. *Environmental Chemistry, 12*(5), 539−550.

Dubaish, F., & Liebezeit, G. (2013). Suspended microplastics and black carbon particles in the Jade system, southern North Sea. *Water, Air, and Soil Pollution, 224*(2), 1352.

Eerkes-Medrano, D., Thompson, R. C., & Aldridge, D. C. (2015). Microplastics in freshwater systems: A review of the emerging threats, identification of knowledge gaps and prioritisation of research needs. *Water Research, 75*, 63−82.

Eriksen, M., Lebreton, L. C., Carson, H. S., Thiel, M., Moore, C. J., Borerro, J. C., Galgani, F., et al. (2014). Plastic pollution in the world's oceans: More than 5 trillion plastic pieces weighing over 250,000 tons afloat at sea. *PLoS One, 9*(12), e111913.

Filella, M. (2015). Questions of size and numbers in environmental research on microplastics: Methodological and conceptual aspects. *Environmental Chemistry, 12*(5), 527−538.

Fisherscientific (2019a) *Sodium iodide, 99 + %, pure, anhydrous,* ACROS Organics. Retrieved on March 28, 2021 from https://www.fishersci.com/shop/products/sodium-iodide-99-pure-anhydrous-acros-organics-5/AC203180050.

Fisherscientific (2019b) *Sodium chloride,* Fisher BioReagents. Retrieved on March 28, 2021 from https://www.fishersci.com/shop/products/sodium-chloride-fisher-bioreagents-3/BP3581.

Galloway, T. S., Cole, M., & Lewis, C. (2017). Interactions of microplastic debris throughout the marine ecosystem. *Nature Ecology and Evolution, 1*(5), 0116.

Geyer, R., Jambeck, J. R., & Law, K. L. (2017). Production, use, and fate of all plastics ever made. *Science Advances, 3*(7), e1700782.

Gregory, M. R. (2009). Environmental implications of plastic debris in marine settings—entanglement, ingestion, smothering, hangers-on, hitch-hiking and alien invasions. *Philosophical Transactions of the Royal Society B: Biological Sciences, 364*(1526), 2013−2025.

Hammer, J., Kraak, M. H., & Parsons, J. R. (2012). Plastics in the marine environment: The dark side of a modern gift. *Reviews of environmental contamination and toxicology* (pp. 1−44). Springer.

Hidalgo-Ruz, V., Gutow, L., Thompson, R. C., & Thiel, M. (2012). Microplastics in the marine environment: A review of the methods used for identification and quantification. *Environmental Science & Technology, 46*(6), 3060−3075.

Hodges, G. M., Carr, E. A., Hazzard, R. A., & Carr, K. E. (1995). Uptake and translocation of microparticles in small intestine. *Digestive Diseases and Sciences, 40*(5), 967−975.

Imhof, H. K., Ivleva, N. P., Schmid, J., Niessner, R., & Laforsch, C. (2013). Contamination of beach sediments of a subalpine lake with microplastic particles. *Current Biology, 23*(19), R867−R868.

Imhof, H. K., Schmid, J., Niessner, R., Ivleva, N. P., & Laforsch, C. (2012). A novel, highly efficient method for the separation and quantification of plastic particles in sediments of aquatic environments. *Limnology and Oceanography: Methods, 10*(7), 524−537.

Ivleva, N. P., Wiesheu, A. C., & Niessner, R. (2017). Microplastic in aquatic ecosystems. *Angewandte Chemie International Edition, 56*(7), 1720−1739.

Jabeen, K., Su, L., Li, J., Yang, D., Tong, C., Mu, J., & Shi, H. (2017). Microplastics and mesoplastics in fish from coastal and fresh waters of China. *Environmental Pollution (Barking, Essex: 1987), 221*, 141−149.

Jambeck, J. R., Geyer, R., Wilcox, C., Siegler, T. R., Perryman, M., Andrady, A., Narayan, R., et al. (2015). Plastic waste inputs from land into the ocean. *Science (New York, N.Y.), 347*(6223), 768−771.

Jiang, C., Yin, L., Wen, X., Du, C., Wu, L., Long, Y., Liu, Y., et al. (2018). Microplastics in sediment and surface water of west dongting lake and south dongting lake: Abundance, source and composition. *International Journal of Environmental Research and Public Health, 15*(10), 2164.

Kamboj, M. (2016). Degradation of plastics for clean environment. *International Journal of Advanced Research in Engineering and Applied Sciences, 5*(3), 10−19.

Khan, F. R., Syberg, K., Shashoua, Y., & Bury, N. R. (2015). Influence of polyethylene microplastic beads on the uptake and localization of silver in zebrafish (Danio rerio). *Environmental Pollution (Barking, Essex: 1987), 206*, 73−79.

Klein, S., Worch, E., & Knepper, T. P. (2015). Occurrence and spatial distribution of microplastics in river shore sediments of the Rhine-Main area in Germany. *Environmental Science & Technology, 49*(10), 6070−6076.

Kumar, R., & Sharma, P. (2021a). Microplastics pollution pathways to groundwater in India. *Current Science, 120*, 249.

Kumar, R., & Sharma, P. (2021b). Recent developments in extraction, identification, and quantification of microplastics from agricultural soil and groundwater. In P. K. Gupta, & R. N. Bharagava (Eds.), *Fate and transport of subsurface pollutants* (pp. 125−143). Singapore: Springer. Available from https://doi.org/10.1007/978-981-15-6564-9_7.

Kumar, R., Sharma, P., & Bandyopadhyay, S. (2021a). Evidence of microplastics in wetlands: extraction and quantification in freshwater and coastal ecosystems. *Journal of Water Process Engineering, 40*, 101966.

Kumar, R., Sharma, P., Manna, C., & Jain, M. (2021b). Abundance, interaction, ingestion, ecological concerns, and mitigation policies of microplastic pollution in riverine ecosystem: A review. *The Science of the Total Environment, 782*, 146695. Available from https://doi.org/10.1016/j.scitotenv.2021.146695.

Kumar, R., Sharma, P., Verma, A, Jha, P. K., Singh, P., Gupta, P. K., Chandra, R., & Vara Prasad, P. V. (2021c). Effect of physical characteristics and hydrodynamic conditions on transport and deposition of microplastics in riverine ecosystem. *Water, 13*(19)2710, In press.

Kumar, R., Verma, A., Shome, A., Sinha, S., Jha, P. K., Kumar, R., Kumar, P., Trivedi, S., Das, S., Sharma, P., & Vara Prasad, P. V. (2021d). Impacts of plastic pollution on ecosystem services, sustainable development goals and need to focus on circular economy and policy interventions. *Sustainability, 13*(17)9963.

Kunwar, B., Cheng, H., Chandrashekaran, S. R., & Sharma, B. K. (2016). Plastics to fuel: A review. *Renewable & Sustainable Energy Reviews, 54*, 421−428.

Law, K. L. (2017). Plastics in the marine environment. *Annual Review of Marine Science, 9*, 205−229.

Li, J., Yang, D., Li, L., Jabeen, K., & Shi, H. (2015). Microplastics in commercial bivalves from China. *Environmental Pollution (Barking, Essex: 1987), 207*, 190−195.

Liebezeit, G., & Liebezeit, E. (2013). Non-pollen particulates in honey and sugar. *Food Additives & Contaminants: Part A, 30*(12), 2136−2140.

Liebezeit, G., & Liebezeit, E. (2014). Synthetic particles as contaminants in German beers. *Food Additives & Contaminants: Part A, 31*(9), 1574−1578.

Lusher, A., Mchugh, M., & Thompson, R. (2013). Occurrence of microplastics in the gastrointestinal tract of pelagic and demersal fish from the English Channel. *Marine Pollution Bulletin, 67*(1−2), 94−99.

Lusher, A. L., Tirelli, V., O'Connor, I., & Officer, R. (2015). Microplastics in Arctic polar waters: The first reported values of particles in surface and sub-surface samples. *Scientific Reports, 5*, 14947.

Maes, T., Jessop, R., Wellner, N., Haupt, K., & Mayes, A. G. (2017). A rapid-screening approach to detect and quantify microplastics based on fluorescent tagging with Nile Red. *Scientific Reports, 7*, 44501.

Moore, C. J. (2008). Synthetic polymers in the marine environment: A rapidly increasing, long-term threat. *Environmental Research, 108*(2), 131−139.

Nel, H., & Froneman, P. (2015). A quantitative analysis of microplastic pollution along the south-eastern coastline of South Africa. *Marine Pollution Bulletin, 101*(1), 274−279.

Nicholson, J. W. (1997). *Kinetics of spatial structures in the photodegradation of polymers* (2nd ed.). The Royal Society of Chemistry.

Ogata, Y., Takada, H., Mizukawa, K., Hirai, H., Iwasa, S., Endo, S., Mato, Y., et al. (2009). International pellet watch: Global monitoring of persistent organic pollutants (POPs) in coastal waters. 1. Initial phase data on PCBs, DDTs, and HCHs. *Marine Pollution Bulletin, 58* (10), 1437−1446.

Pandey P. (2013) *Patna generates 900MT garbage daily*. Times of India. Retrieved on March, 28, 2021 from https://timesofindia.indiatimes.com/city/patna/patna-generates-900mt-garbage-daily/articleshow/19587311.cms.

PlasticEurope (2016) *Plastics − the Facts 2016: An analysis of European plastics production, demand and waste data*. Retrieved on March 28, 2021 from https://www.plasticseurope.org/en/resources/publications/3-plastics-facts-2016.

PlasticEurope (2017) *Plastics − the Facts 2017: Analysis of European plastics production, demand and waste data*. Retrieved on March 28, 2021 from https://www.plasticseurope.org/en/resources/publications/274-plastics-facts-2017.

PlasticEurope (2018) *Plastics − the Facts 2018: An analysis of European plastics production, demand and waste data*. Retrieved on March 28, 2021 from https://www.plasticseurope.org/en/resources/publications/619-plastics-facts-2018.

Powell, J. J., Faria, N., Thomas-McKay, E., & Pele, L. C. (2010). Origin and fate of dietary nano-particles and microparticles in the gastrointestinal tract. *Journal of Autoimmunity, 34*(3), J226−J233.

Quinn, B., Murphy, F., & Ewins, C. (2017). Validation of density separation for the rapid recovery of microplastics from sediment. *Analytical Methods, 9*(9), 1491−1498.

Rieux, A., Ragnarsson, E. G., Gullberg, E., Préat, V., Schneider, Y. J., & Artursson, P. (2005). Transport of nanoparticles across an in vitro model of the human intestinal follicle associated epithelium. *European Journal of Pharmaceutical Sciences Sci., 25*(4−5), 455−465.

Rocha-Santos, T., & Duarte, A. C. (2015). A critical overview of the analytical approaches to the occurrence, the fate and the behavior of microplastics in the environment. *Trends in Analytical Chemistry, 65*, 47−53.

Ryan, P. G. (2015). A brief history of marine litter research. *Marine anthropogenic litter* (pp. 1−25). Cham: Springer.

Sarkar, D. J., Sarkar, S. D., Das, B. K., Manna, R. K., Behera, B. K., & Samanta, S. (2019). Spatial distribution of meso and microplastics in the sediments of river Ganga at eastern India. *The Science of the Total Environment, 694*, 133712.

Scheurer, M., & Bigalke, M. (2018). Microplastics in Swiss floodplain soils. *Environmental Science & Technology, 52*(6), 3591−3598.

Singh, N., Mondal, A., Bagri, A., Tiwari, E., Khandelwal, N., Monikh, F. A., & Darbha, G. K. (2021). Characteristics and spatial distribution of microplastics in the lower Ganga River water and sediment. *Marine Pollution Bulletin, 163*, 111960.

Sruthy, S., & Ramasamy, E. (2017). Microplastic pollution in Vembanad Lake, Kerala, India: The first report of microplastics in lake and estuarine sediments in India. *Environmental Pollution (Barking, Essex: 1987)*, *222*, 315–322.

Stolte, A., Forster, S., Gerdts, G., & Schubert, H. (2015). Microplastic concentrations in beach sediments along the German Baltic coast. *Marine Pollution Bulletin*, *99*(1–2), 216–229.

Teuten, E. L., Saquing, J. M., Knappe, D. R., Barlaz, D. R., Jonsson, M. A., Björn, S., Rowland, A., et al. (2009). Transport and release of chemicals from plastics to the environment and to wildlife. *Philosophical Transactions of the Royal Society B: Biological Sciences*, *364*(1526), 2027–2045.

Thompson, R. C., Olsen, Y., Mitchell, R. P., Davis, A., Rowland, S. J., John, A. W., McGonigle, D., et al. (2004). Lost at sea: Where is all the plastic? *Science (New York, N.Y.)*, *304*(5672), 838-838.

Toxicslink. (2014). *Plastics and the environment assessing the impact of the complete ban on plastic carry bag*. New Delhi: Central Pollution Control Board.

Wang, W., Ndungu, A. W., Li, Z., & Wang, J. (2017). Microplastics pollution in inland freshwaters of China: A case study in urban surface waters of Wuhan, China. *The Science of the Total Environment*, *575*, 1369–1374.

WHO (2018a) *14 of world's most polluted 15 cities in India, Kanpur tops WHO list*, India Today: Retrieved on March 18, 2019 from https://www.indiatoday.in/education-today/gk-currentaffairs/story/14-worlds-most-polluted-15-cities-india-kanpur-tops-who-list-1224730-201805-02.

WHO (2018b) *14 of world's most polluted 15 cities in India, Kanpur tops WHO list*, India Today. Retrieved from https://www.indiatoday.in/education-today/gk-currentaffairs/story/14-worlds-most-polluted-15-cities-india-kanpur-tops-who-list-1224730-201805-02.

Wright, S. L., & Kelly, F. J. (2017). Plastic and human health: A micro issue? *Environmental Science & Technology*, *51*(12), 6634–6647.

Wright, S. L., Thompson, R. C., & Galloway, T. S. (2013). The physical impacts of microplastics on marine organisms: A review. *Environmental Pollution (Barking, Essex: 1987)*, *178*, 483–492.

Zbyszewski, M., & Corcoran, P. L. (2011). Distribution and degradation of freshwater plastic particles along the beaches of Lake Huron, Canada. *Water, Air, and Soil Pollution*, *220*(1–4), 365–372.

Ziajahromi, S., Kumar, A., Neale, P. A., & Leusch, F. D. (2017). Impact of microplastic beads and fibers on waterflea (*Ceriodaphnia dubia*) survival, growth, and reproduction: Implications of single and mixture exposures. *Environmental Science & Technology*, *51*(22), 13397–13406.

Ziajahromi, S., Neale, P. A., & Leusch, F. D. (2016). Wastewater treatment plant effluent as a source of microplastics: Review of the fate, chemical interactions and potential risks to aquatic organisms. *Water Science and Technology*, *74*(10), 2253–2269.

CHAPTER 11

Assessment of long-term groundwater variation in India using GLDAS reanalysis

Swatantra Kumar Dubey[1], Preet Lal[2], Pandurang Choudhari[3], Aditya Sharma[4] and Aditya Kumar Dubey[5]

[1]Department of Geology, Sikkim University, Gangtok, India, [2]Department of Geoinformatics, Central University of Jharkhand, Ranchi, India, [3]Department of Geography, University of Mumbai, Mumbai, India, [4]Department of Atmospheric Science, School of Earth Sciences, Central University of Rajasthan, Ajmer, India, [5]Department of Earth and Environmental Sciences, Indian Institute of Science Education and Research Bhopal, Bhopal, India

11.1 Introduction

Water is one of the essential natural resources for human beings and other living organisms. It is unevenly distributed across the planet Earth. Groundwater quality is an essential factor for water suitability for various purposes (Li et al., 2015). Groundwater is one of the most valuable natural resources on the earth's surface and is utilized for drinking, industrial activities, and agricultural activities. The global groundwater depletion between 1900 and 2008 was estimated to be 4500 km^3 (or $41.4 \text{ km}^3 \text{ year}^{-1}$), which represents a significant threat to global water security, thus potentially causing a decline in agricultural productivity and energy production (Frappart & Ramillien, 2018). The groundwater fluctuations are driven by natural (rainfall, vegetation, soil types) and anthropogenic (socioeconomic concerns, land use/land cover change, damming) processes with complex, nonlinear interactions between them (Lambin et al., 2003; Saikia et al., 2020). Thus one can expect to see a wide range of variability in yearly groundwater storage, with some areas facing unexpected shortages or flooding (Lal, Prakash, Kumar, 2020; Lal, Prakash, Kumar, Srivastava, et al., 2020). The area experiencing high interannual groundwater storage variability faces a higher risk of water supply shortages, even if there is little to no net loss of groundwater (Ouma et al., 2015). Various methods have been used by researchers worldwide for qualitative and quantitative evaluation of groundwater. Groundwater can be optimally utilized and sustained only when the quantity and quality are adequately assessed (Sadat-Noori et al., 2014). Groundwater storage changes

Advances in Remediation Techniques for Polluted Soils and Groundwater. DOI: https://doi.org/10.1016/B978-0-12-823830-1.00018-3

within the watershed are playing a vital role in the global water cycle as well as its variability with the aquifer system (Güntner et al., 2007) and climatic variability and change affect groundwater systems both through groundwater use changes directly and indirectly by recharge groundwater system (Green, 2016). Many factors are responsible for the occurrence and movement of groundwater, that is, geography, rocky area, weathering, elevation, drainage structure, land use—land cover, climatic conditions, and the interrelationship between these factors (Dubey & Sharma, 2018). Thus it is essential to observe and analyze groundwater storage variations in space and time for long-term sustainability (Singaraja, 2017).

Further, the groundwater fluctuation study also helps in understanding the effect of climate change. Some groundwater investigations require trained manpower, are time-consuming, and are very costly (Mukherjee et al., 2018). In hydrological sciences, one of the emerging technologies is geospatial technology that focuses on collecting, estimating, and managing resources at different scales (Ashtekar & Mohammed-Aslam, 2019; Oulidi et al., 2020). The in situ measurement of groundwater level data is costly and not available in many parts of India. However, much progress has been made with groundwater storage estimation using remotely sensed data (Kumar et al., 2021; Lee et al., 2020). Nowadays, remote-sensing technologies have been used by many researchers as handy tools in the field of hydrogeology (Patra et al., 2016). A remote sensor cannot identify groundwater directly as the presence of groundwater is derived from different indicators from satellite imagery, that is, formation of land, land use and land cover (LULC), water bodies, and topography. The geographic information system technique has been utilized efficiently in the last decade or so for different groundwater quality assessment purposes (Anbazhagan & Nair, 2004). According to Hoffmann and Sander (2007), "Remote sensing is an excellent tool for hydrologists and geologists to understand the perplexing problems of groundwater exploration better." However, of all the hydrological applications of remote sensing, the hydrogeological analysis of aerial photographs and satellite imagery is one of the most difficult because "groundwater by its very nature is not available for direct observation" (Kumar & Pandey, 2016). Although a variety of remotely sensed techniques such as sonar, downhole televiewer, deep-well current meters, and other geophysical and mechanical logging devices are used in groundwater hydrology, it is usual to confine the discussion of remote sensing to those techniques that involve the recording of reflected or emitted electromagnetic radiation by satellite or airborne sensors (Boulding, 1993). This covers various techniques that include conventional photography, visible, near-, middle- and thermal-infrared scanner imagery, and radar that can be used to guide, speed, and reduce the costs of groundwater investigations. Remote sensing for hydrogeological research has its main applications in groundwater exploration, although it can also be used in hazard monitoring and geothermal energy exploration (Avtar et al., 2019; Waters et al., 1990).

Over the past few decades, satellites have provided much useful information about conditions around the globe at a spatial and temporal resolution of practical significance (Degbelo & Kuhn, 2018). The two techniques for spatial modeling of the ground surface, that is, the Gravity Recovery and Climate Experiment (GRACE) satellite and the Global Land Data Assimilation System (GLDAS), provide information related to underground water storage changes (Frappart & Ramillien, 2018; Ouma et al., 2015). The GRACE satellite data measures the anomalies of gravity variation across the Earth (Rodell et al., 2004), which has been utilized in the GLDAS model for reanalysis groundwater storage data at higher spatial resolution. In March 2002 the collaborative project and mission of the GRACE of NASA and the German Aerospace Center (Deutsches Zentrum für Luft - DLR) had launched the first space—the time-based discovery of terrestrial water storage (TWS) at a large scale to measure the storage of water (Tang et al., 2010). The satellite observations from the GRACE sensor are a measure of changes in earth gravity. More recently, researchers used satellite-based interferometric synthetic aperture radar data for the deformation of image surface associated with groundwater extraction and replenishment (Vasco et al., 2019). Also, land surface models are a useful tool for investigating and predicting the spatio-temporal variations in TWS and other hydrological variables (Yin et al., 2020).

The largest groundwater consumer in the world is India, with more than 230 km^3 annual withdrawals. For drinking and domestic requirements, more than 90% of the rural and 30% of the urban population depend on groundwater as per the Ministry of Water Resources, India. India is now facing a severe water shortage problem in many states due to the uneven rainfall distribution. The unplanned exploitation of groundwater has depleted this vital resource in many parts of India and various parts of the world. Due to over-exploitation of water for agriculture, the groundwater resources in India are declining rapidly (Pathak & Dodamani, 2019). Therefore it seems essential to understand the ways and methods for groundwater estimation and measurement using satellite datasets at different scales (national, regional, and local). Hence, this chapter exemplifies the investigation and importance of satellite data for groundwater use in India.

11.2 Data used and methodology

In this study the precipitation and groundwater storage data from Global Land Data Assimilation System Version 2 (GLDAS-2) reanalysis for 2003—19 was used. The reanalysis GLDAS-2 includes three components as GLDAS-2.0, GLDAS-2.1, and GLDAS-2.2, whereas GLDAS-2.0 is forced entirely with the Princeton meteorological forcing input data and provides a temporally consistent series from 1948 through 2014. GLDAS-2.1 is forced with a combination of model and observation data from 2000 to 2020. The GLDAS v2.2 reanalysis data were developed based on v2.0 daily catchment model simulations and forced with the operation European Centre for Medium-Range Weather Forecasts for precipitation data, but TWS was assimilated using GRACE satellite observation.

For time series analysis, Theil−Sen median trend analysis and Mann−Kendall test were used for the change and significance test of groundwater storage and precipitation in India. Theil−Sen median trend analysis is a robust nonparametric statistical trend analysis method that divided the data into $u(u-1)/2$ pairs of combinations, and calculating the median slope of them with the following equation:

$$P = median \frac{Index_m - Index_n}{m - n} \tag{11.1}$$

where $index_m$ and $index_n$ are index values of year m and n, respectively, where $P > 0$ indicates an increasing trend of the series; otherwise, it shows a decreasing trend.

The nonparametric Mann−Kendall test is a statistical significance test that does not require data in a particular distribution manner. The equation is as follows:

For a set of time series $Index_i$, the value z is expressed as

$$Z = \begin{cases} \frac{S-1}{\sqrt{S(S)}}, S > 0 \\ \frac{0}{\sqrt{s(S)}}, S < 0 \end{cases} \tag{11.2}$$

where $P = \sum_{m=1}^{p-1} \sum_{m=n+1}^{p} significance \ (Index_m - Index_n)$,

$$significance \ (Index_m - Index_n)$$

$$= \begin{cases} 1 & Index_m - Index_n & > 0 \\ 0 & Index_m - Index_n & = 0, \quad s(S) = \frac{u(u-1)(2u-5)}{18} \\ -1 & Index_m - Index_n & < 0 \end{cases} \tag{11.3}$$

where $Index_m$ and $Index_n$ are index values of year m and n, respectively and u is the length of the time series. The value of $z = -$ infinity to $+$ infinity, whereas significance is generally considered 0.05.

11.3   Results and discussion

11.3.1   Rainfall variations

The monthly rainfall variation was assessed by the selected years (2003–19) to find the monthly changes in the study area. The maximum rainfall was observed in the monsoon season, that is, June, July, August, and September, which can be observed in Fig. 11.1. The highest rainfall was observed in the eastern part of India and low on the western side. December, January, February, March, and April months show the least rainfall due to the seasonal changes, but the eastern and northern parts show some rainfall in these months. The maximum rainfall was observed in July and August months in the selected study period.

11.3.1.1   Groundwater variations

The groundwater variation was observed using the GLDAS-2.2 data for the period of 2003–19. Fig. 11.2 shows the monthly groundwater changes using satellite data. The maximum groundwater is observed in August, September, October, and November months in the selected period. In July the maximum rainfall was observed, but the groundwater was observed low because of the less percolation and high runoff during this month. The groundwater depletion was observed from November to June due to more groundwater extraction for agriculture and industrial and drinking purposes. The maximum groundwater depletion was observed in May and June in most of the parts of India. The deficit was primarily observed in the Gangetic plain areas in most of the months. This area is agriculture dominant and uses groundwater for crop cultivation in the Rabi and Kharif seasons. The central part of India shows the minimum groundwater depletion compared to other parts of India because Central India comes under the monsoon core zone, and the depletion of groundwater was low.

11.3.2   Seasonal variations in rainfall and groundwater

The Indian monsoon was divided into four seasons, that is, December, January, February (DJF); March, April, May (MAM); June, July, August (JJA); and September, October, November (SON) months. The maximum rainfall was observed in the seasonal months, that is, JJA, as shown in Fig. 11.3. The DJF season shows the least rainfall in the months, but MAM months have rainfall in the eastern part of India. In the SON season the rainfall is maximum in India's eastern and southern parts (Fig. 11.3). Their rainfall variation affects the groundwater level in different parts of the country.

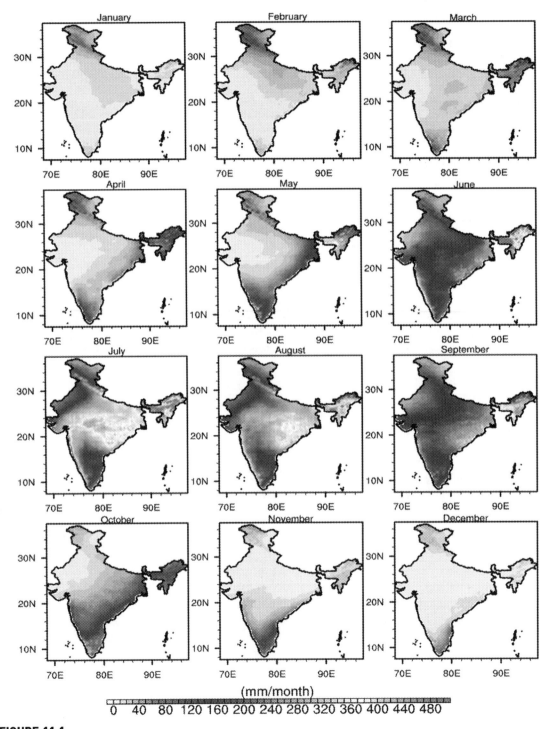

FIGURE 11.1

Monthly rainfall variations on the period (2003—19).

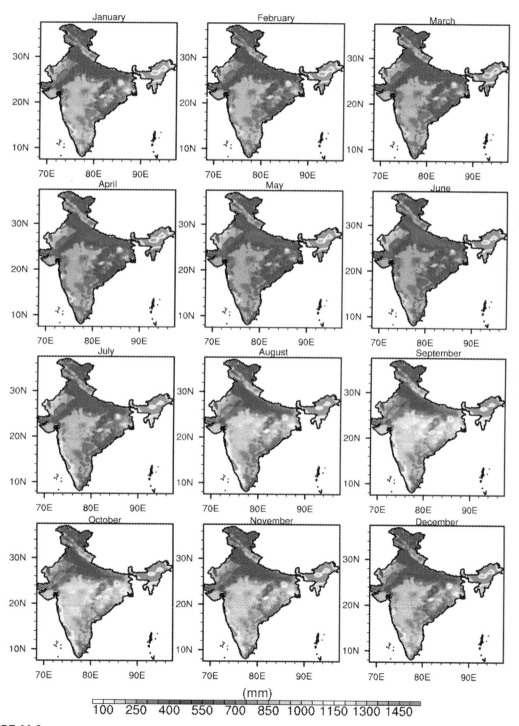

FIGURE 11.2

Monthly groundwater variations of period 2003–19.

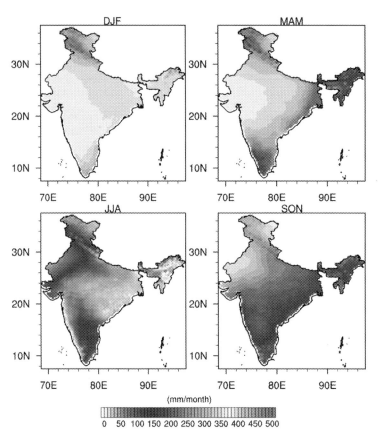

FIGURE 11.3
Seasonal rainfall variations of period 2003–19.

The maximum seasonal variation in groundwater was observed in the SON season, and it has shown the maximum groundwater during the period 2003–19. But the monsoon season, that is, JJA has the least groundwater, but the rainfall was high in the JJA season. The maximum groundwater depletion was observed in India's Gangetic plain and western part, as shown in Fig. 11.4. The maximum groundwater depletion exhibited in the MAM because of the less or no rainfall during these months.

11.3.3   Trend analysis of precipitation and groundwater

The precipitation trend analysis shows the negative and significance in the Gangetic plain area in most seasons. The maximum negative trend was

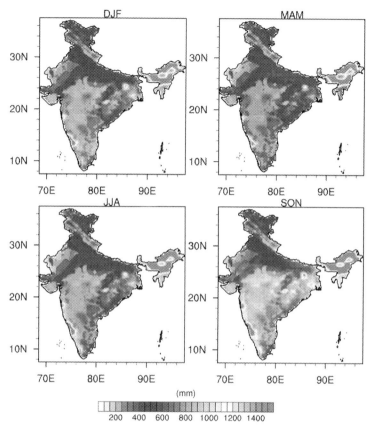

FIGURE 11.4

Seasonal groundwater variations of period 2003−19.

observed in the DJF season, followed by the MAM, JJA, and SON seasons. A positive and significant trend was observed in the Central India region, and most of the part showed a positive trend but not significance. The trend analysis observed that the negative trend and significance show less rainfall in the Gangetic region, as shown in Fig. 11.5. In the groundwater trend analysis, the southern part of India showed a negative trend and significance during 2003−19. A negative significance trend was observed in the MAM, JJA, and SON seasons in the southern part of India in the seasonal analysis. The JJA shows a positive significance trend in Central India due to maximum rainfall in the monsoon season. The eastern part also has shown a positive significance trend during the analysis period in the monsoon

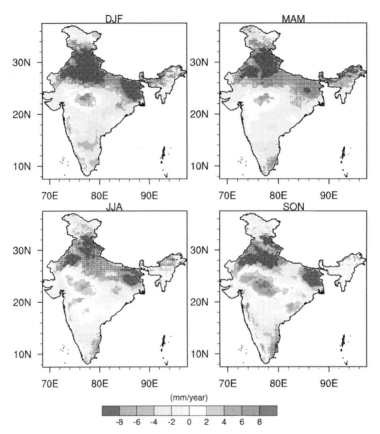

FIGURE 11.5
Trend analysis of precipitation of period 2003—19.

season. In the DJF season the groundwater has a positive trend but is not significant in most places in India. The Gangetic plain shows the negative trend in all the seasons, but it was not shown the significance; it observed that the groundwater withdraws high in this area during the all-season (Fig. 11.6).

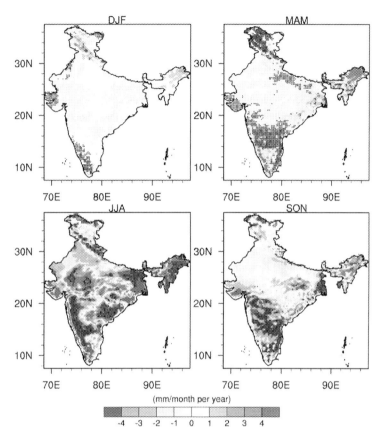

FIGURE 11.6
Trend analysis of groundwater of period 2003–19.

11.4 Conclusion

The spatiotemporal analysis of groundwater level for India utilizes the GLDAS-2.2 information and a factual method to discover the groundwater variance pattern investigation. The groundwater-level data were employed in this study to understand the groundwater fluctuation over space and time. According to the spatial interpolation of seasonal changes, groundwater declination has a severe rate from 2003 to 2019. This decline is observed in all seasons and mainly witnessed in southern, southeastern parts of the study area. However, it has overcome naturally in the forthcoming years due to sufficient rainfall occurrence, reflected in the groundwater-level fluctuation. This analysis indicates that the decrease of rainfall increases groundwater depletion and increases drought, which leads the aquifer to stress in many parts of India.

The study concludes that the groundwater level is declined in the western and northern parts of India during all the monsoon seasons. More surface runoff, less infiltration, and overexploitation of groundwater are the leading causes of high groundwater fluctuation. The results also indicate that the overexploitation of groundwater for irrigation purposes in the north region leads to more groundwater fluctuation during the summer season, and the aquifer comes under stress. These variations in groundwater levels for space, time, and depth mainly affect the farmers. Therefore it is necessary to enhance the groundwater resources in the affected areas through artificial recharge techniques by constructing suitable structures.

References

Anbazhagan, S., & Nair, A. M. (2004). Geographic information system and groundwater quality mapping in Panvel Basin, Maharashtra, India. *Environmental Geology, 45*, 753–761. Available from https://doi.org/10.1007/s00254-003-0932-9.

Ashtekar, A. S., & Mohammed-Aslam, M. A. (2019). Geospatial technology application for groundwater prospects mapping of sub-upper Krishna Basin, Maharashtra. *Journal of the Geological Society of India, 94*, 419–427. Available from https://doi.org/10.1007/s12594-019-1331-5.

Avtar, S., Aggarwal, C., Kharrazi, Y., & Dou, K. (2019). Exploring renewable energy resources using remote sensing and GIS—A review. *Resources, 8*, 149. Available from https://doi.org/10.3390/resources8030149.

Boulding, J. R. (1993). *Use of airborne, surface, and borehole geophysical techniques at contaminated sites: A reference guide (no. PB-94-123825/XAB)*. Lexington, MA: Eastern Research Group, Inc.

Degbelo, A., & Kuhn, W. (2018). Spatial and temporal resolution of geographic information: An observation-based theory. *Open Geospatial Data, Software and Standards, 3*, 1–22. Available from https://doi.org/10.1186/s40965-018-0053-8.

Dubey, S. K., & Sharma, D. (2018). Assessment of climate change impact on yield of major crops in the Banas River Basin, India. *The Science of the Total Environment, 635*, 10–19. Available from https://doi.org/10.1016/j.scitotenv.2018.03.343.

Frappart, F., & Ramillien, G. (2018). Monitoring groundwater storage changes using the Gravity Recovery and Climate Experiment (GRACE) satellite mission: A review. *Remote Sensing, 10*, 829. Available from https://doi.org/10.3390/rs10060829.

Green, T. R. (2016). Linking climate change and groundwater. In A. J. Jakeman, O. Barreteau, R. J. Hunt, J.-D. Rinaudo, & A. Ross (Eds.), *Integrated groundwater management: Concepts, approaches and challenges* (pp. 97–141). Cham: Springer International Publishing. Available from https://doi.org/10.1007/978-3-319-23576-9_5.

Güntner, A., Schmidt, R., & Döll, P. (2007). Supporting large-scale hydrogeological monitoring and modelling by time-variable gravity data. *Hydrogeology Journal, 15*, 167−170.

Hoffmann, J., & Sander, P. (2007). Remote sensing and GIS in hydrogeology. *Hydrogeology Journal, 15*, 1−3. Available from https://doi.org/10.1007/s10040-006-0140-2.

Kumar, A., Kumar, S., Lal, P., Saikia, P., Srivastava, P. K., & Petropoulos, G. P. (2021). Chapter 1 − Introduction to GPS/GNSS technology. In G. p Petropoulos, & P. K. Srivastava (Eds.), *GPS and GNSS technology in geosciences* (pp. 3−20). Elsevier. Available from https://doi.org/10.1016/B978-0-12-818617-6.00001-9.

Kumar, A., & Pandey, A. C. (2016). Geoinformatics based groundwater potential assessment in hard rock terrain of Ranchi urban environment, Jharkhand state (India) using MCDM−AHP techniques. *Groundwater for Sustainable Development, 2−3*, 27−41. Available from https://doi.org/10.1016/j.gsd.2016.05.001.

Lal, P., Prakash, A., & Kumar, A. (2020). Google Earth Engine for concurrent flood monitoring in the lower basin of Indo-Gangetic-Brahmaputra plains. *Natural Hazards, 104*, 1947−1952. Available from https://doi.org/10.1007/s11069-020-04233-z.

Lal, P., Prakash, A., Kumar, A., Srivastava, P. K., Saikia, P., Pandey, A. C., Srivastava, P., & Khan, M. L. (2020). Evaluating the 2018 extreme flood hazard events in Kerala, India. *Remote Sensing Letters, 11*, 436−445. Available from https://doi.org/10.1080/2150704X.2020.1730468.

Lambin, E. F., Geist, H. J., & Lepers, E. (2003). Dynamics of land-use and land-cover change in tropical regions. *Annual Review of Environment and Resources, 28*, 205−241. Available from https://doi.org/10.1146/annurev.energy.28.050302.105459.

Lee, S., Hyun, Y., Lee, S., & Lee, M.-J. (2020). Groundwater potential mapping using remote sensing and GIS-based machine learning techniques. *Remote Sensing, 12*, 1200. Available from https://doi.org/10.3390/rs12071200.

Li, B., Rodell, M., & Famiglietti, J. S. (2015). Groundwater variability across temporal and spatial scales in the central and northeastern U.S. *Journal of Hydrology, 525*, 769−780. Available from https://doi.org/10.1016/j.jhydrol.2015.04.033.

Mukherjee, S., Aadhar, S., Stone, D., & Mishra, V. (2018). Increase in extreme precipitation events under anthropogenic warming in India. *Weather and Climate Extremes, 20*, 45−53. Available from https://doi.org/10.1016/j.wace.2018.03.005.

Oulidi, H. J., Fadil, A., & Semane, N. E. (Eds.), (2020). *Geospatial technology: Application in water resources management, advances in science, technology & innovation*. Springer International Publishing. Available from https://doi.org/10.1007/978-3-030-24974-8.

Ouma, Y. O., Aballa, D. O., Marinda, D. O., Tateishi, R., & Hahn, M. (2015). Use of GRACE time-variable data and GLDAS-LSM for estimating groundwater storage variability at small basin scales: A case study of the Nzoia River Basin. *International Journal of Remote Sensing, 36*, 5707−5736. Available from https://doi.org/10.1080/01431161.2015.1104743.

Pathak, A. A., & Dodamani, B. M. (2019). Trend analysis of groundwater levels and assessment of regional groundwater drought: Ghataprabha river basin, India. *Natural Resources Research, 28*, 631−643. Available from https://doi.org/10.1007/s11053-018-9417-0.

Patra, H. P., Adhikari, S. K., & Kunar, S. (2016). Remote sensing in groundwater studies. In H. P. Patra, S. K. Adhikari, & S. Kunar (Eds.), *Groundwater prospecting and management* (pp. 7−45). Singapore: Springer Hydrogeology. Springer. Available from https://doi.org/10.1007/978-981-10-1148-1_2.

Rodell, M., Houser, P., Jambor, U., Gottschalck, J., Mitchell, K., Meng, C.-J., Arsenault, K., Cosgrove, B., Radakovich, J., Bosilovich, M., ... Toll, D. (2004). The global land data assimilation system. *Bulletin of the American Meteorological Society, 85*, 381−394.

Sadat-Noori, S. M., Ebrahimi, K., & Liaghat, A. M. (2014). Groundwater quality assessment using the Water Quality Index and GIS in Saveh-Nobaran aquifer, Iran. *Environmental Earth Sciences, 71*, 3827−3843. Available from https://doi.org/10.1007/s12665-013-2770-8.

Saikia, P., Kumar, A., Diksha., Lal, P., Nikita., & Khan, M. L. (2020). Ecosystem-based adaptation to climate change and disaster risk reduction in Eastern Himalayan forests of Arunachal Pradesh, Northeast India. In S. Dhyani, A. K. Gupta, & M. Karki (Eds.), *Nature-based solutions for resilient ecosystems and societies, disaster resilience and green growth* (pp. 391–408). Singapore: Springer. Available from https://doi.org/10.1007/978-981-15-4712-6_22.

Singaraja, C. (2017). Relevance of water quality index for groundwater quality evaluation: Thoothukudi District, Tamil Nadu, India. *Applied Water Science, 7*, 2157–2173. Available from https://doi.org/10.1007/s13201-017-0594-5.

Tang, Q., Gao, H., Yeh, P., Oki, T., Su, F., & Lettenmaier, D. P. (2010). Dynamics of terrestrial water storage change from satellite and surface observations and modeling. *Journal of Hydrometeorology, 11*, 156–170. Available from https://doi.org/10.1175/2009JHM1152.1.

Vasco, D. W., Farr, T. G., Jeanne, P., Doughty, C., & Nico, P. (2019). Satellite-based monitoring of groundwater depletion in California's Central Valley. *Scientific Reports, 9*, 16053. Available from https://doi.org/10.1038/s41598-019-52371-7.

Waters, P., Greenbaum, D., Smart, P. L., & Osmaston, H. (1990). Applications of remote sensing to groundwater hydrology. *Remote Sensing Reviews, 4*, 223–264. Available from https://doi.org/10.1080/02757259009532107.

Yin, K., Xu, S., Zhao, Q., Huang, W., Yang, K., & Guo, M. (2020). Effects of land cover change on atmospheric and storm surge modeling during typhoon event. *Ocean Engineering, 199*, 106971.

CHAPTER 12

Emerging contaminants in subsurface: sources, remediation, and challenges

Anuradha Garg and Shachi

Research Scholar, Department of Hydrology, IIT Roorkee, Roorkee, India

12.1 Introduction

An "emerging" concern among the scientific community around the world is the newly detected range of anthropogenic contaminants occurring in the air, water, food, and soil, etc., as a result of reckless use and careless disposal of chemicals (Cabeza et al., 2012). The previously unknown micropollutants found mainly in the aqueous environment can or have the potential to cause risks to life sustenance. Since some of them are persistent in nature, they attract more attention from scientists, researchers, managers, citizens, and other stakeholders in the direct/indirect use of contaminated resources. This new category of contaminants consists of pharmaceutical compounds, personal care products (PCPs), nanomaterials, microplastics, various drugs, pesticides, UV filters, insect repellents, fire retardants, etc. Another class of compounds is endocrine disrupting chemicals or compounds (EDCs) that disturb the functioning of the endocrine system found in humans, animals, birds, fish, crustaceans, snails, and in most other living species (Diamanti-Kandarakis et al., 2009). These compounds are not categorized based on any physical or chemical property but by the disruption they cause to the endocrine system. They are the chemicals that have been found responsible for causing an interruption in the glands producing hormones for reproduction, growth, development, metabolism, tissue function, sleep, and mood regulation. The endocrine system is sensitive to even trace concentrations of such chemicals, making most animal species vulnerable to this type of contamination (Snyder et al., 2004). Emerging contaminants (ECs), thus, include, but are not limited to a range of EDCs, pharmaceuticals, PCPs, and some other drugs, nanomaterials. The list is "ever-growing" as the contaminants are still being discovered. While some "new" compounds are being added, there are many "old" contaminants that were discovered earlier but whose health and environmental risks were unidentified. Hence, the term "emerging" is used for newly identified compounds and the existing compounds, contamination potential of which has only been known recently (Gogoi et al., 2018).

Advances in Remediation Techniques for Polluted Soils and Groundwater. DOI: https://doi.org/10.1016/B978-0-12-823830-1.00014-6

The fresh list of contaminants is not fixed as new compounds keep on adding regardless of how and where they are identified. They could be present widely like PCPs or remain limited to a few sites. This class of compounds has not been studied comprehensively, and scientists still need to understand the risks associated with their exposure to the environment and humans. Thus the treatment of contaminated sites remains a far picture when even the level of pollution risk is not fully explained. The advanced and highly sensitive instruments have facilitated the detection of these micropollutants, due to which new concerns have surfaced. However, their concentration has still not been studied deeply to decide upon permissible limits.

Studies in the past decades have found traces of many of these ECs in groundwater (Bartelt-Hunt et al., 2011; Cabeza et al., 2012). Since most of them have industrial, urban, medicinal, and agricultural origins, their pathways and fate into and from a system become important to understand. People studying ECs in the source—pathway—receptor model have suggested that the contaminants have high chances of ending up in subsurface, which causes a threat to living species using it for drinking and other consumption purposes.

12.2 Sources of emerging contaminants in groundwater

ECs have been found in trace concentrations in soil, surface water, and groundwater ranging from few nanograms to micrograms per liter. Sources of these contaminants are varied, which majorly fall into two category types: point and nonpoint sources (Table 12.1). Industrial effluent discharge to soil and surface water, municipal sewage treatment plants, landfills, and other waste disposal sites, resource extraction units, small workshops, hospital water discharge, etc. are the main point sources of emerging compounds' pollution (Bedding et al., 1982; Ritter et al., 2002). These sources release pollutants at discrete places, after which they are transported to other systems/resources through various travel mechanisms and reactions taking place. The other category of sources, that is, nonpoint sources are responsible for the diffusion of contaminants over a geographically broader area. Agricultural, urban, and stormwater runoff, infiltration from agricultural fields, atmospheric deposition of pollutants, leakages from automobiles/trains carrying chemical goods, etc. are some of the common sources. ECs through these sources may enter either or all of the soil, surface, and ground water resources. These chemicals, however, may ultimately reach groundwater resources through infiltration and recharge creating bigger problems. Being an invisible resource, it is more difficult to track and eliminate pollutants from groundwater than from surface water. The following diagram explains how these pollutants may end up in the groundwater while entering from different sources in the environment.

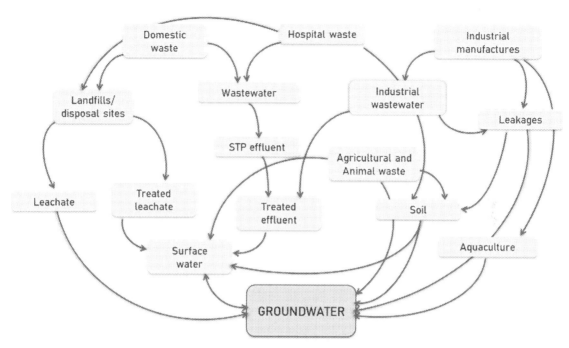

FIGURE 12.1

Schematic diagram showing sources and pathways of emerging contaminants in groundwater.

As can be seen from Fig. 12.1, groundwater is vulnerable to contamination through infiltration from many sources (Calvo-Flores et al., 2018). Liquid leaching out from waste disposal sites, soils exposed to the industrial waste discharge, infiltration from septic systems, effluents from wastewater treatment plants (WWTPs) and agricultural fields, etc. collectively put together extreme pollution pressure on groundwater resources (Gogoi et al., 2018). Even surface and groundwater interactions exchange a large number of chemicals that are difficult to account for. Though groundwater has mostly been found to be comparatively less contaminated than surface water bodies, the lack of accountability of sources and concentration makes the situation more alarming (Loos et al., 2010; Vulliet & Cren-olivé, 2011; Schaider et al., 2014). Also, a low microbial population and lesser redox activity increase the residence time of these contaminants in groundwater (Luo et al., 2014). ECs existing in groundwater are a major concern as the resource is most used for direct drinking and agricultural purposes, especially in developing countries, which directs its path to the human body majorly through oral ingestion. Considering the extent of the problem, their occurrence has not yet been adequately characterized in most countries, except for some parts in Europe and North America (Calvo-Flores et al., 2018).

Table 12.1 Point and nonpoint sources of emerging contaminants (ECs) in groundwater.

Source	Route	ECs released	References
Point sources			
Wastewaters	Sewage waters, industrial waste waters, contaminated surface water sources, hospital wastewater, WWTPs, etc.	Pharmaceuticals and PCPs, therapeutic drugs, lifestyle products, industrial chemicals, surfactants, etc.	Putschew et al. (2000), Ternes and Hirsch (2000), Sacher et al. (2001), Drewes et al. (2003), Clara et al. (2004), Heberer and Adam (2004), Kreuzinger et al. (2004), Mansell and Drewes (2004), Snyder et al. (2004), Glassmeyer et al. (2005), Grünheid et al. (2005), Rabiet et al. (2006), Díaz-Cruz and Barceló (2008), Gasser et al. (2010), Katz et al. (2009), Pecoraino et al. (2008), Schulz et al. (2008)
Landfills	Domestic waste landfills, industrial waste landfills	Pharmaceuticals like antibiotics and antiinflammatories; industrial compounds like detergents, fire retardants, plasticizers, and antioxidants; and lifestyle products like caffeine, nicotine, and its metabolite cotinine	Ahel et al. (1998), Barnes et al. (2004), Buszka et al. (2009), Eckel et al. (1993), Holm et al. (1995)
Septic tanks	Rural and semi urban areas with shallow groundwater tables and septic tanks	Pharmaceuticals like ibuprofen, acetaminophen, PCPs, nicotine, cotinine, etc.	Godfrey et al. (2007), Swartz et al. (2006), Verstraeten et al. (2005)
CAFOs	Leakage from waste lagoons	Veterinary antibiotics, endogenous estrogen E2 and estrone, androgens testosterone and androstenedione, etc.	Bartelt-Hunt et al. (2011), Bradford et al. (2008), Chomycia et al. (2008), Kolodziej et al. (2004), Thorne (2007)
Nonpoint sources			
Agricultural soils	Manure and biosolids application to agricultural soils	Halogenated hydrocarbons, perfluorochemicals, and polychlorinated alkanes, veterinary antibiotics, antimicrobials, saccharin	Buerge et al. (2011), Clarke and Smith (2011), Hu et al. (2010), Sarmah et al. (2006)
Surface waters and MAR	Surface water and groundwater exchange processes, artificial recharge of contaminated water to aquifers	Pharmaceuticals, PCPs, lifestyle products, many industrial chemicals, surfactants, etc.	Barnes et al. (2008), Buerge et al. (2009), Drewes (2009), Focazio et al. (2008), Lapworth et al. (2009), Lewandowski et al. (2011)
Leaky sewage systems	Leaky or underrepair sewage systems	Pharmaceuticals, caffeine, nicotine and its metabolites, surfactants, PCPs, etc.	Ellis (2006), Morris and Cunningham (2007), Morris et al. (2005), Rueedi et al. (2009)
Aerial deposition	Agricultural, industrial, vehicular transport deposition, etc.	Pesticides, industrial chemicals, etc.	Leister and Baker (1994), Villanneau et al. (2011)

CAFOs, *Concentrated animal feeding operations;* MAR, *Managed aquifer recharge;* PCP, *Personal care product;* WWTPs, *Waste water treatment plants.*

12.3   Detection and analysis

The advances in analytical methods have enabled the detection of chemicals that are present in trace concentrations yet are significant from a health perspective. Medicinal and personal care products, pesticides, and other industrial chemicals that fall in the category of ECs have been existing for years in the environment yet went unnoticed due to their concentrations that have always been below the detection limits. With efficient modern technologies the detection of emerging pollutants and analyses of their toxicity levels have become possible than ever. With the help of new analytical tools, all kinds of matrices, including soil, sediments, surface water, sewage, groundwater, and even plant samples, can be potentially tested for low levels of contamination. The most crucial and laborious step in the detection of ECs is the process of sample preparation (Dimpe & Nomngongo, 2016). The very low concentration of these compounds in the samples, which also sometimes does not remain constant through the analysis period, makes it difficult to accurately detect and evaluate the samples (Calvo-Flores et al., 2018). Thus to properly analyze the samples, they need to be extracted and cleaned prior to the analysis. This step is the longest in the entire process of EC's detection and analysis.

The extraction process can be done using any suitable method from the list of the latest available technologies, depending on the type of sample and the contaminant to be detected. Groundwater samples are most commonly extracted using these methods: solid-phase extraction, solid-phase microextraction (SPME), dispersive SPME, and liquid- phase microextraction. The purpose of extraction is to clean, recover, and prepare the sample with the right concentration required for quantitative analysis (Calvo-Flores et al., 2018). After the sample is ready, it is subjected to analyte isolation using chromatographic techniques. Liquid chromatography (LC) is the most popularly used technique to isolate the target product from the groundwater samples selectively (Petrovic & Barcelo, 2006).

The analysis of the samples is done using the highly selective and sensitive mass spectrometry (Ms) technique that tells about the compound's molecular structure. A number of methods such as gas chromatography/Ms, LC electrospray ionization Ms, or high-performance LC Ms are used for the analysis and characterization of emerging pollutants. The advanced versions of these methods provide high sensitivity toward trace compounds by also offering reduced analysis time. Modern Ms technologies such as time-of-flight—Ms or linear ion trap—Ms offer the analysis of the broader spectrum of polar and nonpolar organic metabolites with increased mass accuracy and high sensitivity (Petrovic & Barcelo, 2006).

12.4 Types of emerging contaminants

12.4.1 Pesticides

The term "pesticides" is used for the substances used to kill pests and control weeds and diseases in plants. A large variety of pesticides are used for agricultural purposes today which may be biochemically active and may be toxic to the environment. With agricultural runoff and leaching, most of these pesticides end up in groundwater and remain there for a fairly long time period. These chemicals may transform into metabolites after being released into the environment. The transformation may take place through various processes (biotic and abiotic). Studies have suggested that some of these products are more mobile, polar, and persistent than their parent compounds and, thus, can be found at higher concentration and frequency (Glassmeyer et al., 2005; Kolpin et al., 2000, 2004; Holtze et al., 2008; Stuart et al., 2011). The problem, however, lies with their toxicity levels and not persistence, which also is higher than the parent compounds in some cases (Sinclair & Boxall, 2003; Stuart et al., 2011). Since the current knowledge about pesticide metabolites is less, its detection and analysis get complex. The formation, breakdown, movement, accumulation, and transformation of pesticide compounds depend on physicochemical properties like leachability, solubility, and persistence, which also decide their ultimate fate.

12.4.2 Pharmaceutical products

Pharmaceutical products are compounds utilized for treating medical illnesses in humans and animals. They include a whole range of medicines: antibiotics, antiinflammatories, steroids, analgesics, alcoholic drugs, lipid regulators, and some illegal drugs that raise many biological and environmental concerns (Fatta-Kassinos et al., 2011; Gogoi et al., 2018). Bioaccumulation of pharmaceutical compounds and their transformation products (TPs) may have serious undesirable consequences (Arnold et al., 2013). Pharmaceutical industries manufacturing these drugs are responsible for discharging their waste in surface and/or groundwater, resulting in higher bioaccumulation. Other potential routes are human excretion, disposal of expired/unused medicines, veterinary excretion, etc. These compounds travel from landfills, soil, and surface water and may reach groundwater bodies. Though the concentration of pharmaceuticals in groundwater has mostly been found very low, their ecotoxicological impacts are little understood with the present knowledge. Some of the most commonly detected compounds of this class are ibuprofen, acetaminophen, salicylic acid, phenazone, trimipramine, coumarin, sulfamethoxazole propyphenazone, sulfamethazine, carbamazepine, triclosan, diclofenac, fenofibrate, ketoprofen, azithromycin, etc. (Gogoi et al., 2018; Stuart et al., 2011).

12.4.3 Personal care products

This is another class of emerging compounds that are enormously used by humans for personal care and hygiene purpose. The major fraction comes from skin care and cosmetic products like sunscreens, shampoos, soaps, perfumes, moisturizers, talcum powders, lipsticks, beauty creams, dietary supplements, cleaning products, and the list goes on. The major chemicals that are particularly problematic are methyl and propyl paraben, oxybenzone, carbamodithioic acid, dimethyl-, methyl ester, isopropyl myristate, octocrylene, 2-phenoxy-ethanol, drometrizole, ethylhexyl methoxycinnamate, propranolol, caffeine, and polycyclic musks like galaxolide (HHCB), tonalide (AHTN), HHCB-lactone, 2-amino-musk ketone, and so on (Martin et al., 2007; Matamoros et al., 2009; Stuart et al., 2011; Sui et al., 2015).

12.4.4 Industrial chemicals

Industrial chemicals that are of emerging concern include a range of plasticizers, flame retardants, polycyclic aromatic hydrocarbons, volatile organic compounds, surfactants, antioxidants, corrosion inhibitors, gasoline additives, and so on. Some of these chemicals fall under this category due to the unknown effects of their TPs. These derivatives are usually released from WWTPs, domestic effluents, landfills, and underground chemical storage spills.

The most common plasticizers falling in this category are bisphenol-A (BPA), phthalates, and N-butylbenzenesulfonamide. These chemicals are ubiquitous in the environment and in groundwater as well. They are persistent and have significant endocrine disrupting activities on human bodies. For example, BPA, which is used in manufacturing a series of resins, polymers, adhesives, building materials, paints, etc., is a common contaminant in groundwater bodies (Calvo-Flores et al., 2018). It disrupts the endocrine system by binding to α- and β-estrogen receptors and competing with the natural hormones like E2. Several studies have also presented the effects of BPA on the reproductive system, thyroid, breast cancer, obesity, etc. (Diamanti-Kandarakis et al., 2009; Vandenberg et al., 2010). Corrosion inhibitors like benzotriazoles and benzothiazoles used in aircrafts, antifreeze liquids, engine coolants, and in dye production, fungicides, herbicides, etc. are increasingly found in groundwater samples all over the world. They are found to have estrogenic effects and may also be carcinogenic to humans (Richardson, 2012).

12.4.5 Lifestyle products

The synthetic organic compounds that exist because of our lifestyle choice are coming up as problematic chemicals. These mostly include various types of food additives, supplements, sweeteners, preservatives, stimulants, etc. Caffeine and nicotine are among the most pervasively found ECs in groundwater (Cabeza et al., 2012; Estévez et al., 2012; Teijon et al., 2010). The concentration levels are very high in some regions of the world depending on their "lifestyle" and hence, the consumption. Although these compounds have not been studied enough to establish severe impacts on the ecology and environment, their ever-increasing higher concentration in the environment especially in groundwater is becoming a serious concern. Different types of food additives, including nutrients, taste enhancers, stabilizers, preservatives, and food colors, fall in this category, omnipresence of which in the environment raises research questions. Some of these compounds are considered to be endocrine disrupters and oxidants (Jobling et al., 1995) and are enlisted as high-priority emerging pollutants (Calvo-Flores et al., 2018), while research is still going on to carefully understand their impacts.

12.4.6   Surfactants

Popularly known for use in the manufacturing of products like laundry detergents, dishwashing liquids and soaps, shampoos, surface cleaners, and bathing soaps surfactants are well-identified environmental pollutants. They are also widely used in a variety of industries founding applications in textile processing, wastewater treatment, petroleum recovery, and also in lubricants, pesticides, paper, leather, metals, etc. These chemicals constitute two parts (hydrophilic head and hydrophobic tail), which helps in removing the oil and dirt from the surface. This long-chain hydrophobic part is what causes toxicity to the aquatic environments. Surfactants and their metabolites can be found in higher concentrations in urban groundwater, mainly due to domestic and industrial discharge (Corada-Fernández et al., 2011; Eadsforth et al., 2006). All of the anionic, cationic, amphoteric, and nonionic surfactants may be found in various quantities in the environment; however, detailed research has only been conducted for anionic surfactants like Linear alkylbenzene sulfonate (LAS) and nonionic surfactants like nonylphenol ethoxylates (NPEOs) (Ahel et al., 1994; Ding et al., 1999; Eichhorn et al., 2002; Ferguson et al., 2001; González-Mazo et al., 1998, 2002; Isobe & Takada, 2004; Jonkers et al., 2003; Takada & Ogura, 1992). Surfactants that may potentially persist in aquatic environments and are difficult to eliminate (Montgomery-Brown & Reinhard, 2003; Soares et al., 2008). The most common types of surfactants that are in use are alkylphenol ethoxylates, alkyl ethoxysulfates, alkyl sulfates, α-olefin sulfonates, NPEOs, LAS, fluorosurfactants, etc. (Fig. 12.2).

12.5   Fate of emerging contaminants in groundwater

Since the categorization of emerging chemicals is not on the usual basis of their properties, the class of compounds consists of a large number of chemicals with very diverse physicochemical nature. This is the reason for some of

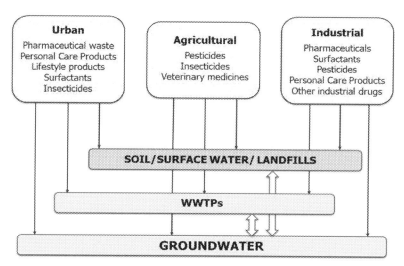

FIGURE 12.2

Diagram showing major types of emerging contaminants released from urban, agricultural, and industrial sources.

them being easily attenuated under natural conditions, while some behaving persistently in the environment. As these contaminants enter the subsurface environment, their fate depends on their physicochemical properties like water solubility and residence time; the properties of the subsurface system; and other environmental conditions (Lapworth et al., 2012; Sedlak & Pinkston, 2003; Wells, 2006). The hydraulic properties of the overlying soil and the aquifer where the chemical of interest is present affect its fate (Worrall & Dana, 2004). Aquifer confinement is one of the major factors as unconfined or shallow aquifers may generally be more closely exposed to the contaminant source (Fram & Belitz, 2011). Shallow alluvial aquifers and karst aquifers offer shorter residence time and, thus, restrict the process of natural attenuation (Bruchet et al., 2005; Katz et al., 2009; Nakada et al., 2008; Osenbrück et al., 2007; Rabiet et al., 2006). The former is also under constant exchange with surface waters and, hence, changing the concentration levels and other desirable conditions. The transport, accumulation, and degradation of chemicals also depend on their sorption potential on the surface matrix. Ionic molecules get adsorbed on oppositely charged surfaces, which is also affected by the surrounding pH conditions (Lapworth et al., 2012). Redox conditions of the aquifer environment affect the attenuation of ECs as aerobic degradation is generally faster than anaerobic (Barnes et al., 2004; Godfrey et al., 2007; Watanabe et al., 2010). The presence of microorganisms in the unsaturated zone also has a role in the attenuation of chemicals during their residence time (Alvarez & Illman, 2005; Semple et al., 2007; Wick et al., 2007).

Hence, the series of studies conducted so far to understand the fate of ECs in groundwater has found both easily attenuable and long-persisting compounds, largely depending on the properties of the chemical and the surrounding aquifer characteristics and conditions.

12.6   Potential risks associated with emerging contaminants

While the increasing levels of ECs in the groundwater are being detected, information on their behavior, fate, and effects on the environment, ecology, and humans has not yet been reported adequately (Gavrilescu et al., 2015). Some of these chemicals are comparatively benign, while others may arouse serious environmental and biological health concerns. Although the complete range of repercussions appearing due to ECs has not been documented yet, a large number of these chemicals have the ability to persist in the environment due to low degradation potential. Some of them may accumulate, magnify, and/or transform into other products after entering the biological species systems. Depending on their properties, different types of ECs have been found to be neurotoxic, carcinogenic, genotoxic, endocrine disruptors, possessing reproductive toxicity, mutagenic toxicity, immune toxicity, and also causing allergies, intolerance (Balducci et al., 2012; Pereira et al., 2015).

Table 12.2 provides information on the adverse effects of various types of ECs on human health. This information is entirely based on the little number of studies conducted to derive correlations among the presence of these contaminants and their potential health effects. However, the current number of investigations is not enough to draw a profile of the possible health hazards associated with a certain type of contaminant. There is a huge need for wide and deep investigation to establish accurate and reliable relationships between the contaminant and the potential effects.

Fig. 12.3 represents the possible pathways of any EC in the human body. The body can get exposed to the contaminant present in groundwater through three possible routes: oral (drinking and cooking water), dermal (bathing, cleaning, and washing water), and respiratory (breathing in bathing showers, etc.). Once the contaminant enters the body, it can get mixed up in the blood, through which it gets distributed to other body organs. The biologically effective dose of the contaminant may interact with biomolecules of the body, for example, proteins, lipids, and DNA and make changes. Later, the metabolism of the parent compound into biotransformation products takes place. The EC or its metabolites may accumulate in any of the body organs depending on their association and/or may get eliminated from the body. Once the contaminant is out of the system, it reenters the environment with the chance of ending up in groundwater again.

Table 12.2 Health effects of emerging contaminants (ECs).

Type	Common ECs	Health effects	References
Pharmaceuticals and veterinary antibiotics	Propranolol, triclosan, etc.	Microbial resistance, skin irritations, endocrine disruption, resistance to antibiotics in animals	Brooks et al., (2003), Dhillon et al. (2015), Esiobu et al. (2002), Huggett et al. (2002), Kumar et al. (2009)
Perfluorinated compounds	PFOS, PFOA, PFAAs, PFASs, PFNA, etc.	Cancer in liver, bladder, prostate, pancreas, and breasts, decrease in fecundity, reproductive dysfunctions, thyroid, and gene modification	Eriksen et al. (2009), Bonefeld-Jorgensen et al. (2011), Fletcher et al. (2013), Hardell et al. (2014), Joensen et al. (2009), Melzer et al. (2010), Saikat et al. (2013), Shrestha et al. (2015)
Personal care and lifestyle products	Salicylic acid, caffeine, UV filters like BP-3, etc.	Gastrointestinal toxicity, ulceration, stomach bleeding, endocrine disruption, HPT, reproductive and developmental functions disruption, dermatitis, etc.	IARC (1997), Frederiksen et al. (2014), Krause et al. (2012)
Pesticides	All major pesticides, including dithiocarbamates, dipyridyl derivatives, and malathion	Endocrine disruption, neurotoxicity (Parkinson's disease), genotoxicity, cytotoxicity, carcinogenicity, and mutagenicity	Bolognesi (2003), Chomycia et al. (2008), Colborn et al. (1993), Ruiz and Marzin (1997); Sailaja et al. (2006), Simonelli et al. (2007), USEPA (2006), Franco et al. (2010), George and Shukla (2011), Ji et al. (2008), McKinlay et al. (2008), Yan et al. (2010)
Plasticizers	BPA, phthalates, triphenyl phosphate, etc.	Neurotoxicity, cytotoxicity, hepatotoxicity, mutagenicity, carcinogenicity, oxidative stress, immunotoxicity, reproductive toxicity, endocrine disruption, insidious behavior in adulthood, etc.	Adamakis et al. (2013), Benjamin et al. (2015), Hassan et al. (2012), Michałowicz (2014), Rosado-Berrios et al. (2011), Rosenfeld (2015)
Flame retardants	Tetrabromodiphenyl ether—BDE 47, pentabromodiphenyl ether—BDE 100, hexabromodiphenyl ether—BDE 154, etc.	Genotoxicity, endocrine disruption, thyroid hormone disruption, neurotoxicity, hepatotoxicity, and carcinogenicity	He et al. (2008), Huang et al. (2009), Madia et al. (2004), Macaulay et al. (2015), Pereira et al. (2014), Pereira et al. (2013), Pazin et al. (2015)

BPA, Bisphenol-A; HPT, hypothalamic–pituitary–thyroid axis; PFOS, perfluorooctanesulfonic acid; PFOA, perfluorooctanoic acid; PFAAs, perfluoroalkyl acids; PFASs, perfluoroalkyl substances; PFNA, perfluorononanoic acid.

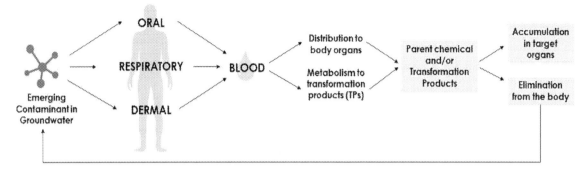

FIGURE 12.3

Diagram showing the adsorption, distribution, metabolism, and accumulation of emerging contaminants in the human body. The exposure routes may be oral (drinking and cooking water), dermal (bathing and washing water, etc.), and respiratory (bathing showers).

Table 12.3 Remediation technologies for emerging contaminants (ECs).

Method	Technique	Type of EC	References
Physical	Adsorption on activated carbon, membrane filtration	Organic micropollutants, PhACs	Schäfer et al. (2011), Jermann et al. (2009), Radjenović et al. (2009)
Chemical	Oxidation process: ozonation and fenton/H_2O_2	Naproxen, carbamazepine (pharmaceuticals), atrazine (pesticide)	Gerrity et al. (2011), Hernández-Leal et al. (2011)
Biological	Use of biofilms: microorganisms, polymers, and particulate material	EDCs	Spring et al. (2007), Gavrilescu et al. (2015), Gavrilescu (2010)

EDC, *Endocrine disrupting compound*; PhACs, *pharmaceutically active compounds*.

12.7   Remediation of emerging contaminants

The accumulation and persistence of ECs in the environment have resulted in several ecological risks and health hazards. The remediation of these contaminants has so far not been adequately studied because of their complex biochemical properties, transformation reactions, and extremely low concentrations in the environment. It is very difficult to define methods that can be appropriately applied to all or most of the chemicals. Based on the available information and limited research literature, few methods have been found working efficiently in filtering and/or removing the pollutants. However, it is to be noted that no single method is suitable for all the contaminants of emerging concern and, thus, each method works selectively based on the properties of the compound (Table 12.3, Fig. 12.4).

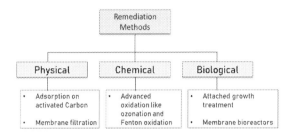

FIGURE 12.4

Schematic diagram showing methods to remove emerging contaminants.

1. *Physical methods*: Methods that are effective in physically eliminating the ECs include adsorption on activated carbon and membrane filtration. The adsorption process on activated carbon (powder or granules) is efficient in wiping out nonbiodegradable, persistent organic compounds. Membrane filters use a semipermeable selective membrane to retain the target compounds and letting the fluid (water) flow through depending on the pore size. Specific membrane may be used for a specific type of chemical, also depending on the operating conditions. Improved efficiency is achieved through the adsorption of the target compound onto the filtration membrane (Jermann et al., 2009; Schäfer et al., 2011). Membrane filtration is sometimes integrated with biological reactors (called membrane bioreactors or MBRs) for the biological degradation of organic matter, to potentially remove a broader range of low concentration compounds (Ngo et al., 2012; Radjenović et al., 2009).

2. *Chemical methods*: The most effective chemical method in removing organic contaminants is oxidation process like ozonation and Fenton/ H_2O_2 oxidation where the contaminant can be oxidized by either ozone or HO radical or both depending on its chemical properties (Gerrity et al., 2011; Hernández-Leal et al., 2011). ECs like naproxen, carbamazepine, meprobamate (pharmaceuticals), and atrazine (pesticide) can be effectively removed using this method.

3. *Biological methods*: Attached growth treatment processes use biofilms made of microorganisms, polymers, and particulate material providing higher efficiency than the activated sludge process in the removal of organic compounds (Loupasaki & Diamadopoulos, 2013; Guo et al., 2012). The operational costs are low and improved efficiency is achieved through better oxygen supply and large biomass concentrations. The bioreactors can work in both fixed bed and moving bed systems.

MBRs are used where physical separation is done using a selective membrane, and biological removal is achieved by allowing a strain of microorganisms adsorbed on the membrane surface. This biofilter is able to biologically transform and degrade the contaminant, thus allowing complete removal of the parent compound (Reungoat et al., 2011). This method provides promising results specially for EDCs, as it combines both biotic (microbes) and abiotic (membrane filter) processes to transform and immobilize the target compound (Gavrilescu, 2010; Gavrilescu et al., 2015; Gavrilescu & Chisti, 2005; Spring et al., 2007; Sørensen et al., 2009).

12.8   Challenges and scope

ECs are problematic as they are present in almost every aspect of our everyday life. Hundreds of chemical products that we use today release these compounds in the environment, most of which are still undetected or unidentified in terms of toxicity and risks. The use of products that release ECs is not new to the world; however, they went unnoticed for years and were only recently detected in air, water, soil, and biological samples. The current contaminants of emerging concern in the environment may present health risks like carcinogenicity, genotoxicity, neurotoxicity, and endocrine disruption to humans and other living species as well. Some of them may do their job at extremely low concentrations. Their increasing presence in our surroundings is a concern not only to humans but also to other biological species. A large number of potential target species are out there in the wild with different physiological mechanisms, which have not been studied even remotely. If any, these impacts may have large and deteriorating consequences, disturbing the entire ecosystem. And hence, the present level of information is insufficient to conclude on the permissible limits, toxicity levels, and other dynamics of these pollutants in the environment.

Biomonitoring studies are applied to understand the association and establish cause−effect relationship between contaminants and the biota. However, these studies are less popular and require more attention to address the problem to greater length and depth. And, thus, with the current incomplete knowledge and understanding, it is very difficult to plan and implement remediation actions. The research also needs to focus on the production of toxic metabolites that might become "the real problem" in the long term. The study of ECs presents a multidisciplinary approach where several disciplines like chemistry, engineering, environmental science, biology, microbiology, molecular biology, and biomedical engineering need to work independently as well as in integration to fill the current knowledge gap and provide future solutions.

12.9   Discussion and conclusion

Until recently, scientists' major focus has only been on the compounds found in larger concentrations in the environment (Pereira et al., 2015). ECs have now started to add up to the concern as they can pose toxicity at trace concentrations. The environment has been exposed to the contamination for years; nonetheless, the detection of emerging compounds has become possible only with the advancement in analytical tools and technologies. These compounds display low degradation potential and are persistent in the environment. They can also get transformed to a range of products that might be more persistent and toxic to the environment. Many of these compounds end up in the subsurface, and due to lower oxidation and microbial activities in the subsurface environment, their attenuation gets difficult. Their long residence time in the groundwater allows them to enter the human bodies through multiple exposure routes, leading to various health hazards. The available information regarding ECs is less, and so, requires more research to understand the problem and find solutions. Thus enhancing knowledge about the existence and impacts of ECs is important to ensure effective management and reduce future risks.

References

Adamakis, I.-D. S., Panteris, E., Cherianidou, A., & Eleftheriou, E. P. (2013). Effects of bisphenol A on the microtubule arrays in root meristematic cells of *Pisum sativum* L. *Mutation Research – Genetic Toxicology and Environmental Mutagenesis, 750*(1–2), 111–120. Available from https://doi.org/10.1016/j.mrgentox.2012.10.012.

Ahel, M., Mikac, N., Cosovic, B., Prohic, E., & Soukup, V. (1998). The impact of contamination from a municipal solid waste landfill (Zagreb, Croatia) on underlying soil. *Water Science and Technology, 37*(8), 203–210. Available from https://doi.org/10.1016/S0273-1223(98)00260-1.

Ahel, M., Giger, W., & Schaffner, C. (1994). Behaviour of alkylphenol polyethoxylate surfactants in the aquatic environment—II. Occurrence and transformation in rivers. *Water Research, 28*(5), 1143–1152.

Alvarez, P. J. J., & Illman, W. A. (2005). *Biodegradation principle, Bioremediation and natural attenuation: Process fundamentals and mathematical models* (3, pp. 49–114). John Wiley and Sons. Inc.

Arnold, S. M., Clark, K. E., Staples, C. A., Klecka, G. M., Dimond, S. S., Caspers, N., & Hentges, S. G. (2013). Relevance of drinking water as a source of human exposure to bisphenol A. *Journal of Exposure Science and Environmental Epidemiology, 23*(2), 137–144. Available from https://doi.org/10.1038/jes.2012.66.

Balducci, C., Perilli, M., Romagnoli, P., & Cecinato, A. (2012). New developments on emerging organic pollutants in the atmosphere. *Environmental Science and Pollution Research, 19*(6), 1875–1884.

Barnes, K. K., Christenson, S. C., Kolpin, D. W., Focazio, M. J., Furlong, E. T., Zaugg, S. D., Meyer, M. T., & Barber, L. B. (2004). Pharmaceuticals and other organic waste water contaminants within a leachate plume downgradient of a municipal landfill. *Ground Water Monitoring and Remediation, 24*(2), 119−126.

Barnes, K. K., Kolpin, D. W., Furlong, E. T., Zaugg, S. D., Meyer, M. T., & Barber, L. B. (2008). A national reconnaissance of pharmaceuticals and other organic wastewater contaminants in the United States—I Groundwater. *Science of the Total Environment, 402*(2−3), 192−200.

Bartelt-Hunt, S., Snow, D. D., Damon-Powell, T., & Miesbach, D. (2011). Occurrence of steroid hormones and antibiotics in shallow groundwater impacted by livestock waste control facilities. *Journal of Contaminant Hydrology, 123*(3−4), 94−103. Available from https://doi.org/10.1016/j.jconhyd.2010.12.010.

Bedding, N. D., McIntyre, A. E., Perry, R., & Lester, J. N. (1982). Organic contaminants in the aquatic environment I. Sources and occurrence. *Science of the Total Environment, 25*, 143−167.

Benjamin, S., Pradeep, S., Josh, M. S., Kumar, S., & Masai, E. (2015). A monograph on the remediation of hazardous phthalates. *Journal of Hazardous Materials, 298*, 58−72. Available from https://doi.org/10.1016/j.jhazmat.2015.05.004.

Bolognesi, C. (2003). Genotoxicity of pesticides: A review of human biomonitoring studies. *Mutation Research − Reviews in Mutation Research, 543*(3), 251−272.

Bonefeld-Jorgensen, E. C., Rossana Bossi, M. L., Ayotte, P., Asmund, G., Krüger, T., Ghisar, M., Mulvad, G., Kern, P., Nzulumiki, P., & Dewailly, E. (2011). Perfluorinated compounds are related to breast cancer risk in Greenlandic Inuit: A case control study. *Environmental Health, 10*(1), 1−16.

Bradford, S. A., Segal, E., Zheng, W., Wang, Q., & Hutchins, S. R. (2008). Reuse of concentrated animal feeding operation wastewater on agricultural lands. *Journal of Environmental Quality, 37*(S5), S−97-S-115.

Brooks, B. W., Foran, C. M., Richards, S. M., Weston, J., Turner, P. K., Stanley, J. K., Solomon, K. R., Slattery, M., & La Point, T. W. (2003). Aquatic ecotoxicology of fluoxetine. *Toxicology Letters, 142*(3), 169−183.

Bruchet, A., Hochereau, C., Picard, C., Decottignies, V., Rodrigues, J. M., & Janex-Habibi, M. L. (2005). Analysis of drugs and personal care products in French source and drinking waters: The analytical challenge and examples of application. *Water Science and Technology, 52*(8), 53−61.

Buerge, I. J., Buser, H. R., Kahle, M., Muller, M. D., & Poiger, T (2009). Ubiquitous occurrence of the artificial sweetener acesulfame in the aquatic environment: An ideal chemical marker of domestic wastewater in groundwater. *Environmental Science and Technology, 43*(12), 4381−4385.

Buerge, I. J., Keller, M., Buser, H. R., Muller, M. D., & Poiger, T. (2011). Saccharin and other artificial sweeteners in soils: Estimated inputs from agriculture and households, degradation, and leaching to groundwater. *Environmental Science and Technology, 45*(2), 615−621.

Buszka, P. M., Yeskis, D. J., Kolpin, D. W., Furlong, E. T., Zaugg, S. D., & Meyer, M. T. (2009). Waste-indicator and pharmaceutical compounds in landfill-leachate-affected ground water near Elkhart, Indiana, 2000−2002. *Bulletin of Environmental Contamination and Toxicology, 82*(6), 653−659.

Cabeza, Y., Candela, L., Ronen, D., & Teijon, G. (2012). Monitoring the occurrence of emerging contaminants in treated wastewater and groundwater between 2008 and 2010. The Baix Llobregat (Barcelona, Spain). *Journal of Hazardous Materials, 239−240*, 32−39. Available from https://doi.org/10.1016/j.jhazmat.2012.07.032.

Calvo-Flores, F. G., Isac-Garcia, J., & Dobado, J. A. (2018). *Emerging pollutants*. In Handbook of environmental analysis, Weinheim: Wiley-VCH.

Chomycia, J. C., Hernes, P. J., Harter, T., & Bergamaschi, B. A. (2008). Land management impacts on dairy-derived dissolved organic carbon in ground water. *Journal of Environmental Quality*, *37*(2), 333–343.

Clara, M., Strenn, B., & Kreuzinger, N. (2004). Carbamazepine as a possible anthropogenic marker in the aquatic environment: Investigations on the behaviour of carbamazepine in wastewater treatment and during groundwater infiltration. *Water Research*, *38*(4), 947–954.

Clarke, B. O., & Smith, S. R. (2011). Review of 'emerging' organic contaminants in biosolids and assessment of international research priorities for the agricultural use of biosolids. *Environment International*, *37*(1), 226–247. Available from https://doi.org/10.1016/j.envint.2010.06.004.

Colborn, T., Vom Saal, F. S., & Soto, A. M. (1993). Developmental effects of endocrine-disrupting chemicals in wildlife and humans. *Environmental Health Perspectives*, *101*(5), 469–489.

Corada-Fernández, C., Lara-Martín, P. A., Candela, L., & González-Mazo, E. (2011). Tracking sewage derived contamination in riverine settings by analysis of synthetic surfactants. *Journal of Environmental Monitoring*, *13*(7), 2010–2017.

Dhillon, G. S., Kaur, S., Puricharla, R., Brar, S. K., Cledon, M., Verma, M., & Surampalli, R. Y. (2015). Triclosan: Current status, occurrence, environmental risks and bioaccumulation potential. *International Journal of Environmental Research and Public Health*, *12*(5), 5657–5684.

Diamanti-Kandarakis, E., Bourguignon, J. P., Giudice, L. C., Hauser, R., Prins, G. S., Soto, A. M., Zoeller, T., & Gore, A. C. (2009). Endocrine-disrupting chemicals: An endocrine society scientific statement. *Endocrine Reviews*, *30*(4), 293–342.

Díaz-Cruz, M. S., & Barceló, D. (2008). Trace organic chemicals contamination in ground water recharge. *Chemosphere*, *72*(3), 333–342.

Dimpe, K. M., & Nomngongo, P. N. (2016). Current sample preparation methodologies for analysis of emerging pollutants in different environmental matrices. *TrAC — Trends in Analytical Chemistry*, *82*, 199–207. Available from https://doi.org/10.1016/j.trac.2016.05.023.

Ding, W.-h, Tzing, S.-h, & Lo, J.-h (1999). Occurrence and concentrations of aromatic surfactants and their degradation products in river waters in Taiwan. *Chemosphere*, *38*(11), 2597–2606.

Drewes, J. E. (2009). Ground water replenishment with recycled water — Water quality improvements during managed aquifer recharge. *Ground Water*, *47*(4), 502–505.

Drewes, J. E., Heberer, T., Rauch, T., & Reddersen, K. (2003). Fate of pharmaceuticals during ground water recharge. *Groundwater Monitoring and Remediation*, *23*(3), 64–72.

Eadsforth, C. V., Sherren, A. J., Selby, M. A., Toy, R., Eckhoff, W. S., McAvoy, D. C., & Matthijs, E. (2006). Monitoring of environmental fingerprints of alcohol ethoxylates in Europe and Canada. *Ecotoxicology and Environmental Safety*, *64*(1), 14–29.

Eckel, W. P., Ross, B., & Isensee, R. K. (1993). Pentobarbital found in ground water. *Ground Water*, *31*(5), 801–804.

Eichhorn, P., Rodrigues, S. V., Baumann, W., & Knepper, T. P. (2002). Incomplete degradation of linear alkylbenzene sulfonate surfactants in Brazilian surface waters and pursuit of their polar metabolites in drinking waters. *Science of the Total Environment*, *284*(1–3), 123–134.

Ellis, J. B. (2006). Pharmaceutical and personal care products (PPCPs) in urban receiving waters. *Environmental Pollution*, *144*(1), 184–189.

Eriksen, K. T., Sørensen, M., McLaughlin, J. K., Lipworth, L., Tjønneland, A., Overvad, K., & Raaschou-Nielsen, O. (2009). Perfluorooctanoate and perfluorooctanesulfonate plasma levels and risk of cancer in the general Danish population. *Journal of the National Cancer Institute*, *101*(8), 605–609.

Esiobu, N., Armenta, L., & Ike, J. (2002). Antibiotic resistance in soil and water environments. *International Journal of Environmental Health Research*, *12*(2), 133–144.

Estévez, E., Cabrera, M. D. C., Molina-Díaz, A., Robles-Molina, J., Palacios-Díaz, M. D. P., et al. (2012). Screening of emerging contaminants and priority substances (2008/105/EC) in reclaimed water for irrigation and groundwater in a volcanic aquifer (Gran Canaria, Canary Islands, Spain). *Science of the Total Environment, 433*, 538–546. Available from https://doi. org/10.1016/j.scitotenv.2012.06.031.

Fatta-Kassinos, D., Meric, S., & Nikolaou, A. (2011). Pharmaceutical residues in environmental waters and wastewater: Current state of knowledge and future research. *Analytical and Bioanalytical Chemistry, 399*(1), 251–275.

Ferguson, P. L., Iden, C. R., & Brownawell, B. J. (2001). Distribution and fate of neutral alkylphenol ethoxylate metabolites in a sewage-impacted urban estuary. *Environmental Science and Technology, 35*(12), 2428–2435.

Fletcher, T., Galloway, T. S., Melzer, D., Holcroft, P., Cipelli, R., Pilling, L. C., Mondal, D., Luster, M., & Harries, L. W. (2013). Associations between PFOA, PFOS and changes in the expression of genes involved in cholesterol metabolism in humans. *Environment International, 57–58*, 2–10. Available from https://doi.org/10.1016/j.envint.2013.03.008.

Focazio, M. J., Kolpin, D. W., Barnes, K. K., Furlong, E. T., Meyer, M. T., Zaugg, S. D., Barber, L. B., & Thurman, M. E. (2008). A national reconnaissance for pharmaceuticals and other organic wastewater contaminants in the United States − II Untreated drinking water sources. *Science of the Total Environment, 402*(2–3), 201–216.

Fram, M. S., & Belitz, K. (2011). Occurrence and concentrations of pharmaceutical compounds in groundwater used for public drinking-water supply in California. *Science of the Total Environment, 409*(18), 3409–3417. Available from https://doi.org/10.1016/j.scitotenv.2011.05.053.

Franco, R., Li, S., Rodriguez-Rocha, H., Burns, M., & Panayiotidis, M. I. (2010). Molecular mechanisms of pesticide-induced neurotoxicity: Relevance to Parkinson's disease. *Chemico-Biological Interactions, 188*(2), 289–300. Available from https://doi.org/10.1016/j.cbi.2010.06.003.

Frederiksen, H., Jensen, T. K., Jørgensen, N., Kyhl, H. B., Husby, S., Skakkebæk, N. E., Main, K. M., Juul, A., & Andersson, A. M. (2014). Human urinary excretion of non-persistent environmental chemicals: An overview of Danish data collected between 2006 and 2012. *Reproduction, 147*(4), 555–565.

Gasser, G., Rona, M., Voloshenko, A., Shelkov, R., Tal, N., Pankratov, I., Elhanany, S., & Lev, O. (2010). Quantitative evaluation of tracers for quantification of wastewater contamination of potable water sources. *Environmental Science and Technology, 44*(10), 3919–3925.

Gavrilescu, M., Demnerová, K., Aamand, J., Agathos, S., & Fava, F. (2015). Emerging pollutants in the environment: Present and future challenges in biomonitoring, ecological risks and bioremediation. *New Biotechnology, 32*(1), 147–156.

Gavrilescu, M., & Chisti, Y. (2005). Biotechnology − A sustainable alternative for chemical industry. *Biotechnology Advances, 23*(7–8), 471–499.

Gavrilescu, M. (2010). Environmental biotechnology: Achievements, opportunities and challenges. *Dynamic Biochemistry, Process Biotechnology and Molecular Biology, 4*(May), 1–36.

George, J., & Shukla, Y. (2011). Pesticides and cancer: Insights into toxicoproteomic-based findings. *Journal of Proteomics, 74*(12), 2713–2722. Available from https://doi.org/10.1016/j. jprot.2011.09.024.

Gerrity, D., Gamage, S., Holady, J. C., Mawhinney, D. B., Quiñones, O., Trenholm, R. A., & Snyder, S. A. (2011). Pilot-scale evaluation of ozone and biological activated carbon for trace organic contaminant mitigation and disinfection. *Water Research, 45*(5), 2155–2165.

Glassmeyer, S. T., Furlong, E. T., Kolpin, D. W., Cahill, J. D., Zaugg, S. D., Werner, S. L., Meyer, M. T., & Kryak, D. D. (2005). Transport of chemical and microbial compounds from known wastewater discharges: Potential for use as indicators of human fecal contamination. *Environmental Science and Technology, 39*(14), 5157−5169.

Godfrey, E., Woessner, W. W., & Benotti, M. J. (2007). Pharmaceuticals in on-site sewage effluent and ground water, western Montana. *Ground Water, 45*(3), 263−271.

Gogoi, A., Mazumder, P., Tyagi, V. K., Chaminda, G. G. T., Kyoungjin An, A., & Kumar, M. (2018). Occurrence and fate of emerging contaminants in water environment: A review. *Groundwater for Sustainable Development, 6*(September 2017), 169−180. Available from https://doi.org/10.1016/j.gsd.2017.12.009.

González-Mazo, E., María Forja, J., & Gómez-Parra, A. (1998). Fate and distribution of linear alkylbenzene sulfonates in the littoral environment. *Environmental Science and Technology, 32*(11), 1636−1641.

González-Mazo, E., León, V. M., Sáez, M., & Gómez-Parra, A. (2002). Occurrence and distribution of linear alkylbenzene sulfonates and sulfophenylcarboxylic acids in several Iberian littoral ecosystems. *Science of the Total Environment, 288*(3), 215−226.

Grünheid, S., Amy, G., & Jekel, M. (2005). Removal of bulk dissolved organic carbon (DOC) and trace organic compounds by bank filtration and artificial recharge. *Water Research, 39*(14), 3219−3228.

Guo, W., Hao Ngo, H., & Vigneswaran, S. (2012). *Enhancement of membrane processes with attached growth media. Membrane technology and environmental applications* (pp. 603−634). USA: American Society of Civil Engineers (ASCE).

Hardell, E., Kärrman, A., Bavel, B. V., Bao, J., Carlberg, M., & Hardell, L. (2014). Case-control study on perfluorinated alkyl acids (PFAAs) and the risk of prostate cancer. *Environment International, 63*, 35−39. Available from https://doi.org/10.1016/j.envint.2013.10.005.

Hassan, Z. K., Elobeid, M. A., Virk, P., Omer, S. A., ElAmin, M., Daghestani, M. H., & AlOlayan, E. M. (2012). Bisphenol a induces hepatotoxicity through oxidative stress in rat model. *Oxidative Medicine and Cellular Longevity, 2012*.

He, P., He, W., Wang, A., Xia, T., Xu, B., Zhang, M., & Chen, X. (2008). PBDE-47-induced oxidative stress, DNA damage and apoptosis in primary cultured rat hippocampal neurons. *NeuroToxicology, 29*(1), 124−129.

Heberer, T., & Adam, M. (2004). Transport and attenuation of pharmaceutical residues during artificial groundwater replenishment. *Environmental Chemistry, 1*(1), 22−25.

Hernández-Leal, L., Temmink, H., Zeeman, G., & Buisman, C. J. N. (2011). Removal of micropollutants from aerobically treated grey water via ozone and activated carbon. *Water Research, 45*(9), 2887−2896.

Holm, J. V., Rugge, K., Bjerg, P. L., & Christensen, T. H. (1995). Occurrence and distribution of pharmaceutical organic compounds in the groundwater downgradient of a landfill (Grindsted, Denmark). *Environmental Science & Technology, 29*(5), 1415−1420.

Holtze, M. S., Sørensen, S. R., Sørensen, J., & Aamand, J. (2008). Microbial degradation of the benzonitrile herbicides dichlobenil, bromoxynil and ioxynil in soil and subsurface environments − Insights into degradation pathways, persistent metabolites and involved degrader organisms. *Environmental Pollution, 154*(2), 155−168.

Hu, X., Zhou, Q., & Luo, Y. (2010). Occurrence and source analysis of typical veterinary antibiotics in manure, soil, vegetables and groundwater from organic vegetable bases, Northern China. *Environmental Pollution, 158*(9), 2992−2998. Available from https://doi.org/10.1016/j.envpol.2010.05.023.

Huang, S. C., Giordano, G., & Costa, L. G. (2009). Comparative cytotoxicity and intracellular accumulation of five polybrominated diphenyl ether congeners in mouse cerebellar granule neurons. *Toxicological Sciences, 114*(1), 124−132.

Huggett, D. B., Brooks, B. W., Peterson, B., Foran, C. M., & Schlenk, D. (2002). Toxicity of select beta adrenergic receptor-blocking pharmaceuticals (B-blockers) on aquatic organisms. *Archives of Environmental Contamination and Toxicology, 43*(2), 229−235.

IARC. (1997). *Non*-steroidal anti-inflammatory drugs (vol. 1). International Agency for Research on Cancer.

Isobe, T., & Takada, H. (2004). Determination of degradation products of alkylphenol poly-ethoxylates in municipal wastewaters and rivers in Tokyo, Japan. *Environmental Toxicology and Chemistry, 23*(3), 599−605.

Jermann, D., Pronk, W., Boller, M., & Schäfer, A. I. (2009). The role of NOM fouling for the retention of estradiol and ibuprofen during ultrafiltration. *Journal of Membrane Science, 329*(1−2), 75−84.

Ji, F., Zhao, L., Yan, W., Feng, Q., & Lin, J. M. (2008). Determination of triazine herbicides in fruits and vegetables using dispersive solid-phase extraction coupled with LC−MS. *Journal of Separation Science, 31*(6−7), 961−968.

Jobling, S., Reynolds, T., White, R., Parker, M. G., & Sumpter, J. P. (1995). A variety of environmen-tally persistent chemicals, including some phthalate plasticizers, are weakly estrogenic. *Environmental Health Perspectives, 103*(6), 582−587.

Joensen, U. N., Bossi, R., Leffers, H., Jensen, A. A., Skakkebaek, N. E., & Jørgensen, N. (2009). Do perfluoroalkyl compounds impair human semen quality? *Environmental Health Perspectives, 117*(6), 923−927.

Jonkers, N., Laane, R. W. P. M., & de Voogt, P. (2003). Fate of nonylphenol ethoxylates and their metabolites in two Dutch estuaries: Evidence of biodegradation in the field. *Environmental Science & Technology, 37*, 321−327.

Katz, B. G., Griffin, D. W., & Davis, J. H. (2009). Groundwater quality impacts from the land application of treated municipal wastewater in a large karstic spring basin: Chemical and microbiological indicators. *Science of the Total Environment, 407*(8), 2872−2886.

Kolodziej, E. P., Harter, T., & Sedlak, D. L. (2004). Dairy wastewater, aquaculture, and spawning fish as sources of steroid hormones in the aquatic environment. *Environmental Science and Technology, 38*(23), 6377−6384.

Kolpin, D. W., Schnoebelen, D. J., & Thurman, E. M. (2004). Degradates provide insight to spa-tial and temporal trends of herbicides in ground water. *Ground Water, 42*(4), 601−608.

Kolpin, D. W., Barbash, J. E., & Gilliom, R. J. (2000). Pesticides in ground water of the United States, 1992−1996. *Ground Water, 38*(6), 858−863.

Krause, M., Klit, A., Jensen, M. B., Søeborg, T., Frederiksen, H., Schlumpf, M., Lichtensteiger, W., Skakkebaek, N. E., & Drzewiecki, K. T. (2012). Sunscreens: Are they beneficial for health? An overview of endocrine disrupting properties of UV-filters. *International Journal of Andrology, 35* (3), 424−436.

Kreuzinger, N., Clara, M., Strenn, B., & Vogel, B. (2004). Investigation on the behaviour of selected pharmaceuticals in the groundwater after infiltration of treated wastewater. *Water Science and Technology, 50*(2), 221−228.

Kumar, V., Chakraborty, A., Kural, M. R., & Roy, P. (2009). Alteration of testicular steroidogene-sis and histopathology of reproductive system in male rats treated with triclosan. *Reproductive Toxicology, 27*(2), 177−185.

Lapworth, D. J., Gooddy, D. C., Allen, D., & Old, G. H. (2009). Understanding groundwater, sur-face water, and hyporheic zone biogeochemical process in a Chalk catchment using fluores-cence properties of dissolves and colloidal organic matter. *Journal of Geophysical Research: Biogeosciences, 114*(G3), 1−10.

Lapworth, D. J., Baran, N., Stuart, M. E., & Ward, R. S. (2012). Emerging organic contaminants in groundwater: A review of sources, fate and occurrence. *Environmental Pollution*, *163*, 287–303. Available from https://doi.org/10.1016/j.envpol.2011.12.034.

Leister, D. L., & Baker, J. E. (1994). Atmospheric deposition of organic contaminants to the Chesapeake Bay. *Atmospheric Environment*, *28*(8), 1499–1520.

Lewandowski, J., Putschew, A., Schwesig, D., Neumann, C., & Radke, M. (2011). Fate of organic micropollutants in the hyporheic zone of a eutrophic lowland stream: Results of a preliminary field study. *Science of the Total Environment*, *409*(10), 1824–1835. Available from https://doi.org/10.1016/j.scitotenv.2011.01.028.

Loos, R., Locoro, G., Comero, S., Contini, S., Schwesig, D., Werres, F., Balsaa, P., Gans, O., Weiss, S., Blaha, L., Bolchi, M., & Gawlik, B. M. (2010). Pan-European survey on the occurrence of selected polar organic persistent pollutants in ground water. *Water Research*, *44*(14), 4115–4126. Available from https://doi.org/10.1016/j.watres.2010.05.032.

Loupasaki, E., & Diamadopoulos, E. (2013). Attached growth systems for wastewater treatment in small and rural communities: A review. *Journal of Chemical Technology and Biotechnology*, *88*(2), 190–204.

Luo, Y., Guo, W., Ngo, H. H., Nghiem, L. D., Hai, F. I., Zhang, J., Liang, S., & Wang, X. C. (2014). A review on the occurrence of micropollutants in the aquatic environment and their fate and removal during wastewater treatment. *Science of the Total Environment*, *473–474*, 619–641. Available from https://doi.org/10.1016/j.scitotenv.2013.12.065.

Macaulay, L. J., Bailey, J. M., Levin, E. D., & Stapleton, H. M. (2015). Persisting effects of a PBDE metabolite, 6-OH-BDE-47, on larval and juvenile Zebrafish swimming behavior. *Neurotoxicology and Teratology*, *52*, 119–126. Available from https://doi.org/10.1016/j.ntt.2015.05.002.

Madia, F., Giordano, G., Fattori, V., Vitalone, A., Branchi, I., Capone, F., & Costa, L. G. (2004). Differential in vitro neurotoxicity of the flame retardant PBDE-99 and of the PCB Aroclor 1254 in human astrocytoma cells. *Toxicology Letters*, *154*(1–2), 11–21.

Mansell, J., & Drewes, J. E. (2004). Fate of steroidal hormones during soil-aquifer treatment. *Ground Water Monitoring and Remediation*, *24*(2), 94–101.

Martin, C., Moeder, M., Daniel, X., Krauss, G., & Schlosser, G. (2007). Biotransformation of the polycyclic musks HHCB and AHTN and metabolite formation by fungi occurring in freshwater environments. *Environment Science & Technology*, *41*(15), 5395–5402.

Matamoros, V., Jover, E., & Bayona, J. M. (2009). Advances in the determination of degradation intermediates of personal care products in environmental matrixes: A review. *Analytical and Bioanalytical Chemistry*, *393*(3), 847–860.

McKinlay, R., Plant, J. A., Bell, J. N. B., & Voulvoulis, N. (2008). Calculating human exposure to endocrine disrupting pesticides via agricultural and non-agricultural exposure routes. *Science of the Total Environment*, *398*(1–3), 1–12.

Melzer, D., et al. (2010). Association between serum perfluorooctanoic acid (PFOA) and thyroid disease in the U.S. National Health and Nutrition Examination Survey. *Environmental Health Perspectives*, *118*(5), 686–692.

Michałowicz, J. (2014). Bisphenol A — Sources, toxicity and biotransformation. *Environmental Toxicology and Pharmacology*, *37*(2), 738–758. Available from https://doi.org/10.1016/j.etap.2014.02.003.

Montgomery-Brown, J., & Reinhard, M. (2003). Occurrence and behavior of alkylphenol polyethoxylates in the environment. *Environmental Engineering Science*, *20*(5), 471–486.

Morris, B., & Cunningham, J. (2007). Suburbanisation of important aquifers in England and Wales: Estimating its current extent. *Water and Environment Journal*, *22*(2), 88–99.

Morris, B. L., Darling, W. G., Gooddy, D. C., Litvak, R. G., Neumann, I., Nemaltseva, E. J., & Poddubnaia, I. (2005). Assessing the extent of induced leakage to an urban aquifer using environmental tracers: An example from Bishkek, capital of Kyrgyzstan, central Asia. *Hydrogeology Journal, 14*(1−2), 225−243.

Nakada, N., Kiri, K., Shinohara, H., Harada, A., Kuroda, K., Takizawa, S., & Takada, H. (2008). Evaluation of pharmaceuticals and personal care products as water-soluble molecular markers of sewage. *Environmental Science and Technology, 42*(17), 6347−6353.

Ngo, H. H., Guo, W., & Vigneswaran, S. (2012). *Membrane processes for water reclamation and reuse. Membrane technology and environmental applications* (pp. 239−275). USA: American Society of Civil Engineers (ASCE).

Osenbrück, K., Glaser, H. R., Knoller, K., Weise, S. M., Moder, M., Wennrich, R., Schirmer, M., Reinstorf, F., Busch, W., & Strauch, G. (2007). Sources and transport of selected organic micropollutants in urban groundwater underlying the city of Halle (Saale), Germany. *Water Research, 41*(15), 3259−3270.

Pazin, M., Cristina Pereira, L., & Junqueira Dorta, D. (2015). Toxicity of brominated flame retardants, BDE-47 and BDE-99 stems from impaired mitochondrial bioenergetics. *Toxicology Mechanisms and Methods, 25*(1), 34−41.

Pecoraino, G., Scalici, L., Avellone, G., Ceraulo, L., Favara, R., Candela, E. G., Provenzano, M. C., & Scaletta, C. (2008). Distribution of volatile organic compounds in Sicilian groundwaters analysed by head space-solid phase micro extraction coupled with gas chromatography mass spectrometry (SPME/GC/MS). *Water Research, 42*(14), 3563−3577.

Pereira, L. C., de Souza, A. O., Bernardes, M. F. F., Pazin, M., Tasso, M. J., Pereira, P. H., & Dorta, D. J. (2015). A perspective on the potential risks of emerging contaminants to human and environmental health. *Environmental Science and Pollution Research, 22*(18), 13800−13823. Available from https://doi.org/10.1007/s11356-015-4896-6.

Pereira, L. C., Oliveira De Souza, A., & Junqueira Dorta, D. (2013). Polybrominated diphenyl ether congener (BDE-100) induces mitochondrial impairment. *Basic and Clinical Pharmacology and Toxicology, 112*(6), 418−424.

Pereira, L. C., Felippe Cabral Miranda, L., Oliveira De Souza, A., & Junqueira Dorta, D. (2014). BDE-154 induces mitochondrial permeability transition and impairs mitochondrial bioenergetics. *Journal of Toxicology and Environmental Health − Part A: Current, 1−3*(77), 24−36.

Petrovic, M., & Barcelo, D. (2006). Application of liquid chromatography/quadrupole time-of-flight mass spectrometry (LC-QqTOF-MS) in the environmental analysis. *Journal of Mass Spectrometry, 41*, 1259−1267.

Putschew, A., Wischnack, S., & Jekel, M. (2000). Occurrence of triiodinated X-ray contrast agents in the aquatic environment. *Science of the Total Environment, 255*, 129−134.

Rabiet, M., Togola, A., Brissaud, F., Seidel, J. L., Budzinski, H., & Elbaz-Poulichet, F. (2006). Consequences of treated water recycling as regards pharmaceuticals and drugs in surface and ground waters of a medium-sized Mediterranean catchment. *Environmental Science and Technology, 40*(17), 5282−5288.

Radjenović, J., Petrović, M., & Barceló, D. (2009). Fate and distribution of pharmaceuticals in wastewater and sewage sludge of the conventional activated sludge (CAS) and advanced membrane bioreactor (MBR) treatment. *Water Research, 43*(3), 831−841. Available from https://doi.org/10.1016/j.watres.2008.11.043.

Reungoat, J., Escher, B. I., Macova, M., & Keller, J. (2011). Biofiltration of wastewater treatment plant effluent: Effective removal of pharmaceuticals and personal care products and reduction of toxicity. *Water Research, 45*(9), 2751−2762. Available from https://doi.org/10.1016/j.watres.2011.02.013.

Richardson, S. D. (2012). Environmental mass spectrometry: Emerging contaminants and current issues. *Analytical Chemistry, 84*(2), 747−778.

Ritter, L., Solomon, K., Sibley, P., Hall, K., Keen, P., Mattu, G., & Linton, B. (2002). Sources, pathways, and relative risks of contaminants in surface water and groundwater: A perspective prepared for the Walkerton inquiry. *Journal of Toxicology and Environmental Health − Part A, 65*, 1−142.

Rosado-Berrios, C. A., Vélez, C., & Zayas, B. (2011). Mitochondrial permeability and toxicity of diethylhexyl and monoethylhexyl phthalates on TK6 human lymphoblasts cells. *Toxicology in Vitro, 25*(8), 2010−2016. Available from https://doi.org/10.1016/j.tiv.2011.08.001.

Rosenfeld, C. S. (2015). Bisphenol A and phthalate endocrine disruption of parental and social behaviors. *Frontiers in Neuroscience, 9*(Mar), 1−15.

Rueedi, J., Cronin, A. A., & Morris, B. L. (2009). Estimation of sewer leakage to urban groundwater using depth-specific hydrochemistry. *Water and Environment Journal, 23*, 134−144.

Ruiz, M. J., & Marzin, D. (1997). Genotoxicity of six pesticides by salmonella mutagenicity test and SOS chromotest. *Mutation Research − Genetic Toxicology and Environmental Mutagenesis, 390*(3), 245−255.

Sacher, F., Lange, F. T., Brauch, H.-J., & Blankenhorn, I. (2001). Pharmaceuticals in groundwaters analytical methods and results of a monitoring program in Baden-Wurttemberg, Germany. *Journal of Chromatography A, 938*(1−2), 199−210.

Saikat, S., Kreis, I., Davies, B., Bridgman, S., & Kamanyire, R. (2013). The impact of PFOS on health in the general population: A review. *Environmental Sciences: Processes and Impacts, 15*(2), 329−335.

Sailaja, N., Chandrasekhar, M., Rekhadevi, P. V., Mahboob, M., Rahman, M. F., Vuyyuri, S. B., Danadevi, K., Hussain, S. A., & Grover, P. (2006). Genotoxic evaluation of workers employed in pesticide production. *Mutation Research − Genetic Toxicology and Environmental Mutagenesis, 609*(1), 74−80.

Sarmah, A. K., Meyer, M. T., & Boxall, A. B. A. (2006). A global perspective on the use, sales, exposure pathways, occurrence, fate and effects of veterinary antibiotics (VAs) in the environment. *Chemosphere, 65*(5), 725−759.

Schäfer, A. I., Akanyeti, I., & Semião, A. J. C. (2011). Micropollutant sorption to membrane polymers: A review of mechanisms for estrogens. *Advances in Colloid and Interface Science, 164*(1−2), 100−117.

Schaider, L. A., Rudel, R. A., Ackerman, J. M., Dunagan, S. C., & Brody, J. G. (2014). Pharmaceuticals, perfluorosurfactants, and other organic wastewater compounds in public drinking water wells in a shallow sand and gravel aquifer. *Science of the Total Environment, 468−469*, 384−393. Available from https://doi.org/10.1016/j.scitotenv.2013.08.067.

Schulz, M., Löffler, D., Wagner, M., & Ternes, T. A. (2008). Transformation of the X-ray contrast medium iopromide in soil and biological wastewater treatment. *Environmental Science and Technology, 42*(19), 7207−7217.

Sedlak, D. L., & Pinkston, K. E. (2003). Factors affecting the concentrations of pharmaceuticals released to the aquatic environment. *Health (San Francisco)*, 56−64.

Semple, K. T., Kieron, J. D., Wick, L. Y., & Harms, H. (2007). Microbial interactions with organic contaminants in soil: Definitions, processes and measurement. *Environmental Pollution, 150*(1), 166−176.

Shrestha, S., Bloom, M. S., Yucel, R., Seegal, R. F., Wu, Q., Kannan, K., Rej, R., & Fitzgerald, E. F. (2015). Perfluoroalkyl substances and thyroid function in older adults. *Environment International, 75*, 206−214. Available from https://doi.org/10.1016/j.envint.2014.11.018.

Simonelli, A., Basilicata, P., Miraglia, N., Castiglia, L., Guadagni, R., Acampora, A., Sannolo, N., et al. (2007). Analytical method validation for the evaluation of cutaneous occupational exposure to different chemical classes of pesticides. *Journal of Chromatography B: Analytical Technologies in the Biomedical and Life Sciences, 860*(1), 26−33.

Sinclair, C. J., & Boxall, A. B. A. (2003). Assessing the ecotoxicity of pesticide transformation products. *Environmental Science and Technology, 37*(20), 4617−4625.

Snyder, S. A., Leising, J., Westerhoff, P., Yoon, Y., Mash, H., & Vanderford, B. (2004). Biological and physical attenuation of endocrine disruptors and pharmaceuticals: Implications for water reuse. *Ground Water Monitoring and Remediation, 24*(2), 108−118.

Soares, A., Guieysse, B., Jefferson, B., Cartmell, E., & Lester, J. N. (2008). Nonylphenol in the environment: A critical review on occurrence, fate, toxicity and treatment in wastewaters. *Environment International, 34*(7), 1033−1049.

Sørensen, S. R., Simonsen, A., & Aamand, J. (2009). Constitutive mineralization of low concentrations of the herbicide linuron by a *Variovorax* sp. strain. *FEMS Microbiology Letters, 292*(2), 291−296.

Spring, A. J., Bagley, D. M., Andrews, R. C., Lemanik, S., & Yang, P. (2007). Removal of endocrine disrupting compounds using a membrane bioreactor and disinfection. *Journal of Environmental Engineering and Science, 6*(2), 131−137.

Stuart, M. E., Manamsa, K., Talbot, J. C., & Crane, E. J. (2011). *Emerging contaminants in water*. British Geological Survey.

Sui, Q., Cao, X., Lu, S., Zhao, W., Qiu, Z., & Yu, G. (2015). Occurrence, sources and fate of pharmaceuticals and personal care products in the groundwater: A review. *Emerging Contaminants, 1*(1), 14−24. Available from https://doi.org/10.1016/j.emcon.2015.07.001.

Swartz, C. H., Reddy, S., Benotti, M. J., Yin, H., Barber, L. B., Brownawell, B. J., & Rudel, R. A. (2006). Steroid estrogens, nonylphenol ethoxylate metabolites, and other wastewater contaminants in groundwater affected by a residential septic system on cape cod, MA. *Environmental Science and Technology, 40*(16), 4894−4902.

Takada, H., & Ogura, N. (1992). Removal of linear alkylbenzenesulfonates (LAS) in the Tamagawa Estuary. *Marine Chemistry, 37*(3−4), 257−273.

Teijon, G., Candela, L., Tamoh, K., Molina-Díaz, A., & Fernández-Alba, A. R. (2010). Occurrence of emerging contaminants, priority substances (2008/105/CE) and heavy metals in treated wastewater and groundwater at Depurbaix facility (Barcelona, Spain). *Science of the Total Environment, 408*(17), 3584−3595. Available from https://doi.org/10.1016/j.scitotenv.2010.04.041.

Ternes, T. A., & Hirsch, R. (2000). Occurrence and behavior of X-ray contrast media in sewage facilities and the aquatic environment. *Environmental Science and Technology, 34*(13), 2741−2748.

Thorne, P. S. (2007). Environmental health impacts of concentrated animal feeding operations: Anticipating hazards − Searching for solutions. *Environmental Health Perspectives, 115*(2), 296−297.

USEPA. (2006). *Triazine cumulative risk assessment*. USEPA.

Vandenberg, L. N., Chahoud, I., Heindel, J. J., Padmanabhan, V., Paumgartten, F. J. R., & Schoenfelder, G. (2010). Urinary, circulating, and tissue biomonitoring studies indicate widespread exposure to bisphenol A. *Environmental Health Perspectives, 118*(8), 1055−1070.

Verstraeten, I. M., Fetterman, G. S., Meyer, M. J., Bullen, T., & Sebree, S. K. (2005). Use of tracers and isotopes to evaluate vulnerability of water in domestic wells to septic waste. *Ground Water Monitoring and Remediation, 25*(2), 107−117.

Villanneau, E. J., Saby, N. P. A., Marchant, B. P., Jolivet, C. C., Boulonne, L., Caria, G., Barriuso, E., Bispo, A., Briand, O., & Arrouays, D. (2011). Which persistent organic pollutants can we map in soil using a large spacing systematic soil monitoring design? A case study in Northern France. *Science of the Total Environment, 409*(19), 3719−3731. Available from https://doi.org/10.1016/j.scitotenv.2011.05.048.

Vulliet, E., & Cren-olivé, C. (2011). Screening of pharmaceuticals and hormones at the regional scale, in surface and groundwaters intended to human consumption. *Environmental Pollution, 159*(10), 2929−2934. Available from https://doi.org/10.1016/j.envpol.2011.04.033.

Watanabe, N., Bergamaschi, B. A., Loftin, K. A., Meyer, M. T., & Harter, T. (2010). Use and environmental occurrence of antibiotics in freestall dairy farms with manured forage fields. *Environmental Science and Technology, 44*(17), 6591−6600.

Wells, M. J. M. (2006). Log DOW: Key to understanding and regulating wastewater-derived contaminants. *Environmental Chemistry, 3*(6), 439−449.

Wick, L. Y., Remer, R., Wurz, B., Reichenbach, J., Braun, S., Schafer, F., & Harms, H. (2007). Effect of fungal hyphae on the access of bacteria to phenanthrene in soil. *Environmental Science and Technology, 41*(2), 500−505.

Worrall, F., & Dana, W. K. (2004). Aquifer vulnerability to pesticide pollution − Combining soil, land-use and aquifer properties with molecular descriptors. *Journal of Hydrology, 293*(1−4), 191−204.

Yan, S., Subramanian, B., Tyagi, R. D., Surampalli, R. Y., & Zhang, T. C. (2010). Emerging contaminants of environmental concern: Source, transport, fate, and treatment. *Practice Periodical of Hazardous, Toxic, and Radioactive Waste Management, 14*(1), 2−20.

CHAPTER 13

Selenium and naturally occurring radioactive contaminants in soil–water systems

Pankaj Kumar Gupta[1], Gaurav Saxena[2] and Basant Yadav[3]

[1]Wetland Hydrology Research Laboratory, Faculty of Environment, University of Waterloo, Waterloo, ON, Canada, [2]Laboratory for Microbiology, Department of Microbiology, Baba Farid Institute of Technology, Dehradun, India, [3]Department of Water Resources Development and Management (WRD&M), Indian Institute of Technology Roorkee, Roorkee, India

13.1 Introduction

From the Bay of Bengal delta regions to the central Gangetic plain, arsenic pollution has been increasing day by day (Kumar et al., 2018). Similarly, the groundwater system of the Western-North regions is highly affected by fluoride (Hussain et al., 2012), salinity, and nitrate (CGWB, 2010). Selenium (Se) and uranium (U) contaminations are a major growing issue in the northern states Punjab and Haryana. These contaminants in soil–water system may cause long-term health hazards, including cancer, genetic toxicity to a large population of the affected area. In this situation, providing safe drinking water to the world's largest population is a key challenge for policymakers.

Selenium (Se) occurs in volcanic and sedimentary (coal, shale, and uranium) lithological formations (Winkel et al., 2012). Some anthropogenic sources are pigments used in plastics, paints, enamels, inks, rubber, and coal burning. It occurs in soils in several forms, according to its possible oxidation states: selenides (Se^{2-}), amorphous or polymeric elemental selenium (Se^0), selenites (Se^{4+}), and selenates (Se^{6+}). Inorganic selenites are reduced to selenium (Se^0) in high acidic and reducing subsurface, while alkaline and oxidizing conditions favor the formation of selenates (Se^{6+}). Thus well-aerated alkaline soils that favor its oxidation start leaching of selenium, where selenites and selenates are soluble in water. While selenium is water-insoluble under reducing conditions and be like to retain in wet. Its concentration in alkaline soil is available for the plant as uptake mechanisms. At the same time, the availability of selenium in acidic soils tends to be limited by the adsorption of selenites and selenates to

259

Advances in Remediation Techniques for Polluted Soils and Groundwater. DOI: https://doi.org/10.1016/B978-0-12-823830-1.00020-1

iron and aluminum oxide. Winkel et al. (2012) presented a schematic global cycle of selenium, especially highlighting the environmental pathway of Se.

There are few theories of high uranium concentration in South-West Punjab. One of them states that the percolation of irrigation water dissolved CO_2 from plant respiration and microbial oxidation in the root zone (as the area is agriculture-dominated) and forms carbonic acid. Such carbonic acid dissolves the calcium carbonate (calcareous soil) to produce bicarbonate which leaches uranium from soils to the groundwater (Alrakabi et al., 2012). Phosphate fertilizers (containing uranium) are another possible source of soil/groundwater in these regions. Acidic magmatites, metamorphites are reported primary natural sources of uranium, originate soluble U^{6+} by dissolving to groundwater, while insoluble U^{4+} get deposited under redox conditions. The Sivalik sedimentary formations host significant uranium mineralization in these regions. Region of high alkalinity/salt is driving the uranium solubility in Tamil Nadu. Thivya et al. (2014) found high U and 222Rn levels in granite formations followed by Fissile hornblende biotite gneiss and Charnockite formation. The main objective of this short review update is to present the current status of the geographical distribution of selenium and uranium in the soil−water system. Thus an extensive literature review has been performed to discuss the current knowledge of the environmental fate, distribution, future scenarios, and remedial measures for both pollutants. This review will help environmental scientists, geochemists, and policy makers frame management and remediation plans for Se- and U-affected areas.

13.2   Selenium: distribution in Indian soil−water systems

A high background concentration of selenium has been reported in most North-West India locations, especially Punjab. Dhillon and Dhillon (2003a, 2003b) checked groundwater samples' quality for selenium in the seleniferous region of Punjab. It was observed that the selenium concentration found to be ranged between 0.25 and 69.5 $\mu g\,L^{-1}$ with an average value of 4.7 $\mu g\,L^{-1}$. This study also highlights that the shallow tube wells contain tow/three times more selenium than deep tube wells. After that, Bajaj et al. (2011) investigated selenium concentration in soil and groundwater in Punjab and Haryana. This study reports a high concentration in shallow (73 m deep) groundwater, for example, 45−341 $\mu g\,L^{-1}$, in two villages named Jainpur and Barwa villages in Punjab. Again, Dhillon and Dhillon (2016) collected about 750 groundwater samples from different locations of Punjab and reported a selenium concentration of range 0.01−35.6 $\mu g\,L^{-1}$. Most sites in and around Northeastern Sivalik foothill zone (NSFZ), Central, and Southwestern zone of state have high Se concentration than average.

Similarly, Lapworth et al. (2017) conducted an extensive field investigation to explore the selenium in the Bist-Doab region, covering a 9000-km^2 area between the River Sutlej and River Beas and the Sivalik Hills in Punjab. In this study, selenium concentration was found in a shallow aquifer with a range of 0.01−40 μg L^{-1}. It is also reported that more mobile Se is leached under oxidizing conditions. Virk (2018) recently investigated the Se concentration in groundwater samples from the Majha belt (Amritsar and Gurdaspur District, Punjab). This study listed the values of Se concentration observed at the different villages of the study area with a depth of water samples. The highest concentration, for example, 0.076 mg L^{-1}, was observed in groundwater samples collected from the hand pump of Abadi Harijan Basti of Tarn Taran district. A high concentration of selenium (133−931 mg kg^{-1}; dry weight) was also observed in wheat and Indian mustard, grown in a seleniferous area in Punjab by Eiche (2015) and Eiche et al. (2015). In this regard, few studies (Dhillon & Dhillon, 2003a, 2003b; Bajaj et al., 2011; Eiche, 2015; Eiche et al., 2015) support that irrigation with Se rich water is the main driving factor of elevated Se concentration North-West Regions of India. Another explanation is that selenium-containing media was originating from nearby hills of the Sivalik range was transported with floodwater (Dhillon and Dhillon, 2003a, 2003b).

Rodell et al. (2009) reported that groundwater in North India, including the seleniferous region of Punjab, is being depleted at the rate of 4.0 ± 1 cm year^{-1}. Likewise, Tiwari et al. (2009) reported a loss of groundwater with the rate of 54 ± 9 km year^{-1} between April 2002 and June 2008. A similar rate of groundwater depletion was also reported by MacDonald et al. (2016). These studies highlighted the highest rate of loss of groundwater from this region in India over the years. A depletion in groundwater may increase the selenium storage in a longer unsaturated zone where oxidizing conditions will be favorable. While irrigation may act as soil washing and increases the Se concentration in deep groundwater zone, increasing precipitation (Mishra et al., 2014) may also increase the Se load to groundwater in the future.

13.3 Naturally occurring radioactive material: distribution in Indian soil—water systems

Like selenium, naturally occurring radioactive material (NORM), especially ^{238}U, ^{232}Th, ^{40}K, and ^{137}Cs, has been reported in soil and groundwater in Punjab, India (Srivastava et al., 2014). Singh et al. (2009) reported ~0.9−63 ppb uranium concentrations in the groundwater samples near Amritsar—Bathinda. Likewise, Kumar et al. (2011) reported 73.1 ppb, five times higher than the permissible limit prescribed by the World Health Organization . Kumar et al. (2011) analyzed the subsurface water samples for uranium concentrations collected from different locations in Bathinda, Mansa, Faridkot, and Firozpur of Punjab. This study reported a range between <2 and $644\ \mu g\ L^{-1}$ with a mean value of $73.1\ \mu g\ L^{-1}$ concentrations of uranium in this study area. Alrakabi et al. (2012) reported a maximum concentration of 212 ppb of uranium in groundwater samples collected from Baluana village near Bathinda city, Punjab. Other than Punjab, high uranium and radon (^{222}Rn) concentrations have been reported in Nalgonda district, Andhra Pradesh, by Keesari et al. (2014). In Karnataka, high uranium in the groundwater of the Kolar district has been reported by Babu et al. (2008). Likewise, Thivya et al. (2014) reported high concentrations of uranium and radon (^{222}Rn) in Madurai district, Central Tamil Nadu. Das et al. (2018) reported cooccurrence of uranium in sediments of Brahmaputra River floodplain, which this study was not analyzed uranium concentration in groundwater. In depth soil-water studies are are require to understand the physical and biogeochemical processes (Gupta, 2020; Gupta et al., 2020; Gupta & Sharma, 2018; Gupta & Yadav, 2019, 2020; Kumar et al., 2021; Kumari et al., 2019; Basu et al., 2020; Gupta & Bhargava, 2020; Gupta & Yadav, 2019; Gupta & Yadav, 2017) and to understand it's implications (Amrit et al., 2019; Dhami et al., 2018; Himanshu et al., 2018, 2021; Pandey et al., 2016).

As mentioned in the section about future scenarios of Punjab regions, it is expected that precipitation will increase in the future. A study by Kooperman et al. (2018) suggested that flooding events (expected to increase in the future) will increase CO_2, reduce stomatal conductance and transpiration, and increase soil moisture in the subsurface. High soil moisture content may enhance the dissolution of CO_2 from the root zone and support microbial decompositions at high rates and more forms of carbonic acid in the subsurface. Large carbonic acid will accelerate uranium leaching. As the groundwater table is declining, uranium may get adsorbed in soil.

13.4 Remedial measures

So far, there have been only a few studies of well-established treatment techniques such as pump and treat, air-sparging, and chemical oxidation in the treatment of selenium and uranium-contaminated groundwater resources. Meanwhile, there is a concern of unwanted additional disturbance of the subsurface system, which destroys the ecological function of the system. Thus the use of the in situ techniques, especially the application of nanomaterials, biological agents, and plants, are the best and cost-effective in the case of Se and U decontamination approach. Reduction of Se^{6+} to Se^0 using nanomaterial is one of the possible remediation techniques for selenium-contaminated resources (Ling et al., 2015). Zhou et al. (2016) proved that

Se^{6+} is separated from the water via nZVI by chemical reduction to Se^{2+} and Se^0 and encapsulation in the nanoparticles. Sheng et al. (2016) used nZVI on carbon nanotubes to enhance selenite removal from water. Electrokinetic degradation of Se is another suitable remedial solution for Se removal from groundwater. Microbial bioremediation and phytoremediation are the most appropriate and cost-effective in situ techniques to remediate the selenium from soil and groundwater. El Mehdawi and Pilon-Smits (2012) listed selenium hyperaccumulator species from the genera *Stanleya*, *Astragalus*, *Xylorhiza*, and *Oonopsis*, which can accumulate $1000-15{,}000$ mg Se kg^{-1} DW ($0.1\%-1.5\%$ Se). Some anaerobic bacteria like fumarate reductase, nitrite reductase, hydrogenase, arsenate reductase, and sulfite reductase have been reported to reduce Se^{6+} (He et al., 2018). Contracted wetlands can be effective in selenium removal from agricultural drainage (Bailey, 2017).

Generally, soil washing with an appropriate reagent and absorbents (resin, activated carbon, activated silica, titanium adsorbent) has been reported earlier to decontaminate NORM-polluted sites. Kim et al. (2016) performed soil washing sulfuric acid with and reported adsorption of 90% of the uranium on the S-950 resin and desorbed 87% on S-950 by adding 0.5 M Na_2CO_3 at $60°C$. Phillips et al. (2008) performed batch and column experiments using synthetic resin to remove uranium from groundwater. They reported that Purolite A-520E anion-exchange resins removed more uranium from high pH (pH $= 8$) and low nitrate-containing synthetic groundwater in batch tests than metal-chelating resins (Diphonix and Chelex-100). At the same time, the metal-chelating resins are more effective in acidic (pH $= 5$) and high nitrate-containing groundwater. On the other hand, the application of zVI can be a practical approach for uranium removal from groundwater. Noubactep et al. (2005) reported high removal of uranium from groundwater in batch experiments having ZVI (FeS_2 and MnO_2). Bhalara et al. (2014) reviewed physical remediation techniques generally used to decontaminate the soil−water resources. One can refer to his contribution for more details on physical remediation techniques.

However, physical−chemical techniques need more remediation time and cost as high electricity demand than bioremediation techniques. It is widely reported in the literature that microbial and phytoremediation is the most suitable and cost-effective remediation approach to uranium-contaminated sites (Bhalara et al., 2014). A range of microbial communities like Pseudomonas MGF-48, a Gram-negative, motile, oxidase-negative, catalase-positive, yellow-pigmented bacterium, has been found to accumulate uranium with high efficiency (Malekzadeh et al., 2002). Newsome et al. (2014) highlighted different microbial moderated mechanisms like bioreduction, biomineralization, biosorption, and bioaccumulation of uranium remediation. Khijniak et al. (2005) and Fredrickson et al. (2000) demonstrated the microbial reduction of poorly soluble U(VI) as uramphite using

Thermoterrabacterium ferrireducens and as metaschoepite using *Shewanella putrefaciens* CN32, respectively. Some researchers reported *Pseudomonas* species as effective biomineralization. One can refer to details of uranium bioreduction from soil—water resources reviewed by Newsome et al. (2014).

13.5   Field scale implications and future research

High selenium, NORMs in thr drinking water may cause long-term health hazards to a large population of the north-west states (Punjab—Haryana) of India. This manuscript aims to provide a current update on selenium and uranium pollutants in the soil—water system for environmental scientists, geochemists, and policymakers. In the case of selenium, most of the hilly areas in and around NSFZ, Central, and Southwestern zones of Punjab are reported high Se concentration in groundwater. Se concentration uptake in different plants has also been reported, which may affect the food chain. Likewise, uranium in groundwater is another big issue in the same region of India. High U concentration in soil may cause its leaching due to the interaction of calcareous soil and bicarbonate. One can expect high Se and U concentration load in the near future due to projected increasing precipitation and large unsaturated (oxidative zone) zone due to groundwater table depletion in this region. There is growing literature on the assessment of Se and U distribution for these regions; however, little attention has been given to the fate and transport of Se and U in subsurface media. Additional research could be conducted (1) to examine the distribution of Se and U with soil depth, (2) fate and transport in different soils of the affected area, (3) alternation in subsurface microbial communities, (4) bioremediation of both pollutants, and (5) factors affecting its behavior in subsurface environments.

References

Alrakabi, M., Singh, G., Bhalla, A., Kumar, S., Kumar, S., Srivastava, A., & Mehta, D. (2012). Study of uranium contamination of ground water in Punjab state in India using X-ray fluorescence technique. *Journal of Radioanalytical and Nuclear Chemistry*, 294(2), 221—227.

Amrit, K., Mishra, S. K., Pandey, R. P., Himanshu, S. K., & Singh, S. (2019). Standardized precipitation index-based approach to predict environmental flow condition. *Ecohydrology*, 12(7) e2127.

Babu, M. N. S., Somashekar, R. K., Kumar, S. A., Shivanna, K., Krishnamurthy, V., & Eappen, K. P. (2008). Concentration of uranium levels in groundwater. *International Journal of Environmental Science & Technology*, 5(2), 263—266.

Bailey, R. T. (2017). Selenium contamination, fate, and reactive transport in groundwater in relation to human health. *Hydrogeology Journal*, 25(4), 1191—1217.

Bajaj, M., Eiche, E., Neumann, T., Winter, J., & Gallert, C. (2011). Hazardous concentrations of selenium in soil and groundwater in North-West India. *Journal of Hazardous Materials, 189* (3), 640–646.

Bhalara, P. D., Punetha, D., & Balasubramanian, K. (2014). A review of potential remediation techniques for uranium (VI) ion retrieval from contaminated aqueous environment. *Journal of Environmental Chemical Engineering, 2*(3), 1621–1634.

Basu, S., Yadav, B.K., Mathur, S., & Gupta, P.K. (2020). In situ bioremediation of LNAPL polluted vadose zone: Integrated column and wetland study. *CLEAN - Soil, Air, Water*, 2000118.

CGWB. (2010). *Central Ground Water Board: Groundwater quality in shallow aquifers of India* (p. 117). Faridabad: CGWB.

Das, N., Das, A., Sarma, K. P., & Kumar, M. (2018). Provenance, prevalence and health perspective of co-occurrences of arsenic, fluoride and uranium in the aquifers of the Brahmaputra River floodplain. *Chemosphere, 194*, 755–772.

Dhami, B., Himanshu, S. K., Pandey, A., & Gautam, A. K. (2018). Evaluation of the SWAT model for water balance study of a mountainous snowfed river basin of Nepal. *Environmental Earth Sciences, 77*(1), 1–20.

Dhillon, K. S., & Dhillon, S. K. (2003a). Distribution and management of seleniferous soils. *Advances in Agronomy, 79*(1), 119–184.

Dhillon, K. S., & Dhillon, S. K. (2003b). Quality of underground water and its contribution towards selenium enrichment of the soil–plant system for a seleniferous region of northwest India. *Journal of Hydrology, 272*(1–4), 120–130.

Dhillon, K. S., & Dhillon, S. K. (2016). Selenium in groundwater and its contribution towards daily dietary Se intake under different hydrogeological zones of Punjab, India. *Journal of Hydrology, 533*, 615–626.

Eiche, E. (2015). Microscale distribution and elemental associations of Se in seleniferous soils in Punjab, India. *Environmental Science and Pollution Research, 22*(7), 5425–5436.

Eiche, E., Bardelli, F., Nothstein, A. K., Charlet, L., Göttlicher, J., Steininger, R., & Sadana, U. S. (2015). Selenium distribution and speciation in plant parts of wheat (*Triticum aestivum*) and Indian mustard (*Brassica juncea*) from a seleniferous area of Punjab, India. *Science of the Total Environment, 505*, 952–961.

El Mehdawi, A. F., & Pilon-Smits, E. A. H. (2012). Ecological aspects of plant selenium hyperaccumulation. *Plant Biology, 14*(1), 1–10.

Fredrickson, J. K., Zachara, J. M., Kennedy, D. W., Duff, M. C., Gorby, Y. A., Shu-mei, W. L., & Krupka, K. M. (2000). Reduction of U (VI) in goethite (α-FeOOH) suspensions by a dissimilatory metal-reducing bacterium. *Geochimica et Cosmochimica Acta, 64*(18), 3085–3098.

Gupta, P. K., & Yadav, B. K. (2017). Bioremediation of non-aqueous phase liquids (NAPL$_S$) polluted soil and water resources. In R. N. Bhargava (Ed.), Environmental pollutants and their bioremediation approaches. Florida, USA: CRC Press, Taylor and Francis Group, ISBN 9781138628892.

Gupta, P.K., & Yadav, B.K., (2019). Remediation and management of petrochemical polluted sites under climate change conditions. In: Bhargava, R.N. (Ed.), nvironmental contaminations: Ecological implications and management, Springer Nature Singapore Pte Ltd. ISBN 9789811379048.

Gupta, P. K. (2020). Pollution load on Indian soil-water systems and associated health hazards: A review. *ASCE Journal of Environmental Engineering, 146*(5)03120004.

Gupta, P. K., Kumari, B., Gupta, S. K., & Kumar, D. (2020). Nitrate-leaching and groundwater vulnerability mapping in North Bihar, India. *Sustainable Water Resources Management, 6*, 1–12.

Gupta, P. K., & Sharma, D. (2018). Assessments of hydrological and hydro-chemical vulnerability of groundwater in semi-arid regions of Rajasthan, India. *Sustainable Water Resources Management, 1*(15), 847−861.

Gupta, P. K., & Yadav, B. K. (2019). Subsurface processes controlling reuse potential of treated wastewater under climate change conditions. In R. P. Singh, A. K. Kolok, & L. S. Bartelt-Hunt (Eds.), *Water conservation, recycling and reuse: Issues and challenges.* Singapore: Springer, ISBN: 9789811331787.

Gupta, P. K., & Yadav, B. K. (2020). Three-dimensional laboratory experiments on fate and transport of light NAPL under varying groundwater flow conditions. *ASCE Journal of Environmental Engineering, 146*(4)04020010.

Gupta, P.K., & Bhargava, R.N. (2020). Fate and transport of subsurface pollutants. Springer, eBook ISBN-978-981-15-6564-9. https://www.springer.com/gp/book/9789811565632.

He, Y., Xiang, Y., Zhou, Y., Yang, Y., Zhang, J., Huang, H., & Tang, L. (2018). Selenium contamination, consequences and remediation techniques in water and soils: A review. *Environmental Research, 164,* 288−301.

Himanshu, S. K., Ale, S., Bordovsky, J. P., Kim, J., Samanta, S., Omani, N., & Barnes, E. M. (2021). Assessing the impacts of irrigation termination periods on cotton productivity under strategic deficit irrigation regimes. *Scientific Reports, 11*(1), 1−16.

Himanshu, S. K., Pandey, A., & Patil, A. (2018). Hydrologic evaluation of the TMPA-3B42V7 precipitation data set over an agricultural watershed using the SWAT model. *Journal of Hydrologic Engineering, 23*(4)05018003.

Hussain, I., Arif, M., & Hussain, J. (2012). Fluoride contamination in drinking water in rural habitations of Central Rajasthan, India. *Environmental Monitoring and Assessment, 184*(8), 5151−5158.

Keesari, T., Mohokar, H. V., Sahoo, B. K., & Mallesh, G. (2014). Assessment of environmental radioactive elements in groundwater in parts of Nalgonda district, Andhra Pradesh, South India using scintillation detection methods. *Journal of Radioanalytical and Nuclear Chemistry, 302*(3), 1391−1398.

Khijniak, T. V., Slobodkin, A. I., Coker, V., Renshaw, J. C., Livens, F. R., Bonch-Osmolovskaya, E. A., & Lloyd, J. R. (2005). Reduction of uranium (VI) phosphate during growth of the thermophilic bacterium *Thermoterrabacterium ferrireducens. Applied and Environmental Microbiology, 71*(10), 6423−6426.

Kim, S. S., Han, G. S., Kim, G. N., Koo, D. S., Kim, I. G., & Choi, J. W. (2016). Advanced remediation of uranium-contaminated soil. *Journal of Environmental Radioactivity, 164,* 239−244.

Kooperman, G. J., Fowler, M. D., Hoffman, F. M., Koven, C. D., Lindsay, K., Pritchard, M. S., & Randerson, J. T. (2018). Plant physiological responses to rising CO_2 modify simulated daily runoff intensity with implications for global-scale flood risk assessment. *Geophysical Research Letters, 45*(22), 12−457.

Kumar, R., Sharma, P., Verma, A., Jha, P. K., Singh, P., Gupta, P. K., Chandra, R., & Prasad, P. V. V. (2021). Effect of physical characteristics and hydrodynamic conditions on transport and deposition of microplastics in riverine ecosystem. *Water, 13,* 2710.

Kumar, A., Usha, N., Sawant, P. D., Tripathi, R. M., Raj, S. S., Mishra, M., & Kushwaha, H. S. (2011). Risk assessment for natural uranium in subsurface water of Punjab state, India. *Human and Ecological Risk Assessment, 17*(2), 381−393.

Kumar, M., Ramanathan, A. L., Mukherjee, A., Verma, S., Rahman, M. M., & Naidu, R. (2018). Hydrogeo-morphological influences for arsenic release and fate in the central Gangetic Basin, India. *Environmental Technology & Innovation, 12,* 243−260.

Kumari, B., Gupta, P. K., & Kumar, D. (2019). In-situ observation and nitrate-N load assessment in Madhubani District, Bihar, India. *Journal of Geological Society of India, 93*(1), 113−118.

Lapworth, D. J., Krishan, G., MacDonald, A. M., & Rao, M. S. (2017). Groundwater quality in the alluvial aquifer system of northwest India: New evidence of the extent of anthropogenic and geogenic contamination. *Science of the Total Environment, 599*, 1433–1444.

Ling, L., Pan, B., & Zhang, W. X. (2015). Removal of selenium from water with nanoscale zero-valent iron: Mechanisms of intraparticle reduction of Se (IV). *Water Research, 71*, 274–281.

MacDonald, A. M., Bonsor, H. C., Ahmed, K. M., Burgess, W. G., Basharat, M., Calow, R. C., & Lark, R. M. (2016). Groundwater quality and depletion in the Indo-Gangetic Basin mapped from in situ observations. *Nature Geoscience, 9*(10), 762–766.

Malekzadeh, F., Farazmand, A., Ghafourian, H., Shahamat, M., Levin, M., & Colwell, R. R. (2002). Uranium accumulation by a bacterium isolated from electroplating effluent. *World Journal of Microbiology and Biotechnology, 18*(4), 295–302.

Mishra, V., Shah, R., & Thrasher, B. (2014). Soil moisture droughts under the retrospective and projected climate in India. *Journal of Hydrometeorology, 15*(6), 2267–2292.

Newsome, L., Morris, K., & Lloyd, J. R. (2014). The biogeochemistry and bioremediation of uranium and other priority radionuclides. *Chemical Geology, 363*, 164–184.

Noubactep, C., Meinrath, G., & Merkel, B. J. (2005). Investigating the mechanism of uranium removal by zerovalent iron. *Environmental Chemistry, 2*(3), 235–242.

Pandey, A., Himanshu, S. K., Mishra, S. K., & Singh, V. P. (2016). Physically based soil erosion and sediment yield models revisited. *Catena, 147*, 595–620.

Phillips, D. H., Gu, B., Watson, D. B., & Parmele, C. S. (2008). Uranium removal from contaminated groundwater by synthetic resins. *Water Research, 42*(1–2), 260–268.

Rodell, M., Velicogna, I., & Famiglietti, J. S. (2009). Satellite-based estimates of groundwater depletion in India. *Nature, 460*(7258), 999.

Sheng, G., Alsaedi, A., Shammakh, W., Monaquel, S., Sheng, J., Wang, X., & Huang, Y. (2016). Enhanced sequestration of selenite in water by nanoscale zero valent iron immobilization on carbon nanotubes by a combined batch, XPS and XAFS investigation. *Carbon, 99*, 123–130.

Singh, H., Singh, J., Singh, S., & Bajwa, B. S. (2009). Uranium concentration in drinking water samples using the SSNTDs. *Indian Journal of Physics, 83*(7), 1039–1044.

Srivastava, A., Lahiri, S., Maiti, M., Knolle, F., Hoyler, F., Scherer, U. W., & Schnug, E. W. (2014). Study of naturally occurring radioactive material (NORM) in top soil of Punjab state from the North Western part of India. *Journal of Radioanalytical and Nuclear Chemistry, 302*(2), 1049–1052.

Thivya, C., Chidambaram, S., Tirumalesh, K., Prasanna, M. V., Thilagavathi, R., & Nepolian, M. (2014). Occurrence of the radionuclides in groundwater of crystalline hard rock regions of central Tamil Nadu, India. *Journal of Radioanalytical and Nuclear Chemistry, 302*(3), 1349–1355.

Tiwari, V. M., Wahr, J., & Swenson, S. (2009). Dwindling groundwater resources in northern India, from satellite gravity observations. *Geophysical Research Letters, 36*, 18.

Virk, H. S. (2018). Selenium contamination of groundwater of Majha Belt of Punjab (India). *Research & Reviews: A Journal of Toxicology, 8*(2), 1–7.

Winkel, L. H., Johnson, C. A., Lenz, M., Grundl, T., Leupin, O. X., Amini, M., & Charlet, L. (2012). Environmental selenium research: From microscopic processes to global understanding. *Environmental Science & Technology, 46*, 571–579.

Zhou, Y., Tang, L., Yang, G., Zeng, G., Deng, Y., Huang, B., & Wu, Y. (2016). Phosphorus-doped ordered mesoporous carbons embedded with Pd/Fe bimetal nanoparticles for the dechlorination of 2,4-dichlorophenol. *Catalysis Science & Technology, 6*(6), 1930–1939.

CHAPTER 14

Understanding and modeling the process of seawater intrusion: a review

Lingaraj Dhal and Sabyasachi Swain

Department of Water Resources Development and Management, Indian Institute of
Technology Roorkee, Roorkee, India

14.1 Background

Groundwater is an important component as it is one of the major sources of fresh water available in the earth system. In coastal regions, it is one of the important sources of fresh water for domestic and other uses. Due to seawater intrusion (SI), the availability and utility of the groundwater is restricted in coastal regions. SI is the inward movement of seawater through river, wetlands, or to the coastal aquifers. As the saline water moves inward and mixes with the fresh water, water in the freshwater reservoir gets contaminated. For the effective management of water resources in the coastal regions, the understanding of coastal hydrology is essential. Coastal hydrology includes the following:

1. interaction of surface water and groundwater,
2. submarine groundwater discharge,
3. coastal hydrogeology (i.e., the study of heterogeneity in coastal aquifers),
4. coast line morphology, and
5. behavior and position of fresh water and seawater interface/mixing zone, etc.

More than 50 years of study has been carried out on SI. However, the current understanding is only based on the laboratory experiments, numerical modeling, and some case studies. There is significant advancement for steady-state SI condition but it is difficult to understand the transient flow process. Further, the SI demands a multidisciplinary approach (i.e., hydrology, geology, hydrochemical studies, and biology) to be understood and addressed comprehensively. A simplified conceptual diagram of the SI process is given in Fig. 14.1.

Advances in Remediation Techniques for Polluted Soils and Groundwater. DOI: https://doi.org/10.1016/B978-0-12-823830-1.00009-2

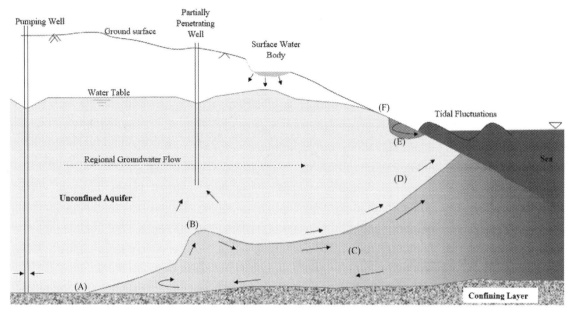

FIGURE 14.1

Conceptual diagram of an unconfined aquifer in coastal area showing (A) toe position of SI wedge, (B) upcoming due to pumping by partially penetrating well, (C) saline water and freshwater mixing zone, (D) fresh submarine groundwater discharge, (E) estuaries influenced by tide, and (F) seepage face near seashore. *SI*, Seawater intrusion.

14.2   Seawater intrusion process

There are different factors that influence the SI processes. These are (1) anthropogenic influences, (2) beach morphology, (3) dispersive mixing, (4) geochemical reaction, (5) geological characteristics (the degree of confinement and aquifer hydraulic and transport properties), (6) paleo-hydrological condition, (7) surface hydrology (i.e., recharge variability, surface–subsurface interaction), (8) tidal effects, (9) tsunami, and (10) unsaturated zone flow and transport process. These factors are briefly discussed next:

1.   Anthropogenic influences: SI is now becoming a global challenge for the coastal water resources managers, especially due to anthropogenic activities. The interference of human activities with the natural groundwater flow is creating complex changes in the saline water and freshwater equilibrium in coastal aquifers (Custodio, 1987). There are some direct and indirect impacts of human activities that are responsible for aggravating the SI process in the coastal aquifers. To meet the growing freshwater demand of different sectors in the coastal area, the freshwater aquifers are being overexploited, leading to reduction in freshwater hydraulic head in the

aquifer and coastal freshwater discharge. It also alters the dynamics of equilibrium and sets a new equilibrium position of seawater—freshwater interface (often with deeper seawater penetration wedge). In most of the cases the wells are dug to tap multiple aquifers for higher yielding. Consequently, the short circuit of the aquifer layers may create a favorable condition for seawater contamination from a seawater infected aquifer to a freshwater aquifer based on the hydraulic gradient (Custodio, 1987). During the execution of a number of civil works like buildings, roads, canals, and subways, a large amount of fresh groundwater are drained to complete the desired works. This leads to reduction in groundwater level in the aquifers and subsequently accelerates the SI. Further, the paving and mismanagement of groundwater recharge area also fuels the SI in coastal areas. In most of the cases the effects of human activities are not visible immediately and may take several years. Therefore to understand the influence of human activities on SI in coastal aquifers, information on the pace of changes along with the continuous data collection programs with rapid response mechanisms are certainly required.

2. Beach morphology: In coastal regions, beach morphology and tide influence the along-shore flow of groundwater and modify the local pore water circulation and salt transport. Further, the bathymetry and hydraulic head of coastal rivers enable farther and more rapid landward intrusion of seawater along the river than coastal aquifers. Due to the landward movement of seawater in the rivers, saline water plumes from near river area, spreads toward inland coastal aquifers, and increases chances of seawater contamination (Zhang et al., 2016).

3. Dispersive mixing: This mechanism influences the thickness of fresh water—saltwater mixing zone. In reality, the SI is a multifactor-influenced complex process. Some studies (Abarca & Clement, 2009; Fahs et al., 2018; Robinson et al., 2016; Takahashi et al., 2018) have tried to understand how SI is influenced by dispersive mixing; however, their inferences do not agree with each other and hence, a generalized conclusion is yet to be reached.

4. Geochemical reaction: Majority of SI studies focus on the hydrogeological systems using traditional density-driven flow and transport processes through experimental and numerical modeling, often ignoring the chemical reactions. The chemical reactions taking place in the aquifer has direct impact on the porosity and permeability of the aquifer material, which ultimately influences the SI process. To improve the SI modeling and better understand the underlying processes, the interdependency of density-driven flow and chemical reactions must be considered. SI may increase by solid matrix dissolution in the seawater mixing zone in the case of carbonate-dominated coastal aquifers. Further, the dissolution of such rocks increases the porosity and permeability of the porous media

(Laabidi & Bouhlila, 2016). The increase in permeability fuels the SI more toward the freshwater aquifer. Boluda-Botella et al. (2008) demonstrated how other geochemical reactions are taking place inside the aquifer during SI.

5. Geological characteristics: It is the main cause of aquifer hydraulic and transport properties, which ultimately governs the groundwater flow and transport processes in the geological formation. The hydraulic conductivity has a monotonous effect on the salt mass on the shore (van Engelen et al., 2021). Arrangement of the aquifers in a coastal region and how these aquifers interact among themselves govern the contamination of seawater from one contaminated aquifer to another. Further, the aquifer hydraulic and transport properties decide the rate of SI.

6. Paleo-hydrological condition: The historically changing boundary conditions are shaping present day distribution of groundwater and its chemical composition. In coastal aquifers the marine transgression is a vital process for seawater contamination (Delsman et al., 2014). Past marine transgression is still reflected in current fresh—saline groundwater distributions in deltas (van Engelen et al., 2021). Therefore paleo-groundwater studies and understanding are prerequisites for current seawater modeling practices.

7. Surface hydrology: Surface water plays a key role in coastal area to mitigate the SI problem. Availability of surface water meets the freshwater demand and reduces the groundwater extraction. Increasing recharge of fresh water to the coastal aquifer increases the submarine freshwater discharge and maintains the equilibrium of saltwater—freshwater mixing near sea and prevents the saline water to intrude toward the freshwater aquifers.

8. Tidal effects: Tide changes the nearshore groundwater flow and transport processes. Tidal fluctuations flatten the brackish freshwater interfaces, widen the dispersion zone, and generate upper saline and freshwater belt (Liu et al., 2017).

9. Tsunami: Tsunami-induced SI causes the vertical saltwater infiltration into coastal aquifers (Liu & Tokunaga, 2019). This results in the unexpected salinization of fresh groundwater. Tsunami inundation height, rainfall recharge rate, and vertical hydraulic conductivity of the soil are the primary factors that decide the severity of tsunami induced SI (Liu & Tokunaga, 2019).

10. Unsaturated zone flow and transport process: The water flow in unsaturated zone can also be vital for SI process. The unsaturated zone flow is significantly influenced by unsaturated hydrodynamics (preferential flow and diffuse flow) and by unsaturated hydrostatics (water content, pressure, energy, and retention) (Nimmo, 2006). The

contamination transport in the coastal aquifer under the influences of SI may be referred from the recent studies (Guo et al., 2020, 2021).

Interaction among these provides a number of possible situations in real-world problem for SI, which makes the study more complicated and challenging for the researchers as well as for groundwater managers for the evaluation and optimal utilization of the groundwater for achieving sustainability. As a lot of influencing factors are associated with SI process, it is difficult to identify the key factors that are primarily responsible for SI in natural system/field conditions. However, in laboratory condition under controlled environment, few influencing factors are taken into consideration and the impact of each factor over SI process is studied. As evident from all the SI studies, the key factors that influence the SI are as follows:

1. the buoyancy forces generated due to the variation in densities due to salt and temperature,
2. advective forces due to submarine fresh groundwater discharge,
3. dispersive mechanism, and
4. hydrological/geometric boundary controls (e.g., aquifer thickness, extent, and properties).

Further, the time lag in response of stresses on aquifer to affect the SI process adds some criticality/complexity in studying and understanding the processes. To understand the SI process, different benchmark problem studies have been carried out. Several researchers (Abarca et al., 2007a; Simpson & Clement, 2004; Younes & Fahs, 2015) modified the standard Henry benchmark problem to study the behavior of saltwater and fresh water when stable density stratification occurs, that is, fresh water remain over the seawater. Dentz et al. (2006) modified the Henry problem by studying the advective and diffusive process considering the P_e (Péclet number). Abarca et al. (2007b) also modified the Henry problem by adding a dimensionless parameter to study the dispersive and anisotropic behavior and asserted the dispersive form of Henry problem to be more realistic. Convective circulation is another crucial component in SI. Since freshwater and seawater mixing zone is an important part of SI study, it is very important to know the position and thickness of the mixing zone for a better management of coastal aquifers. The density and concentration of salt vary across the mixing zone. This is due to the molecular diffusion caused by the density gradient between fresh water and saline water, and mechanical dispersion result of the submarine freshwater discharge. The occurrence of submarine freshwater discharge creates a mobile zone over an immobile zone, which results in the formation of a mixing zone between the freshwater and groundwater zones due to convective circulation. To understand these processes, different laboratory studies (Abarca et al., 2007b; Goswami & Clement, 2007; Jakovovic et al., 2011), numerical modeling along with field experiments (Kim & Yang, 2018; Mansour et al., 2017; Price et al., 2003; Xu et al., 2019) have been conducted. But they did not converge at a single conclusion as the laboratory experiments and numerical models (finer discretization with low dispersive coefficient) give narrow mixing zone. However, in natural conditions, the mixing zone varies from few meters to some kilometers (Paster et al., 2006; Price et al., 2003). Generally, in the steady-state condition, the freshwater—seawater mixing zone is affected by freshwater discharge, density variation between seawater and freshwater, mechanical dispersion (i.e., longitudinal dispersion and transverse dispersion), and molecular diffusion.

For lower freshwater discharge, narrow mixing zone is likely to be there, where dispersion is proportional to velocity. Millero et al. (2008) suggested that local transverse dispersion (i.e., pore scale) is the primary mechanism for steady-state mixing condition in a homogenous aquifer. They also experimentally found the dispersion coefficients to be low, which indicate narrow mixing zone in the abovesaid condition. In reality, the wider mixing zones may not be there because of the local transverse dispersion. Abarca et al. (2007b) demonstrated that both the longitudinal and transverse dispersion coefficients are equally important for controlling the freshwater and saline water mixing zone; and the resultant/effective dispersive coefficient will be the geometric mean of the two in an anisotropic dispersive adaptation of Henry problem. In natural condition the mixing zone is the result of several factors, namely, spatial heterogeneity in hydrogeological characteristics in the geological structure, influence of tide and wave, spatiotemporal variation in groundwater recharge, changes in sea-level position, and pumping activities in the coastal aquifers.

Till date, there is no clear picture how these abovementioned factors affect the SI. Fluctuations in sea level due to climate change and freshwater head influence the saltwater and freshwater mixing zone. Further, the mixing zone is not stationary due to seasonal fluctuations of groundwater level in the coastal aquifers due to groundwater withdrawal and recharge. Anthropogenic activities like pumping from coastal aquifers aggravate the problem especially when it is done from a partially penetrating well in a stratified condition of fresh water and saline water (i.e., fresh water lay over saline water) and the mixing zone then gets wider as the seawater moves upward (Fig. 14.1). This process is called upconing. During the initial phase of upconing, longitudinal dispersion plays key role and when the system gets stabilized then the transverse dispersion takes over and become narrower. Ataie-Ashtiani et al. (2001) studied the effect of tidal forcing in an unconfined aquifer. Effect of single tide may have less influence on the mixing zone but the net influence is the point of concern. During active SI from sea to coastal aquifer, longitudinal dispersion plays a vital role, whereas for slow moving front transverse dispersion plays a vital role.

14.2.1   Heterogeneity in seawater intrusion process

In reality, there are huge variations in the properties of geological settings. The geological properties vary spatially which lead to heterogeneous geological system. When SI study comes in to picture it not only includes the hydrogeological heterogeneities but also includes geochemical heterogeneities. For the study of SI in a coastal aquifer, the scale of heterogeneity also plays an important role. Small-scale heterogeneity generally has a minor influence on mixing zone thickness; however, macroscopic variation/heterogeneity has significant influence on the mixing zone thickness. Further, moderate random heterogeneity has small mixing zone thickness in steady-state conditions. In addition to this, there are differences of heterogeneity which affect the SI process. Preferential flow paths are one among them. These are macroscale heterogeneities which allow the solute and saline water to move faster. Heterogeneity has a negligible effect on transversal mixing, but it influences the direction of the movement of interface. Further, the variation in the aquifer bottom also influences the inward movement of saline water. The transverse strata present in the aquifer also play an important role in SI. If the horizontal stratification is neglected in numerical modeling, the model will overpredict the toe position, if other conditions remain the same. Therefore it is very important in SI study to identify the local features and represent them properly in a numerical model. The circulation of water in a coastal aquifer, the chemical composition of water present in it, and the submarine freshwater discharge zones are highly influenced by heterogeneity.

14.2.2   Sea level fluctuations and seawater intrusion

The sea level influences the SI process (Hawkes, 2013). Different studies on climate change have suggested about the future sea level rise possibilities (Chang et al., 2015; Ercan et al., 2013; Fashae & Onafeso, 2011; Mimura, 2013). In addition, the events like tsunami (Kume et al., 2009; Violette et al., 2009), storm surges are very irregular and unpredictable. Tidal inlets, rivers, estuaries, and tidal wetlands are also influenced by these events and further complicate the SI studies.

14.2.3   Hydrochemical processes involved in seawater intrusion process

The chemical composition of seawater is more or less same all over the world. There is very little difference. However, there is a significant variation in the chemical composition of groundwater. The chemical composition of groundwater mainly depends upon the chemical composition of the aquifer materials and the favorable environment for different chemical reactions. Therefore it is essential to study the hydrochemical properties of groundwater in general and water in coastal aquifers in particular. The seawater and groundwater interface will remain in chemical equilibrium with the aquifer materials. When there is any change/disturbance in the position and thickness of the mixing zone, it ultimately disturbs the chemical equilibrium. This disturbance may cause dissolution of salt present in the aquifer materials, which leads to cavity or pore formation. The cation exchange mechanism is more significant in the mixing zone due to the difference in the ion concentration in the mixing zone. Saline water generally has more sodium (Na^+) and magnesium (Mg^{2+}) ions; however, groundwater has more calcium (Ca^{2+}) ions. During the disturbances in sea water—groundwater interface, the behavior of the media to absorb cations will drive the chemical processes. Due to the chemical reactions, a clear demarcation is left behind by the earlier mixing zone, which is one of the most useful indicators for study the temporal dynamics of the mixing zone.

14.2.4 Upconing

Pumping from a partially penetrating well present in a coastal aquifer creates vertical upward thrust, which develops a cone-shape surface of the underlaying seawater surface (Fig. 14.1) until the upward force is balanced by the density-dependent downward gravitational force. After achieving an equilibrium condition, the cone gets stabilized. This process is called upconing, which increases the thickness of the saline water—freshwater mixing zone. Many factors such as pumping rate, pumping duration, the frequency of pumping, aquifer properties, and initial position of seawater play critical roles in the upconing process. Numerous studies are available in literature that provides a clear understanding of the upconing process (Abdoulhalik & Ahmed, 2018; Dagan & Bear, 1968; Diersch et al., 1984; García-Menéndez et al., 2016; Jakovovic et al., 2011, 2016; Kura et al., 2014).

14.3 Measurement and monitoring of seawater intrusion

Sustainable management of coastal aquifer warrants proper measurement and monitoring of groundwater. For transient groundwater flow studies, sufficient amount and quality of historical data is required. For management and planning at country/regional level, again it becomes a challenging task to understand the system and generalize the result. As SI is a complex process found in complex geological environment, a single measurement method may not be adequate. Therefore multiple measuring and monitoring methods should be used to address the multifaceted SI problem.

14.3.1   Head measurement

To study the subsurface flow in coastal aquifers, generally three types of head measurements are carried out. These are (1) point water head, (2) freshwater head, and (3) environmental head. Point water head is the head when the water inside the well has the same density as the water present in the aquifer. Freshwater head is the head measured during monitoring, if the water filled in the well is fresh water. Environmental head is the head measured during monitoring, if the density variation of water in the well is the same as the density variation of water in the aquifer. In homogenous aquifer, contours are helpful in understanding the groundwater flow in freshwater or saline water zone individually. However, it fails to give the same information in the saline water and freshwater mixing zone. Further, it will give sensible results or information if the exact depth and concentration of water is known. Monitoring wells present in the coastal area connects different layers of single or multiple aquifers. Therefore the water present inside the well may not properly represent the system in its surroundings. To overcome this problem in the measurement and monitoring of SI, proper knowledge of the strata and placement of strainers are necessary.

14.3.2   Geophysical methods

There is a large electrical resistivity difference between fresh water and seawater, as their corresponding values are 0.2 Ωm and greater than 5 Ωm, respectively (Werner et al., 2013). So it is feasible to map SI with the help of electrical resistivity method. A number of well-documented studies are there, where electrical resistivity method is used successfully to map coastal aquifers for SI studies (Bouzaglou et al., 2018; Chen et al., 2018; Hung et al., 2018; Kazakis et al., 2016; Kumar et al., 2015; Martínez-Moreno et al., 2017; Najib et al., 2017). Previously, due to the lack of high computational power and data processing tools, it was restricted to 1D mapping only. However, due to the availability of high computational power and data processing tools nowadays, 2D and 3D mapping by electrical resistivity tomography (ERT) method to map the coastal aquifer heterogeneity can be achieved. With the help of ERT, Acworth & Dasey, (2003) studied the mixing zone between freshwater infiltration and saline water in hyporheic zone below a tidal creek. The submarine groundwater discharge, which is highly influenced by local and small-scale heterogeneity, was also studied (El-Kaliouby & Abdalla, 2015; Martínez-Moreno et al., 2017; Trabelsi et al., 2013; Ziadi et al., 2017). ERT can be used to map the freshwater lances below sea surface. This technique is influenced by resolution and depth. The combination of ERT data with the monitored borehole data adds more sense. Effects of the most influencing environmental factors, that is, rainfall, river stage, and tide on SI, can be studied at different depth and time scale through time-lapsed resistivity by installing permanent electrodes with advanced telemetry system. There is another geophysical method, that is, electromagnetic (EM) method. It is a rapid and cheap method as compared to ERT. It can be used without being in contact with the ground surface. Therefore it is used for airborne mapping of aquifer by mounting the sensors in helicopter or drones. This method helped the coastal freshwater custodians and researchers to map the coastal aquifers even in inaccessible coastal areas. The combination of ERT and EM enhances the efficiency of aquifer mapping. First, the EM survey is executed and the zones are identified where detailed study has to be carried out by ER method. Further, to add more sense and verify the results, the downhole verification is required. During the downhole verification study, the metal casing of borehole should be avoided. For efficient and meaningful SI mapping, a proper calibrated model should be considered that correlates to the salinity and also takes other influencing factors (other than salinity) into account.

14.3.3 Tracer techniques for seawater intrusion studies

There are various radioactive elements that are used as environmental tracer to study SI, that is, $\delta^{18}O$, $\delta^{11}B$, δ^2H, ^3H, ^{14}C, $^{87}Sr/^{86}Sr$ ratios, and Br/Cl ratios (Maurya et al., 2019; Mohanty & Rao, 2019; Yu et al., 2019). Further, with the help of this tracer technique, the modern SI as well as the old SI can be studied (Sukhija et al., 1996). Besides, salinity due to irrigation and the chemicals properties of irrigation return flows can be studied by this technique.

14.4 Seawater intrusion modeling and prediction

There are various models used for groundwater flow and solute transport studies. For SI studies the models that consider the variable density flow are useful, as there is variation in the density of water in the seawater and freshwater mixing zone. The main distinguished feature between the SI and groundwater flow model is the consideration of density of the fluid. SI model consider the density variation of water along with solute transport, whereas groundwater flow models do not consider the same. For SI studies, some researchers examined the sharp interface between the fresh groundwater and the saline seawater, where the fresh water and sea water are considered two immiscible fluids (Kacimov & Obnosov, 2016; Llopis-Albert et al., 2016; Mehdizadeh et al., 2014, 2017; Werner, 2017). In the freshwater and saline water interface, constant and continuous pressure from freshwater side and saline water side are exerted and the system remained in an equilibrium condition. However, others considered the variable density mixing zone between fresh groundwater and seawater. The later is more realistic and the results of the study can be compared with the real-world situations.

14.4.1   Analytical solutions for seawater intrusion problems

For getting analytical solutions to SI problem, one has to solve flow in fresh water and saline water simultaneously. In this condition, hydrostatic pressure and fluxes are normal to the interface from both freshwater and saline water sides. Further, simplification of the problem leads to steady-state interface position, where the flow is usually there in the freshwater zone but the saline water remains stable or in rest. With the help of Dupuit—Forchheimer assumption and Ghyben—Herzberg formula, steady-state analytical solution of SI problem can be solved. The most important assumption in interface flow problem is that the fresh water and saline water are treated as two immiscible fluids. Analytical steady-state solution for a problem of 2D variable density flow in a vertical plane was given by Henry (1964). Later, this problem was modified by other researchers and studied to understand the variable density flow processes (Dentz et al., 2006; Segol, 1994; Voss et al., 2010). Henry problem is a unique problem as it is the only problem that has analytical solution and is useful for SI study. However, it has less utility for real-life situations.

14.4.2   Numerical modeling of seawater intrusion

A very limited number of analytical solutions are available for SI problems. Therefore numerical models play vital roles to understand the SI process. Numerical models are used to understand the natural system and future predictions of SI for the management of coastal aquifers (Lathashri & Mahesha, 2015, 2016; Mao et al., 2006; Xu et al., 2019). These models are used for both local and regional scales. Different processes like tidal fluctuations, hydrodynamics of sea shore, seepage face development, and aquifer heterogeneities are the key factors in SI studies. One of the key challenges in front of the modelers is to represent these real-world processes in the model environment. Most of the numerical codes couple the flow and transport processes to simulate the SI process. Commonly used grid discretization, that is, finite difference method (used in SEAWAT and MODFLOW) and finite element method (used in SUTRA) are handy for flow simulations, however, not that much useful for solute transport. The "method of characteristic" and "total variation diminishing method" are useful for solving transport equations. But these methods require high computational power and time. Therefore a trade-off between accuracy and time has to be done by the researchers during SI study. A list of models used for groundwater flow and transport as well as SI studies is presented in Table 14.1. Different optimization models are used for pumping optimization in coastal aquifers to sustainably manage the water resources in coastal areas (Christelis & Mantoglou, 2019; Singh, 2014).

Table 14.1 List of codes/models used for seawater intrusion studies (Werner et al., 2013).

Sl. no.	Code/model	Basic features
1	FEFLOW	Uses finite element approach Both saturated and unsaturated flow simulation
2	FEMWATER	Uses finite element approach Both saturated and unsaturated flow simulation
3	HYDROGEOSPHERE	Uses finite element approach Both saturated and unsaturated flow simulation
4	MOCDENS3D	Uses finite difference approach Saturated flow simulation
5	SEAWAT	Uses finite difference approach Saturated flow simulation
6	SUTRA	Uses finite element and finite difference approaches Both saturated and unsaturated flow simulation
7	SWI	Uses finite difference approach Saturated flow simulation

14.5 Management of seawater intrusion

Growing demand for water in every sector and the changing climatic conditions are the root causes of spatiotemporal variations in the availability of fresh water, which puts a big challenge in front of water resources managers and environmentalists in general and coastal areas in particular. In coastal regions, groundwater is one of the major freshwater reservoirs to rely on. However, the reliability of the resources is threatened due to SI. The coastal aquifers get contaminated due to SI and become unsuitable for freshwater uses. Proper monitoring and assessment of SI is the call of the day to manage the coastal aquifers in a sustainable ecosystem. But the monitoring and assessment of coastal aquifer is a tedious and expensive affair.

One of the most influencing human-induced activities for SI is overexploitation (specifically in terms of uncontrolled pumping) of coastal aquifer. The optimal use of freshwater in coastal area demands checking the overpumping, which warrants a careful monitoring. Further, SI has considerable impacts on the ecosystem. Some studies have been conducted to assess the impacts of future climate change and sea level rise on SI (Chun et al., 2018; Kalaoun et al., 2018; Mastrocicco et al., 2019), which gives an idea how to manage coastal aquifers in a sustainable way. SI study and management needs a multidisciplinary approach. For the effective management of coastal aquifers and to check SI into the same, certain aspects have to be taken care of. These are as follows:

1. Sufficient knowledge about the hydrogeological system and hydrological processes that directly influence the SI.
2. Continuous and intensive monitoring of coastal aquifers in terms of water quantity and quality, and analysis of the monitored data and interpretation of the results.
3. Consideration of environmental and socioeconomic aspects.
4. The interaction/interplay between these abovementioned three components in the past as well as in future.

The government policies, legislative frameworks, and regulatory authorities have to be formulated for land use and well field management, considering all the previous points. Werner et al. (2011) carried out an extensive study of well field management for controlling SI in coastal aquifers. SI causes reduction in the availability of freshwater resources in the coastal aquifers. Therefore studies on the assessment of vulnerability and resilience are necessary for implementing preventive measures and for a better management of coastal aquifers. When dealing with regional/country level, sometimes the simple methods perform better than the complex numerical models. Some SI vulnerability assessment studies have been carried out using GLADIT (Kazakis et al., 2018; Sophiya & Syed, 2013) and DRASTIC (Kaliraj et al., 2015; Momejian et al., 2019a, 2019b).

For policy-making on the management of coastal aquifers and prevention of SI into the coastal aquifers, various factors like surface water use, groundwater use, SI status, agricultural productivity, water logging conditions, and the economic factors have to be taken into considerations. Qureshi et al. (2008) studied on the management of coastal aquifers considering previous factors. There are certain engineering measures and practices for remediation/mitigation of SI. Conjunctive use of water for agricultural purpose to optimize the crop productivity (economic aspect) and soil salinity (environmental aspect) was a common practice in farm (El-Fadel et al., 2018). Some studies have been conducted regarding the prevention of SI into coastal aquifers by aquifer recharge, and simulation and optimization of pumping in coastal aquifers (Basdurak et al., 2007; Guo et al., 2019; Hussain et al., 2015; Khomine et al., 2011; Nofal et al., 2015; Padilla et al., 1997; Paniconi et al., 2001). Laboratory experiments showed that the injection of water at toe is more effective as compared to other positions in the aquifer for SI remediation. So well injection is more effective measure compared to the surface recharge of aquifer in controlling SI but it is expensive (Luyun et al., 2011). Further, in some cases, the aquifer recharge may not be suitable for the prevention/remediation of SI into coastal aquifers. In such cases, some physiological barriers are used to check SI. Luyun et al. (2011) suggested that these physical barriers are effective when close to sea and in deeper aquifers. More importantly, the height of the barrier plays a vital role in the mixing zone of fresh water and seawater in the seaward side of the barrier (Luyun et al., 2009). Generally, concrete, bentonite clay, etc. are used as physical barrier materials for SI control in the coastal aquifers. For the seasonal prevention of SI, trapped air and pore-filling technique are used to reduce the permeability of the aquifer media. It is practically feasible to inject air to decrease the permeability of media; however, further studies are advocated (Dror et al., 2004).

SI reduces the freshwater availability and alters the chemical composition of water present in the aquifer or other surface water bodies that are influenced by sea, leading to disturbances in the ecosystem. It affects the life of the organisms who depend upon fresh groundwater, for example, stygofauna (i.e., microfauna depends on groundwater). Stygofauna are influenced by the changing chemical composition of aquifer water due to SI. In some cases where surface water bodies are hydraulically connected to aquifer, the saline groundwater discharges of aquifer to surface water bodies remarkably influence the environment there (Simpson et al., 2011). SI influences the microbial activities, which triggers the catalysis of important nutrients transformation in the soil media. One of the major environmental issues is soil salinity due to the application of saline groundwater in agricultural fields, which alters the soil chemistry and reduces soil fertility (Darwish et al., 2005; Qi & Qiu, 2011). Further, the salinization of unsaturated zone due to capillary rise is also a concern for vadose-zone hydrologists. Vegetation present in nontidal zone, which depends on the freshwater availability in coastal aquifers, is also affected by SI. These critical issues urge significant attention toward a proper management of the SI.

14.6 Seawater intrusion, climate change, and sea level rise

Increasing pumping due to dry climate and meeting the growing population demand in coastal aquifers are the main human-induced causes for SI into coastal aquifer. Enhanced frequency of storm surge inundation and marine transgression further aggravates the SI problem. The climate change impacts on SI are sensitive to the boundary conditions of the aquifer system, that is, head- or flux-controlled (Rasmussen et al., 2013; Werner & Simmons, 2009). The regions will be highly vulnerable to future sea level changes, where the water table is governed by the drainage system (Chang et al., 2011). The climate change leading to erratic rainfall and sea level rise can also pose a major challenge for groundwater abstraction and water supply management (Aadhar et al., 2019; Dayal et al., 2019; Pandey & Khare, 2018; Swain et al., 2017a, 2017b, 2020, 2021a, 2021b). A reduction in rainfall results in the reduction of groundwater recharge and, in turn, groundwater head. This causes a significant increase of SI. Moreover, the displacement of groundwater divided toward the sea further intensifies the conditions. Using the regional climate change models, van Roosmalen et al. (2007) estimated the changes in groundwater recharge for the Intergovernmental Panel on Climate Change scenarios. Even the climate change will cause the extreme events to be more frequent and, thus, the probability of flooding along the drainage canals will increase (Swain et al., 2018a, 2018b). Thus it can be inferred that the SI is sensitive to the stage of the drainage canals, groundwater recharge, and sea-level changes. However, the interrelationship between the local anthropogenic activities and the global climate change still remains ambiguous (Swain et al., 2019a, 2019b). Further, climate change causing sea level rise and aggravating SI may also cause the migration of people from the coastal areas (McLeman, 2019; Wrathall et al., 2019). Therefore it is crucial for the coastal managers to devise proper planning and management strategies to protect the environmental and socioeconomic security of the coastal communities. Tariff restructuring, smart metering, incentives for water-saving appliances, and promoting water conservation practices are the steps that can be handy for a sustainable management of water demands (Bahita et al., 2021; Dhami et al., 2018; Himanshu et al., 2019; Swain, 2017). Further, rainwater harvesting, minimizing the losses in water supply network, and water reuse can be the solution, especially in the soaring water-demanding scenarios due to population explosion and high consumption rates (Safi et al., 2018).

14.7 Conclusion

SI has become a major concern, especially when the population explosion and rapid industrialization have led to a drastic increase in water demands. This study provides a detailed description of the SI process and its consequences. A brief review of SI measurement and monitoring practices, modeling and prediction techniques, and salient measures adopted in different regions of world to adapt or mitigate the effects of SI is also presented. The SI possesses remarkable effects over groundwater, especially over the coastal aquifers. Further, the changing climatic conditions leading to sea level rise may have detrimental impacts on SI. The study emphasizes on the adoption of sustainable water management practices to minimize the ill effects of SI.

References

Aadhar, S., Swain, S., & Rath, D. R. (2019). *Application and performance assessment of SWAT hydrological model over Kharun river basin, Chhattisgarh, India. World environmental and water resources congress 2019: Watershed management, irrigation and drainage, and water resources planning and management* (pp. 272−280). American Society of Civil Engineers.

Abarca, E., Carrera, J., Sánchez-Vila, X., & Dentz, M. (2007a). Anisotropic dispersive Henry problem. *Advances in Water Resources, 30*(4), 913–926.

Abarca, E., Carrera, J., Sánchez-Vila, X., & Voss, C. I. (2007b). Quasi-horizontal circulation cells in 3D seawater intrusion. *Journal of Hydrology, 339*(3–4), 118–129.

Abarca, E., & Clement, T. P. (2009). A novel approach for characterizing the mixing zone of a saltwater wedge. *Geophysical Research Letters, 36*(6), 1–5. Available from https://doi.org/10.1029/2008GL036995.

Abdoulhalik, A., & Ahmed, A. A. (2018). Transient investigation of saltwater upconing in laboratory-scale coastal aquifer. *Estuarine, Coastal and Shelf Science, 214,* 149–160.

Acworth, R., & Dasey, G. (2003). Mapping of the hyporheic zone around a tidal creek using a combination of borehole logging, borehole electrical tomography and cross-creek electrical imaging, New South Wales, Australia. *Hydrogeology Journal, 11,* 368–377. doi:10.1007/s10040-003-0258-4.

Ataie-Ashtiani, B., Volker, R. E., & Lockington, D. A. (2001). Tidal effects on groundwater dynamics in unconfined aquifers. *Hydrological Processes, 15*(4), 655–669.

Bahita, T. A., Swain, S., Dayal, D., Jha, P. K., & Pandey, A. (2021). *Water quality assessment of upper ganga canal for human drinking. Climate impacts on water resources in India* (pp. 371–392). Cham: Springer. Available from https://doi.org/10.1007/978-3-030-51427-3_28.

Basdurak, N. B., Onder, H., & Motz, L. H. (2007). Analysis of techniques to limit saltwater intrusion in coastal aquifers. In: *World environmental and water resources congress 2007: Restoring our natural habitat* (pp. 1–8).

Boluda-Botella, N., Gomis-Yagües, V., & Ruiz-Beviá, F. (2008). Influence of transport parameters and chemical properties of the sediment in experiments to measure reactive transport in SI. *Journal of Hydrology, 357*(1–2), 29–41. Available from https://doi.org/10.1016/j.jhydrol.2008.04.021.

Bouzaglou, V., Crestani, E., Salandin, P., Gloaguen, E., & Camporese, M. (2018). Ensemble Kalman filter assimilation of ERT data for numerical modeling of seawater intrusion in a laboratory experiment. *Water (Switzerland), 10*(4), 1–26.

Chang, B., Guan, J., & Aral, M. M. (2015). Scientific discourse: Climate change and sea-level rise. *Journal of Hydrologic Engineering, 20*(1), 1–14.

Chang, S. W., Clement, T. P., Simpson, M. J., & Lee, K. K. (2011). Does sea-level rise have an impact on saltwater intrusion? *Advances in Water Resources, 34*(10), 1283–1291.

Chen, T. T., Hung, Y. C., Hsueh, M. W., Yeh, Y. H., & Weng, K. W. (2018). Evaluating the application of electrical resistivity tomography for investigating seawater intrusion. *Electronics (Switzerland), 7*(7).

Christelis, V., & Mantoglou, A. (2019). Pumping optimization of coastal aquifers using seawater intrusion models of variable-fidelity and evolutionary algorithms. *Water Resources Management, 33*(2), 555–568.

Chun, J. A., Lim, C., Kim, D., & Kim, J. S. (2018). Assessing impacts of climate change and sea-level rise on seawater intrusion in a coastal aquifer. *Water (Switzerland), 10*(4), 1–11.

Custodio, E. (1987). Effects of human activities on salt–fresh water relationships in coastal aquifers. In E. Custodio, & G. A. Bruggeman (Eds.), *Studies and reports in hydrology: Groundwater problems in coastal areas* (pp. 97–117). Paris: UNESCO, [Chapter 4].

Dagan, G., & Bear, J. (1968). Solving the problem of local interface upconing in a coastal aquifer by the method of small perturbations. *Journal of Hydraulic Research, 6*(1), 15–44.

Darwish, T., Atallah, T., El Moujabber, M., & Khatib, N. (2005). Salinity evolution and crop response to secondary soil salinity in two agro-climatic zones in Lebanon. *Agricultural Water Management, 78*(1–2), 152–164.

Dayal, D., Swain, S., Gautam, A. K., Palmate, S. S., Pandey, A., & Mishra, S. K. (2019). *Development of ARIMA model for monthly rainfall forecasting over an Indian River Basin. World environmental and water resources congress 2019: Watershed management, irrigation and drainage, and water resources planning and management* (pp. 264−271). American Society of Civil Engineers.

Delsman, J. R., Hu-A-Ng, K. R. M., Vos, P. C., De Louw, P. G. B., Oude Essink, G. H. P., Stuyfzand, P. J., & Bierkens, M. F. P. (2014). Paleo-modeling of coastal saltwater intrusion during the Holocene: An application to the Netherlands. *Hydrology and Earth System Sciences, 18*(10), 3891−3905. Available from https://doi.org/10.5194/hess-18-3891-2014.

Dentz, M., Tartakovsky, D. M., Abarca, E., Guadagnini, A., Sanchez-Vila, X., & Carrera, J. (2006). Variable-density flow in porous media. *Journal of Fluid Mechanics, 561*, 209−235.

Dhami, B., Himanshu, S. K., Pandey, A., & Gautam, A. K. (2018). Evaluation of the SWAT model for water balance study of a mountainous snowfed river basin of Nepal. *Environmental Earth Sciences, 77*(1).

Diersch, H.-J., Prochnow, D., & Thiele, M. (1984). Finite-element analysis of dispersion-affected saltwater upconing below a pumping well. *Applied Mathematical Modelling, 8*(5), 305−312.

Dror, I., Berkowitz, B., & Gorelick, S. M. (2004). Effects of air injection on flow through porous media: Observations and analyses of laboratory-scale processes. *Water Resources Research, 40*(9).

El-Fadel, M., Deeb, T., Alameddine, I., Zurayk, R., & Chaaban, J. (2018). Impact of groundwater salinity on agricultural productivity with climate change implications. *International Journal of Sustainable Development and Planning, 13*(3), 445−456.

El-Kaliouby, H., & Abdalla, O. (2015). Application of time-domain electromagnetic method in mapping saltwater intrusion of a coastal alluvial aquifer, North Oman. *Journal of Applied Geophysics, 115*, 59−64.

Ercan, A., Bin Mohamad, M. F., & Kavvas, M. L. (2013). The impact of climate change on sea level rise at Peninsular Malaysia and Sabah-Sarawak. *Hydrological Processes, 27*(3), 367−377.

Fahs, M., Koohbor, B., Belfort, B., Ataie-Ashtiani, B., Simmons, C. T., Younes, A., & Ackerer, P. (2018). A Generalized semi-analytical solution for the dispersive Henry problem: Effect of stratification and anisotropy on SI. *Water (Switzerland), 10*(2). Available from https://doi.org/10.3390/w10020230.

Fashae, O. A., & Onafeso, O. D. (2011). Impact of climate change on sea level rise in Lagos, Nigeria. *International Journal of Remote Sensing, 32*(24), 9811−9819.

García-Menéndez, O., Morell, I., Ballesteros, B. J., Renau-Pruñonosa, A., Renau-Llorens, A., & Esteller, M. V. (2016). Spatial characterization of the seawater upconing process in a coastal Mediterranean aquifer (Plana de Castellón, Spain): Evolution and controls. *Environmental Earth Sciences, 75*(9), 1−18.

Goswami, R. R., & Clement, T. P. (2007). Laboratory-scale investigation of saltwater intrusion dynamics. *Water Resources Research, 43*(4), 1−11.

Guo, Q., Huang, J., Zhou, Z., & Wang, J. (2019). Experiment and numerical simulation of seawater intrusion under the influences of tidal fluctuation and groundwater exploitation in coastal multilayered aquifers. *Geofluids, 2019*, 1−17, 2316271.

Guo, Q., Zhang, Y., Zhou, Z., & Hu, Z. (2020). Transport of contamination under the influence of sea level rise in coastal heterogeneous aquifer. *Sustainability, 12*(23), 9838.

Guo, Q., Zhao, Y., Hu, Z., & Li, M. (2021). "Contamination transport in the coastal unconfined aquifer under the influences of seawater intrusion and inland freshwater recharge—Laboratory experiments and numerical simulations. *International Journal of Environmental Research and Public Health, 18*(2), 762.

Hawkes, P. J. (2013). *Sea level change.* Encyclopedia of earth sciences series (pp. 895−900). Cambridge: Cambridge University Press.

Henry, H. R. (1964). Effects of dispersion on salt encroachment in coastal aquifers. *United States geological survey water-supply paper, 1613-C* (pp. C71–C84).

Himanshu, S. K., Pandey, A., Yadav, B., & Gupta, A. (2019). Evaluation of best management practices for sediment and nutrient loss control using SWAT model. *Soil and Tillage Research, 192*, 42–58.

Hung, Y. C., Yeh, Y. H., Hsu, Y. Y., & Weng, K. W. (2018). Investigation of seawater intrusion using resistivity tomography: A case study in Kinmen. In: *Proceedings of 4th IEEE international conference on applied system innovation (ICASI) 2018* (pp. 472–475).

Hussain, M. S., Javadi, A. A., & Sherif, M. M. (2015). Three dimensional simulation of seawater intrusion in a regional coastal aquifer in UAE. *Procedia Engineering, 119*, 1153–1160.

Jakovovic, D., Werner, A. D., de Louw, P. G. B., Post, V. E. A., & Morgan, L. K. (2016). Saltwater upconing zone of influence. *Advances in Water Resources, 94*, 75–86.

Jakovovic, D., Werner, A. D., & Simmons, C. T. (2011). Numerical modelling of saltwater upconing: Comparison with experimental laboratory observations. *Journal of Hydrology, 402* (3–4), 261–273.

Kacimov, A. R., & Obnosov, Y. V. (2016). Size and shape of steady seawater intrusion and sharp-interface wedge: The Polubarinova-Kochina analytical solution to the dam problem revisited. *Journal of Hydrologic Engineering, 21*(8), 1–6.

Kalaoun, O., Jazar, M., & Al Bitar, A. (2018). Assessing the contribution of demographic growth, climate change, and the refugee crisis on seawater intrusion in the Tripoli aquifer. *Water (Switzerland), 10*(8).

Kaliraj, S., Chandrasekar, N., Peter, T. S., Selvakumar, S., & Magesh, N. S. (2015). Mapping of coastal aquifer vulnerable zone in the south west coast of Kanyakumari, South India, using GIS-based DRASTIC model. *Environmental Monitoring and Assessment, 187*(1).

Kazakis, N., Pavlou, A., Vargemezis, G., Voudouris, K. S., Soulios, G., Pliakas, F., & Tsokas, G. (2016). Seawater intrusion mapping using electrical resistivity tomography and hydrochemical data. An application in the coastal area of eastern Thermaikos Gulf, Greece. *Science of the Total Environment, 543*, 373–387.

Kazakis, N., Spiliotis, M., Voudouris, K., Pliakas, F. K., & Papadopoulos, B. (2018). A fuzzy multi-criteria categorization of the GALDIT method to assess seawater intrusion vulnerability of coastal aquifers. *Science of the Total Environment, 621*, 524–534.

Khomine, A., János, S., & Balázs, K. (2011). *Potential solutions in prevention of saltwater intrusion: A modelling approach. Advances in the research of aquatic environment* (pp. 251–257). Berlin, Heidelberg: Springer.

Kim, I. H., & Yang, J. S. (2018). Prioritizing countermeasures for reducing seawater-intrusion area by considering regional characteristics using SEAWAT and a multicriteria decision-making method. *Hydrological Processes, 32*(25), 3741–3757.

Kumar, K. S. A., Priju, C. P., & Prasad, N. B. N. (2015). Study on saline water intrusion into the shallow coastal aquifers of Periyar River basin, Kerala using hydrochemical and electrical resistivity methods. *Aquatic Procedia, 4*, 32–40.

Kume, T., Umetsu, C., & Palanisami, K. (2009). Impact of the December 2004 tsunami on soil, groundwater and vegetation in the Nagapattinam district, India. *Journal of Environmental Management, 90*(10), 3147–3154.

Kura, N. U., Ramli, M. F., Ibrahim, S., Sulaiman, W. N. A., Zaudi, M. A., & Aris, A. Z. (2014). A preliminary appraisal of the effect of pumping on seawater intrusion and upconing in a small tropical island using 2D resistivity technique. *Scientific World Journal, 2014*, 1–10.

Laabidi, E., & Bouhlila, R. (2016). Reactive Henry problem: Effect of calcite dissolution on SI. *Environmental Earth Sciences, 75*(8), 1–15. Available from https://doi.org/10.1007/s12665-016-5487-7.

Lathashri, U. A., & Mahesha, A. (2015). Simulation of saltwater intrusion in a coastal aquifer in Karnataka, India. *Aquatic Procedia, 4*, 700−705.

Lathashri, U. A., & Mahesha, A. (2016). Predictive simulation of seawater intrusion in a tropical coastal aquifer. *Journal of Environmental Engineering, 142*(12)D4015001.

Liu, J., & Tokunaga, T. (2019). Future risks of tsunami-induced SI into unconfined coastal aquifers: Insights from numerical simulations at Nii-jima Island, Japan. *Water Resources Research, 55*(12), 10082−10104. Available from https://doi.org/10.1029/2019WR025386.

Liu, S., Tao, A., Dai, C., Tan, B., Shen, H., Zhong, G., Lou, S., Chalov, S., & Chalov, R. (2017). Experimental study of tidal effects on coastal groundwater and pollutant migration. *Water, Air, and Soil Pollution, 228*(4). Available from https://doi.org/10.1007/s11270-017-3326-4.

Llopis-Albert, C., Merigó, J. M., & Xu, Y. (2016). A coupled stochastic inverse/sharp interface seawater intrusion approach for coastal aquifers under groundwater parameter uncertainty. *Journal of Hydrology, 540*, 774−783.

Luyun, R., Momii, K., & Nakagawa, K. (2009). Laboratory-scale saltwater behavior due to subsurface cutoff wall. *Journal of Hydrology, 377*(3−4), 227−236.

Luyun, R., Momii, K., & Nakagawa, K. (2011). Effects of recharge wells and flow barriers on seawater intrusion. *Ground Water, 49*(2), 239−249.

Mansour, A. Y. S., Baba, A., Gunduz, O., Şimşek, C., Elçi, A., Murathan, A., & Sözbilir, H. (2017). Modeling of seawater intrusion in a coastal aquifer of Karaburun Peninsula, western Turkey. *Environmental Earth Sciences, 76*(22).

Mao, X., Enot, P., Barry, D. A., Li, L., Binley, A., & Jeng, D. S. (2006). Tidal influence on behaviour of a coastal aquifer adjacent to a low-relief estuary. *Journal of Hydrology, 327*(1−2), 110−127.

Martínez-Moreno, F. J., Monteiro-Santos, F. A., Bernardo, I., Farzamian, M., Nascimento, C., Fernandes, J., Casal, B., & Ribeiro, J. A. (2017). Identifying seawater intrusion in coastal areas by means of 1D and quasi-2D joint inversion of TDEM and VES data. *Journal of Hydrology, 552*, 609−619.

Mastrocicco, B., Colombani, V., & Ruberti. (2019). Modelling actual and future seawater intrusion in the Variconi coastal wetland (Italy) due to climate and landscape changes. *Water, 11*(7), 1502.

Maurya, P., Kumari, R., & Mukherjee, S. (2019). Hydrochemistry in integration with stable isotopes ($\delta 18O$ and δD) to assess seawater intrusion in coastal aquifers of Kachchh district, Gujarat, India. *Journal of Geochemical Exploration, 196*, 42−56.

McLeman, R. (2019). International migration and climate adaptation in an era of hardening borders. *Nature Climate Change, 9*(12), 911−918.

Mehdizadeh, S. S., Karamalipour, S. E., & Asoodeh, R. (2017). Sea level rise effect on seawater intrusion into layered coastal aquifers (simulation using dispersive and sharp-interface approaches). *Ocean and Coastal Management, 138*, 11−18.

Mehdizadeh, S. S., Werner, A. D., Vafaie, F., & Badaruddin, S. (2014). Vertical leakage in sharp-interface seawater intrusion models of layered coastal aquifers. *Journal of Hydrology, 519*, 1097−1107.

Millero, F. J., Feistel, R., Wright, D. G., & McDougall, T. J. (2008). The composition of Standard Seawater and the definition of the Reference-Composition Salinity Scale. *Deep-Sea Research Part I: Oceanographic Research Papers, 55*(1), 50−72.

Mimura, N. (2013). Sea-level rise caused by climate change and its implications for society. *Proceedings of the Japan Academy Series B: Physical and Biological Sciences, 89*(7), 281−301.

Mohanty, A. K., & Rao, V. V. S. G. (2019). Hydrogeochemical, seawater intrusion and oxygen isotope studies on a coastal region in the Puri District of Odisha, India. *Catena, 172*(September 2018), 558−571.

Momejian, N., Abou Najm, M., Alameddine, I., & El-Fadel, M. (2019a). Can groundwater vulnerability models assess seawater intrusion? *Environmental Impact Assessment Review, 75,* 13−26.

Momejian, N., Abou Najm, M., Alameddine, I., & El-Fadel, M. (2019b). Groundwater vulnerability modeling to assess seawater intrusion: A methodological comparison with geospatial interpolation. *Water Resources Management, 33*(3), 1039−1052.

Najib, S., Fadili, A., Mehdi, K., Riss, J., & Makan, A. (2017). Contribution of hydrochemical and geoelectrical approaches to investigate salinization process and seawater intrusion in the coastal aquifers of Chaouia, Morocco. *Journal of Contaminant Hydrology, 198,* 24−36.

Nimmo, J. R. (2006). Unsaturated zone flow processes. *Encyclopedia of hydrological sciences.* Available from https://doi.org/10.1002/0470848944.hsa161.

Nofal, E. R., Amer, M. A., El-Didy, S. M., & Fekry, A. M. (2015). Delineation and modeling of seawater intrusion into the Nile Delta Aquifer: A new perspective. *Water Science, 29*(2), 156−166.

Padilla, F., Benavente, J., & Cruz-Sanjuliân, J. (1997). Numerical simulation of the influence of management alternatives of a projected reservoir on a small alluvial aquifer affected by seawater intrusion (Almuñécar, Spain). *Environmental Geology, 33*(1), 72−80.

Pandey, B. K., & Khare, D. (2018). Identification of trend in long term precipitation and reference evapotranspiration over Narmada river basin (India). *Global and Planetary Change, 161,* 172−182.

Paniconi, C., Khlaifi, I., Lecca, G., Giacomelli, A., & Tarhouni, J. (2001). A modelling study of seawater intrusion in the Korba coastal plain, Tunisia. *Physics and Chemistry of the Earth, Part B: Hydrology, Oceans and Atmosphere, 26*(4), 345−351.

Paster, A., Dagan, G., & Guttman, J. (2006). The salt-water body in the Northern part of Yarkon-Taninim aquifer: Field data analysis, conceptual model and prediction. *Journal of Hydrology, 323*(1−4), 154−167.

Price, R. M., Top, Z., Happell, J. D., & Swart, P. K. (2003). Use of tritium and helium to define groundwater flow conditions in Everglades National Park. *Water Resources Research, 39*(9), 1−12.

Qi, S. Z., & Qiu, Q. L. (2011). Environmental hazard from saltwater intrusion in the Laizhou Gulf, Shandong Province of China. *Natural Hazards, 56*(3), 563−566.

Qureshi, M. E., Qureshi, S. E., Bajracharya, K., & Kirby, M. (2008). Integrated biophysical and economic modelling framework to assess impacts of alternative groundwater management options. *Water Resources Management, 22*(3), 321−341.

Rasmussen, P., Sonnenborg, T. O., Goncear, G., & Hinsby, K. (2013). Assessing impacts of climate change, sea level rise, and drainage canals on saltwater intrusion to coastal aquifer. *Hydrology and Earth System Sciences, 17*(1), 421−443.

Robinson, G., Ahmed, A. A., & Hamill, G. A. (2016). Experimental saltwater intrusion in coastal aquifers using automated image analysis: Applications to homogeneous aquifers. *Journal of Hydrology, 538,* 304−313. Available from https://doi.org/10.1016/j.jhydrol.2016.04.017.

Safi, A., Rachid, G., El-Fadel, M., Doummar, J., Abou Najm, M., & Alameddine, I. (2018). Synergy of climate change and local pressures on saltwater intrusion in coastal urban areas: Effective adaptation for policy planning. *Water International, 43*(2), 145−164.

Segol, G. (1994). *Classic groundwater simulations: Proving and improving numerical models.* Prentice Hall.

Simpson, M. J., & Clement, T. P. (2004). Improving the worthiness of the Henry problem as a benchmark for density-dependent groundwater flow models. *Water Resources Research, 40*(1).

Simpson, T. B., Holman, I. P., & Rushton, K. R. (2011). Understanding and modelling spatial drain-aquifer interactions in a low-lying coastal aquifer—The Thurne catchment, Norfolk, UK. *Hydrological Processes, 25*(4), 580−592.

Singh, A. (2014). Optimization modelling for seawater intrusion management. *Journal of Hydrology, 508*, 43–52.

Sophiya, M. S., & Syed, T. H. (2013). Assessment of vulnerability to seawater intrusion and potential remediation measures for coastal aquifers: A case study from eastern India. *Environmental Earth Sciences, 70*(3), 1197–1209.

Sukhija, B. S., Varma, V. N., Nagabhushanam, P., & Reddy, D. V. (1996). Differentiation of palaeomarine and modern seawater intruded salinities in coastal groundwaters (of Karaikal and Tanjavur, India) based on inorganic chemistry, organic biomarker fingerprints and radio-carbon dating. *Journal of Hydrology, 174*(1–2), 173–201.

Swain, S. (2017). Hydrological modeling through soil and water assessment toolin a climate change perspective – A brief review. In: *2017 2nd international conference for convergence in technology (I2CT)* (pp. 358–361). IEEE.

Swain, S., Dayal, D., Pandey, A., & Mishra, S. K. (2019a). *Trend analysis of precipitation and temperature for Bilaspur district, Chhattisgarh, India. World environmental and water resources congress 2019: Groundwater, sustainability, hydro-climate/climate change, and environmental engineering* (pp. 193–204). Reston, VA: American Society of Civil Engineers.

Swain, S., Mishra, S. K., & Pandey, A. (2021a). A detailed assessment of meteorological drought characteristics using simplified rainfall index over Narmada River Basin, India. *Environmental Earth Sciences, 80*, 221. Available from https://doi.org/10.1007/s12665-021-09523-8.

Swain, S., Mishra, S. K., & Pandey, A. (2019b). Spatiotemporal characterization of meteorological droughts and its linkage with environmental flow conditions. In: *AGU fall meeting 2019*. AGU.

Swain, S., Mishra, S. K., & Pandey, A. (2020). Assessment of meteorological droughts over Hoshangabad district, India. In: *IOP conference series: Earth and environmental science* (Vol. 491, No. 1, p. 012012). IOP Publishing.

Swain, S., Mishra, S. K., Pandey, A., & Dayal, D. (2021b). *Identification of meteorological extreme years over central division of Odisha using an index-based approach. Hydrological extremes* (pp. 161–174). Cham: Springer. Available from https://doi.org/10.1007/978-3-030-59148-9_12.

Swain, S., Nandi, S., & Patel, P. (2018a). *Development of an ARIMA model for monthly rainfall forecasting over Khordha district, Odisha, India. Recent findings in intelligent computing technique* (pp. 325–331). Singapore: Springer.

Swain, S., Patel, P., & Nandi, S. (2017a). A multiple linear regression model for precipitation forecasting over Cuttack district, Odisha, India. In: *2017 2nd international conference for convergence in technology (I2CT)* (pp. 355–357). IEEE.

Swain, S., Patel, P., & Nandi, S. (2017b). Application of SPI, EDI and PNPI using MSWEP precipitation data over Marathwada, India. In: *2017 IEEE international geoscience and remote sensing symposium (IGARSS)* (pp. 5505–5507). IEEE.

Swain, S., Verma, M. K., & Verma, M. K. (2018b). *Streamflow estimation using SWAT model over Seonath river basin, Chhattisgarh, India. Hydrologic modeling* (pp. 659–665). Singapore: Springer.

Takahashi, M., Momii, K., & Luyun, R. (2018). Laboratory scale investigation of dispersion effects on saltwater movement due to cutoff wall installation. In: *E3S web of conferences* (Vol. 54, pp. 1–6). https://doi.org/10.1051/e3sconf/20185400038.

Trabelsi, F., Mammou, A., Ben., Tarhouni, J., Piga, C., & Ranieri, G. (2013). Delineation of saltwater intrusion zones using the time domain electromagnetic method: The Nabeul-Hammamet coastal aquifer case study (NE Tunisia). *Hydrological Processes, 27*(14), 2004–2020.

van Engelen, J., Bierkens, M. F. P., Delsman, J. R., & Oude Essink, G. H. P. (2021). Factors determining the natural fresh-salt groundwater distribution in deltas. *Water Resources Research, 57*(1). Available from https://doi.org/10.1029/2020WR027290.

van Roosmalen, L., Christensen, B. S., & Sonnenborg, T. O. (2007). Regional differences in climate change impacts on groundwater and stream discharge in Denmark. *Vadose Zone Journal, 6*(3), 554−571.

Violette, S., Boulicot, G., & Gorelick, S. M. (2009). Tsunami-induced groundwater salinization in southeastern India. *Comptes Rendus − Geoscience, 341*(4), 339−346.

Voss, C. I., Simmons, C. T., & Robinson, N. I. (2010). Three-dimensional benchmark for variable-density flow and transport simulation: Matching semi-analytic stability modes for steady unstable convection in an inclined porous box. *Hydrogeology Journal, 18*(1), 5−23.

Werner, A. D. (2017). Correction factor to account for dispersion in sharp-interface models of terrestrial freshwater lenses and active seawater intrusion. *Advances in Water Resources, 102,* 45−52.

Werner, A. D., Alcoe, D. W., Ordens, C. M., Hutson, J. L., Ward, J. D., & Simmons, C. T. (2011). Current practice and future challenges in coastal aquifer management: Flux-based and trigger-level approaches with application to an Australian case study. *Water Resources Management, 25*(7), 1831−1853.

Werner, A. D., & Simmons, C. T. (2009). Impact of sea-level rise on sea water intrusion in coastal aquifers. *Groundwater, 47*(2), 197−204.

Werner, A. D., Bakker, M., Post, V. E. A., Vandenbohede, A., Lu, C., Ataie-Ashtiani, B., Simmons, C. T., & Barry, D. A. (2013). Seawater intrusion processes, investigation and management: Recent advances and future challenges. *Advances in Water Resources, 51,* 3−26.

Wrathall, D. J., Mueller, V., Clark, P. U., Bell, A., Oppenheimer, M., Hauer, M., Kulp, S., Gilmore, F., Adams, H., Kopp, R., Abel, K., Call, M., Chen, J., deSherbinin, A., Fussell, E., Hay, C., Jones, B., Magliocca, N., Marino, E., ... Warner, K. (2019). Meeting the looming policy challenge of sea-level change and human migration. *Nature Climate Change, 9*(12), 898−901.

Xu, Z., Hu, B. X., Xu, Z., & Wu, X. (2019). Numerical study of groundwater flow cycling controlled by seawater/freshwater interaction in Woodville Karst Plain. *Journal of Hydrology, 579,* 1−12.

Younes, A., & Fahs, M. (2015). Extension of the Henry semi-analytical solution for saltwater intrusion in stratified domains. *Computational Geosciences, 19*(6), 1207−1217.

Yu, H., Ma, T., Du, Y., & Chen, L. (2019). Genesis of formation water in the northern sedimentary basin of South China Sea: Clues from hydrochemistry and stable isotopes (D, 18O, 37Cl and 81Br). *Journal of Geochemical Exploration, 196,* 57−65.

Zhang, Y., Li, L., Erler, D. V., Santos, I., & Lockington, D. (2016). Effects of alongshore morphology on groundwater flow and solute transport in a nearshore aquifer. *Water Resources Research, 52*(2), 990−1008. Available from https://doi.org/10.1002/2015WR017420.

Ziadi, A., Hariga, N. T., & Tarhouni, J. (2017). Use of time-domain electromagnetic (TDEM) method to investigate seawater intrusion in the Lebna coastal aquifer of eastern Cap Bon, Tunisia. *Arabian Journal of Geosciences, 10*(22).

Prioritization of erosion prone areas based on a sediment yield index for conservation treatments: A case study of the upper Tapi River basin

Santosh S. Palmate[1], Kumar Amrit[2], Vikas G. Jadhao[3,4], Deen Dayal[4] and Sushil Kumar Himanshu[5]

[1]Texas A&M AgriLife Research Center at El Paso, El Paso, TX, United States, [2]CSIR-National Environmental Engineering Research Institute, Mumbai, India, [3]Krishi Vigyan Kendra (KVK), Buldana, India, [4]Department of Water Resource Development and Management, Indian Institute of Technology, Roorkee, India, [5]Department of Food, Agriculture and Bioresources, School of Environment, Resources and Development, Asian Institute of Technology, Pathum Thani, Thailand

15.1 Introduction

Prioritization of the basin areas for implementing the conservation measures is time consuming and expensive. Hence, selecting a suitable approach to identify critical basin areas in response to hydrological units is an important step, which would be helpful to manage the existing natural resources more efficiently (Palmate et al., 2021; Pandey and Palmate, 2019). In that case, a criterion of maximum sediment yield of a basin can be used to determine priority for conservation measures. Practically, sedimentation is not of the same rate from each and every unit of a basin. Also, the basin areas contributing a maximum sediment yield should be preferred for immediate action of conservation treatments on a priority basis (Himanshu et al., 2019). Hence, prioritization of hydrological regions needs to be performed. Identifying critical areas is particularly important for prioritization, which could be an issue in data-scarce regions. In this case, the advanced remote sensing and Geographic Information System (GIS) techniques are useful to perform spatial analysis for critical area identification and prioritization (Pandey & Palmate, 2019).

Conventional approaches used for soil erosion and sediment yield estimation are time consuming in the large river basins and even cannot be applied for the silt detention reservoirs. The conventional methods, including the Musgrave equation, the Universal Soil Loss Equation, and Sediment Yield

291

Advances in Remediation Techniques for Polluted Soils and Groundwater. DOI: https://doi.org/10.1016/B978-0-12-823830-1.00019-5

Prediction equations, are generally worked out for a specific set of conditions (Pandey et al., 2016). Given that the erosion factors, thereby, restrict to apply it for other conditions that may not have data availability. Therefore a sediment yield index (SYI) method of universal applicability has been developed based on actual spatial data information. The use of SYI proposed by the All India Soil and Land Use Survey (AISLUS), Government of India, is a widely known criterion prioritizing the basin areas under soil erosion and sedimentation (AISLUS, 1991). In the SYI approach, the soil erosivity factor (weightage value) and transportability of the detached soil (delivery ration value) are taken into account for conceptualizing the sediment delivery into the river water.

Generally, the reservoirs are designed in such a way that they can accumulate the water flows from upstream areas and regulate downstream for many purposes. With the water flows, sediment also gets collected in a large amount over time (Amrit et al., 2019; Himanshu et al., 2017; Kumar, 2021). Hence, the purpose of the reservoir is also designated for the sedimentation rate estimation. However, due to unpredicted climatic events, the actual sedimentation rate always exceeds the presumed estimates used in the reservoir design. Such cases are observed in many areas of India, for instance, the major projects of Hirakud, Maithon, Panchet, Tungabhadra, Ramganga, and Nizamsagar (Bali & Karale, 1977). The difference between the estimated and actual sedimentation rates expresses the need for appropriate conservation management in the upstream basin areas to reduce soil erosion and siltation (Mauget, 2021; Palmate & Pandey, 2021). With this consideration, the Government of India initiated a scheme of soil conservation in several river projects. Since then, several reservoir projects are designed and completed for Indian River basins, which have not yet been studied for critical area prioritization to sustain the reservoirs' useful life.

The approach used in this chapter was tested in a semiarid drainage basin of India, the upper Tapi River basin (UTRB), considered up to the Hatnur dam. The accumulation of sediment is being increased and deposited in the drainage and eventually in the Hatnur dam reservoir (Chandra et al., 2014, 2016; Munoth & Goyal, 2020). This siltation in the reservoir is largely owing to poor conservation practices, barely distributed vegetation, steep slopes, and a generally dry climate with an erratic monsoon rainfall, resulting in high rates of erosion that destroy the productive value of the land (Amrit et al., 2018; Pandey and Palmate, 2018; Ramani et al., 2021; Resmi et al., 2020). Hence, it is a need of an hour to lower this siltation with a further increase in runoff. Identifying critical areas is the first step, and prioritizing critical regions is the second important step to implement the feasible conservation practices necessary for the sustainable development of such a fragile environment. Therefore in this chapter, an attempt has been made to prioritize the basin areas using the SYI approach to implement conservation measures/treatments considering soil erosion and siltation.

15.2   Study area

The Tapi River is the second-largest west-flowing interstate river of Deccan plateau, which covers the parts of Madhya Pradesh and Maharashtra majorly, and some part of Gujrat at the end before joining it to the Arabian Sea. This river flows more or less over the plains of Vidharbha, Khandesh, and Gujarat and is bounded on the three sides by the hill ranges, along with its tributaries. The UTRB area up to the Hatnur dam is about 29,791 km^2. The confluence of Purna with the main Tapi is observed at the Hatnur reservoir. The main Tapi river and its tributaries are monsoon-driven and transport sediment loads from the upstream basin part and deposit them into the Hatnur reservoir (Ramani et al., 2021; Sharma et al., 2019a). The Tapi river originates near Multai in Betul district. Elevation of the study area varies from 206 to 1171 m (Fig. 15.1). The annual rainfall for the UTRB is 935.5 mm. About 90% of the total rainfall receives during the monsoon months, out of which 50% receives during July and August. The temperature of the basin varies from 12°C to 40°C. There are four distinct seasons, that is, winter, summer, monsoon, and postmonsoon, in a water year from June to succeeding May. The agricultural land cover area is dominant accounting for about 66% followed by forest covers (about 25%). Water bodies in this region are available as reservoirs, lakes, barrages, and weirs. The basin consists of deep black soils. The hot semiarid ecoregion with shallow and medium black soils is the major agroecological zone of the study basin (MoWR, 2012).

15.3   Data used

Spatial datasets of digital elevation model (DEM), land use land cover (LULC), soil, and slope obtained from multiple online sources were used in this study. These datasets were further processed in the ArcGIS software environment and extracted important spatial data information and statistics.

15.3.1   Digital elevation model data

The shuttle radar topography mission DEM data of 30 m spatial resolution were obtained from the Earth Explorer website (https://earthexplorer.usgs.gov/). The DEM data were processed to delineate the basin boundary, division into subbasins, and prepare a slope map of the study area (Fig. 15.1). Higher the density of the drainage network at the southeast side of the basin, higher the stream flows contributing more sediment transport (Sargaonkar,

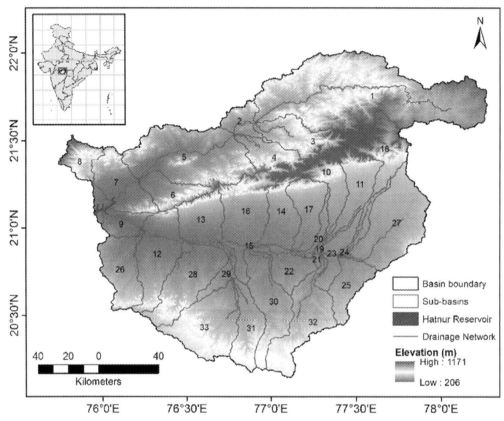

FIGURE 15.1

Location map of the upper Tapi River basin, up to the Hatnur dam reservoir, with delineated subbasins and elevations, obtained using SRTM DEM. *DEM*, digital elevation model; *SRTM*, shuttle radar topography mission.

2006; Sharma et al., 2019a). As per present connectivity of drainage within the study area, the UTRB area was delineated into 33 subbasins. These subbasins are viable for the implementation of conservation measures and other basin development programs.

15.3.2   Land use land cover data

The LULC map of the year 2005 was obtained from a research data product by Roy et al. (2016), from their data website (https://daac.ornl.gov/VEGETATION/guides/Decadal_LULC_India.html; Fig. 15.2). The extracted LULC map shows 10 different classes that cover the total UTRB area. Table 15.1 shows that cropland is a major LULC class covering about 58.9%, and the deciduous broadleaf forest is the second majorly distributed LULC class, covering about 28.3% of the study area.

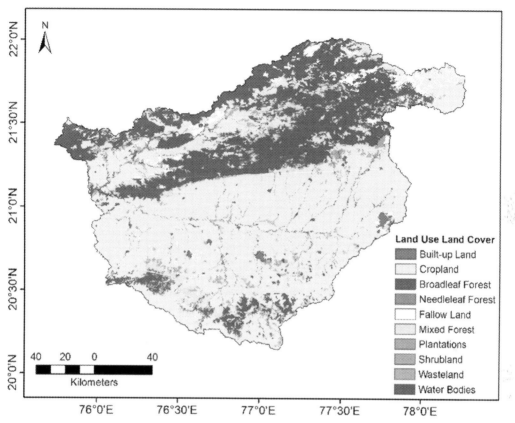

FIGURE 15.2

Land use land cover map of the year 2005. *Source: Roy et al. (2016); https://daac.ornl.gov/VEGETATION/guides/Decadal_LULC_India.html.*

Table 15.1 Area-wise distribution of land use land cover classes in the study area.

Sr. no.	Land use land cover class name	Area (km²)	Area (%)
1	Deciduous broadleaf forest	8437.6	28.3
2	Cropland	17,551.6	58.9
3	Built-up land	168.8	0.6
4	Mixed forest	1080.1	3.6
5	Shrubland	597.7	2.0
6	Fallow land	1136.6	3.8
7	Wasteland	131.4	0.4
8	Water bodies	555.7	1.9
9	Plantations	130.8	0.4
10	Deciduous needleleaf forest	0.6	0.0
	Total	29,790.8	100.0

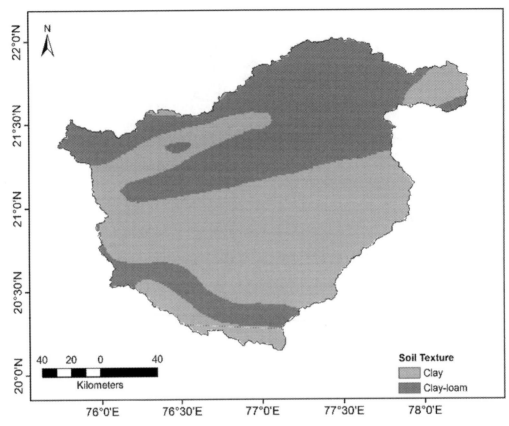

FIGURE 15.3

Soil texture map of the upper Tapi River basin. *Source: HWSD Soil Database http://www.fao.org/soils-portal/data-hub/soil-maps-and-databases/harmonized-world-soil-database-v12/en/.*

15.3.3 Soil data

The soil textures of the study area and their properties were acquired from the Harmonized World Soil Database website (http://www.fao.org/soils-portal/data-hub/soil-maps-and-databases/harmonized-world-soil-database-v12/en/). The study area mainly covers clay and clay loam soil textures at different depths categories (Fig. 15.3). Many factors such as soil texture, effective depth, and soil type need to be considered in erosion intensity mapping. The clay soil covers a major part of the study area of about 48.8% at moderate shallow and 7.3% at deep soil depth categories (Table 15.2). However, clay-loam soil covers about 27.1% at moderate and 16.8% at shallow soil depth categories.

15.3.4 Slope data

The slope map of the study area was generated in the ArcGIS environment using the DEM data (Fig. 15.4). Percent slope is further categorized into five

Table 15.2 Area-wise distribution of soil textural classes and their depth categories in the study area.

Soil depth (mm)	Depth category	Soil texture	Area (km²)	Area (%)
910	Moderate	Clay-loam	8162.4	27.1
630	Shallow	Clay-loam	5068.2	16.8
910	Moderate shallow	Clay	14,702.3	48.8
1000	Deep	Clay	2188.6	7.3
		Total	30,121.5	100.0

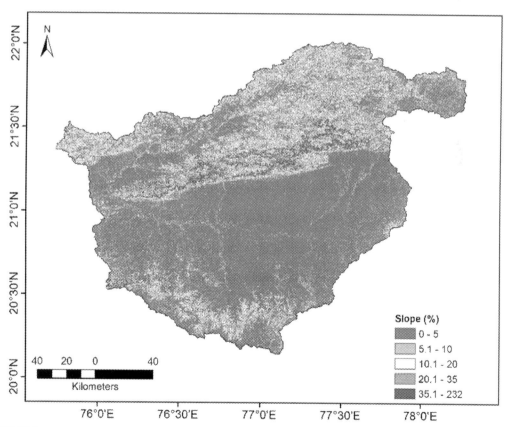

FIGURE 15.4

Percent slope map of the upper Tapi River basin. *Source: Prepared by using SRTM DEM data of 30 m spatial resolution; https://earthexplorer.usgs.gov/.*

Table 15.3 Area-wise distribution of slope classes and their percentages in the study area.

Slope category (%)	Classification	Area (km^2)	Area (%)
0–5	Gentle	18,112.1	61.1
5–10	Moderate	5054.4	17.0
10–20	Moderately steep	3272.4	11.0
20–35	Steep	2167.5	7.3
>35	Very steep	1053.3	3.6
	Total	29,659.7	100.0

different classes that helped to recognize the major variations in the land slopes, which also influence the soil erosion processes within the study area. This analysis shows that the majority of the study area (about 61.1%) is classified as the gentle land slope category, while about 17%, 11%, 7.3%, and 3.6% of the total study areas are classified as moderate, moderately steep, steep, and very steep slope category, respectively (Table 15.3). These percent slopes help to determine the erosion susceptibility of the soil in the study area.

15.4   Methodology

Soil and sediment flow from any area generally occur due to interaction between climatic variables and land surface characteristics. Rainfall is considered an important climate variable that pronounces soil and sediment flow with its more intensity and frequency (Kumar et al., 2019; Amrit et al., 2019; Sharma et al., 2019b). Also, the land characteristics that influence the flows are slope, land use pattern, soil, topography, and contemporary status of erosion and deposition. The SYI approach is based only on land characteristics crucial for determining erosion and sedimentation status in a river basin. Hence, to estimate the SYI value, spatial information of these land characteristics was obtained, processed, and utilized in this study.

15.4.1   Assigning delivery ratio

Sediment detachment, transportation, and deposition are basic phenomena that evidence sediment movement in an area over time. Rainfall acts as an agent for sediment detachment, and the detached sediment may be deposited in the same or an adjacent area after transporting some distance. The sediment deposition is of more significance in reservoirs than the detached sediment from the upstream region. The amount of sediment detached from the upstream region and deposited in the reservoir is defined as the delivery ratio. This delivery ratio depends on different basin characteristics such as

Table 15.4 Assigned delivery ratio to streams according to distance from the reservoir.

Nearest stream distance (km)	Delivery ratio
0−1	1.00
1.1−5	0.95
5.1−10	0.90
10.1−20	0.80
20.1−50	0.70
50.1−100	0.60
100.1−160	0.50

shape and size of the basin, drainage network, land slope, active stream density, detached sediment textural size, and distance from the reservoir. The delivery ratio is an important factor in sediment transportation through existing drainages (Beven et al., 2005). The precise value of the delivery ratio was obtained by considering aggregated effects of these characteristics that influence sediment delivery in the study area. The assigned delivery ratios according to the distance of a stream from the Hatnur reservoir are shown in Table 15.4. These delivery ratios cannot be set without referring to the position of a basin unit in the map. In this study, the delivery ratio value ranged between 0.5 and 1.

15.4.2 Mapping erosion intensity units

A composite erosion intensity map was generated using thematic maps of percent slope, soil, and land use land cover as shown in Figs. 15.1−15.4. This composite map was superimposed on the drainage map with subbasin boundary to extract erosion intensity basin units. These erosion units are further used to assign weightage that indicates the relative erosion intensity.

15.4.3 Assigning weights to basin units

An average weightage value was assigned to the considered basin unit areas for the composite erosion intensity mapping. These weights correspond to the different land characteristics and their aggregated effect on sediment detachment from the upstream region and deposition in the reservoir (Table 15.5). This study combined broadleaf forest, needleleaf forest, and mixed forest classes as one forest class. Water bodies class can have the only deposition, so excluded from the erosion weighing analysis. The low weightage value is given to less or no erosion-producing class, and the high weightage value is given to the more or significant erosion-producing class. It means weighting values are in proportion with the erosion/sediment yields.

Table 15.5 Key factors and their weightage values used for the preparation of a composite erosion intensity map.

Land use land cover class	Weightage	Slope class	Weightage	Soil depth class	Weightage
Forest	1	Gentle	1	Shallow	4
Plantations	2	Moderate	2	Moderate shallow	3
Shrubland	3	Moderately steep	3	Moderate shallow	2
Built-up land	4	Stepp	4	Deep	1
Wasteland	5	Very steep	5		
Cropland	6				
Fallow land	7				

It can also help judge the relative erosion in each basin unit of the study area and prepare a composite erosion intensity map.

A value of 10 is considered as an inertia factor representing a balance between erosion production from the upstream basin part and sediment deposition in the Hatnur dam reservoir. It is also considered as a standard reference for comparison. For a subbasin unit, addition and subtraction from this value indicate a collective effect of different attributed land characteristics in a basin. Any addition to the inertia factor indicates erosion production, and subtraction indicates deposition in proportion to the added factor.

15.4.4 Computation of sediment yield index

The analysis of SYI using different raster datasets was performed in the ArcGIS software environment, using the methodology developed by the AISLUS. The batch processing module of ArcGIS was used to estimate SYI for each subbasin. The SYI is mathematically expressed as follows:

$$SYI = \frac{\sum (A_i \times W_i \times D_i) \times 100}{A_w}, i = 1 \text{ to N} \tag{15.1}$$

where A_i = area of ith the basin unit (also called erosion intensity mapping unit), W_i = weightage value of ith mapping unit, D_i = adjusted delivery ratio assigned to a mapping unit, N = numbers of mapping units, and A_w = total area of a subbasin.

The methodology adopted in this study consists of spatial information based on slope map, LULC map, soil map, and their composite map generated using ArcGIS. All subbasins with their SYI values were further arranged in decreasing order to prioritize the conservation treatments. The priority classes were mainly categorized into very high, high, medium, low, and very low. In this case, specific land characteristics of a subbasin regulate the value of SYI.

Finally, the prioritization was performed based on the distribution of SYI values for all subbasins within the study area.

15.5 Results and discussion

In this study, major result findings follow the erosion intensity mapping and treatment prioritization of the basin areas.

15.5.1 Composite erosion intensity map

Erosion intensities were categorized into five classes, that is, negligible, low (or slight), moderate, severe, and very severe (Fig. 15.5). The analysis result showed that severe erosion occurs in 61.7% of the area, mainly at the

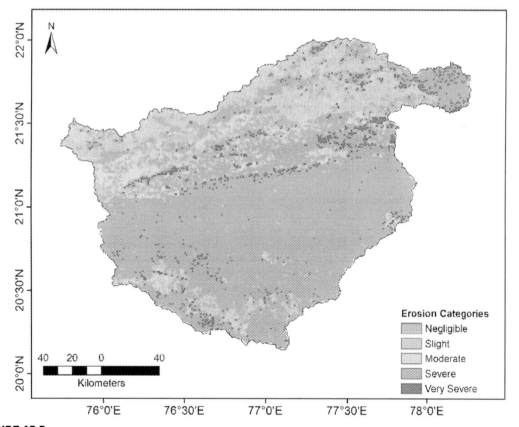

FIGURE 15.5

Composite erosion intensity map of the upper Tapi River basin.

Table 15.6 Erosion intensity classification with output weightages and total areas.

Sr. no.	Sum of weightages	Erosion intensity class	Total area (km²)	Total area (%)
1	0–3	Negligible	243.0	0.8
2	4–5	Low	3291.0	11.1
3	6–6	Moderate	6833.2	23.1
4	9–11	Severe	18276.8	61.7
5	12–16	Very severe	980.9	3.3
		Total	29,624.9	100.0

southern part of the basin due to inadequate land practices. Low and negligible erosion intensity area was only about 12%, and the remaining moderate-to-very severe erosion intensity area was about 26% (Table 15.6). This result revealed that more conservation treatments are needed for moderate to very severe erosion categories. Since cropland is a major land-use class, inappropriate agricultural practices could be an issue of higher erosion in the study area (Ramani et al., 2021). Very severe erosion in the middle-upper basin part is due to a higher slope that significantly influences soil and water flows (Fig. 15.5).

15.5.2   Subbasin wise treatment prioritization

After estimating the SYI, each subbasin was further prioritized for conservation treatments. The priority classes corresponding to SYI values were grouped into very low, low, medium, high, and very high (Fig. 15.6). The priority rating varies from basin to basin, depending on its specific characteristics. In this study, each subbasin was considered to be treated on a priority basis to lower erosion and sedimentation rates in the Hatnur reservoir (Sharma et al., 2019a). The basin unit having an SYI value of more than 1300 was categorized in the very high priority class, and this class was also prone to severe-to-very severe erosion. Other priority categories are shown in Table 15.7.

The results show that 51.6% of the total basin area falls under a medium to very high priority rating. On the other hand, 48.4% of the total basin area falls under low and very low priority classes (Table 15.7). The low and very low priority classes do not need to be considered for the conservation treatments as these classes have less or negligible erosion. Table 15.8 shows 15 subbasins in very low, one subbasin in low, five subbasins in medium, six subbasins in high, and six subbasins in very high priority classes. Also, ranking for each priority treatment has been provided according to the SYI value. Such information can be considered immensely crucial for soil and water

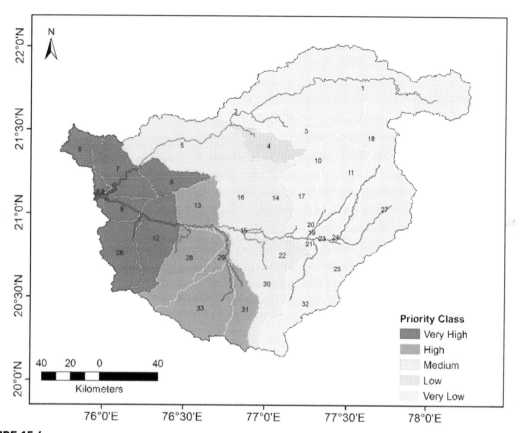

FIGURE 15.6
Treatment prioritization for the upper Tapi River basin.

Table 15.7 Sediment yield index (SYI) value-based priority rating and their coverage areas.

Sr. no.	SYI value	Priority rating	Area (km²)	Area (%)
1	>1300	Very high	4704.5	15.8
2	1200–1299	High	4521.4	15.2
3	1100–1199	Medium	6125.9	20.6
4	1000–1099	Low	667.9	2.2
5	<1000	Very low	13,674.1	46.2
		Total	29,693.8	100.0

Table 15.8 Subbasin-wise priority rating as per the calculated sediment yield index (SYI) value.

Subbasin	Area (km^2)	SYI	Priority	Rank
SB-1	4828.2	704.2	Very low	33
SB-2	367.9	971.2	Very low	19
SB-3	901.7	868.3	Very low	29
SB-4	667.9	1049.8	Low	18
SB-5	2293.8	1119.8	Medium	14
SB-6	587.5	1352.8	Very high	5
SB-7	961.7	1363.6	Very high	4
SB-8	561.9	1346.3	Very high	6
SB-9	395.9	1431.0	Very high	1
SB-10	605.7	927.1	Very low	25
SB-11	848.3	948.3	Very low	23
SB-12	1246.4	1412.6	Very high	2
SB-13	799.9	1270.3	High	7
SB-14	760.5	1105.6	Medium	16
SB-15	3.2	1229.7	High	10
SB-16	1362.6	1107.1	Medium	15
SB-17	737.0	952.4	Very low	22
SB-18	1036.8	759.0	Very low	32
SB-19	34.2	920.8	Very low	26
SB-20	185.0	967.0	Very low	20
SB-21	34.6	904.4	Very low	28
SB-22	585.0	1121.7	Medium	13
SB-23	455.7	966.2	Very low	21
SB-24	23.4	947.8	Very low	24
SB-25	620.0	782.8	Very low	31
SB-26	951.1	1409.5	Very high	3
SB-27	1722.4	791.0	Very low	30
SB-28	1293.5	1240.7	High	9
SB-29	280.0	1257.9	High	8
SB-30	1124.0	1100.5	Medium	17
SB-31	724.6	1203.3	High	12
SB-32	1273.2	918.0	Very low	27
SB-33	1420.2	1221.4	High	11

management policy decisions, river basin planning, fund allocations, and implementation of conservation practices.

15.5.3 Conservation treatments

An important step to planning and managing the critical basin units, inventorying their capacity to sustain the new practice, and averting uses that can harm and degrade the land is implementing conservation treatments (Palmate & Pandey, 2021; Palmate et al., 2021). It is based on soil depth, vegetation cover, and slope steepness important to lower erosion and sediment yields from the basin areas (Himanshu et al., 2019). The subbasin areas with medium to very high erosion intensity units are important in this study. Significant variation in SYI indicates that only critical areas need to prioritize for conservation treatments. The medium, high, and very high prioritized regions are useful to allocate proper soil and water conservation measures regarding the high amount of and intensive rainfall. If the engineering- or nature-based practices are implemented, then a lower erosion results in less sediment yield in the reservoir.

Contour farming, contour bunding, vegetative covers, agricultural zero tillage, reforestation, and gully reclamation are the immediate attentive and adoptive land treatment practices. This would help in soil and water conservation and result in sustainable agricultural land production (Himanshu et al., 2019; Pandey & Palmate, 2019). The southern basin part is with a high drainage density; therefore critical subbasin areas belong to this part may face slightly to moderate sheet and rill erosion and moderate to severe gully erosion. Hence, such critical regions are ideal for soil and water conservation treatments.

15.6 Summary and conclusions

The spatial data information of percent slopes, LULC, and soils were obtained from different sources and were utilized for estimating the SYI for the Hatnur reservoir basin. The aggregated weightage factors between 12 and 20 were assigned to each subbasin depending on the spatial characteristics of each subbasin. Critical area prioritization and raking were performed for the critical subbasin areas. Finally, different soil erosion potentials were assessed to adopt soil and water conservation treatments in each critical subbasin.

Following conclusions are drawn from this study:

1. Higher rate of denudation and removal of soils are observed in subbasins with higher erosion intensity, because of inappropriate land practices and slope that induces more soil and water flows in the study area.
2. A very high SYI value (>1300) was obtained for subbasins SB-9, SB-12, SB-26, SB-7, SB-6, and SB-8. Therefore these critical areas were further prioritized and ranked from 1 to 6 for conservation treatments.
3. High SYI value (1200−1299) was obtained for subbasins SB-13, SB-29, SB-28, SB-15, SB-33, and SB-31, and prioritized with rank from 7 to 12 for implementing conservation treatments.

4. The medium SYI value (1100−1199) was estimated for subbasins SB-22, SB-5, SB-16, SB-14, and SB-30 and prioritized with rank from 13 to 17 for conservation treatments.

5. Overall, the subbasin-wise prioritization is crucial for the immediate implementation of conservation treatments to reduce siltation in the Hatnur reservoir and to maintain full water storage capacity over a long period of time. This could help in sustaining the ecosystem functions of the UTRB.

References

AISLUS. (1991). *Methodology for priority delineation survey*. New Delhi, India: Ministry of Agriculture, Government of India.

Amrit, K., Mishra, S. K., Pandey, R. P., Himanshu, S. K., & Singh, S. (2019). Standardized precipitation index-based approach to predict environmental flow condition. *Ecohydrology, 12*(7), e2127.

Amrit, K., Mishra, S. K., & Pandey, R. P. (2018). Tennant concept coupled with standardized precipitation index for environmental flow prediction from rainfall. *Journal of Hydrologic Engineering, 23*(2), 05017031. Available from https://doi.org/10.1061/(ASCE)HE0.1943-5584.0001605.

Bali, Y. P., & Karale, R. L. (1977). A sediment yield index as a criterion for choosing priority basins. *IAHS-AISH Publication, 122*, 180−188. Available from http://hydrologie.org/redbooks/a122/iahs_122_0180.pdf.

Beven, K., Heathwaite, L., Haygarth, P., Walling, D., Brazier, R., & Withers, P. (2005). On the concept of delivery of sediment and nutrients to stream channels. *Hydrological Processes, 19*(2), 551−556. Available from https://doi.org/10.1002/hyp.5796.

Chandra, P., Patel, P. L., & Porey, P. D. (2016). Prediction of sediment erosion pattern in Upper Tapi Basin, India. *Current Science*, 1038−1049. Available from https://www.jstor.org/stable/24907871.

Chandra, P., Patel, P. L., Porey, P. D., & Gupta, I. D. (2014). Estimation of sediment yield using SWAT model for Upper Tapi basin. *ISH Journal of Hydraulic Engineering, 20*(3), 291−300. Available from https://doi.org/10.1080/09715010.2014.902170.

Himanshu, S. K., Pandey, A., & Shrestha, P. (2017). Application of SWAT in an Indian river basin for modeling runoff, sediment and water balance. *Environmental Earth Sciences, 76*(1), 1−18. Available from https://doi.org/10.1007/s12665-016-6316-8.

Himanshu, S. K., Pandey, A., Yadav, B., & Gupta, A. (2019). Evaluation of best management practices for sediment and nutrient loss control using SWAT model. *Soil and Tillage Research, 192*, 42−58. Available from https://doi.org/10.1016/j.still.2019.04.016.

Kumar, R., Ganapathi, H., & Palmate, S. S (2021). *Wetlands and Water Management: Finding a Common Ground. Water Governance and Management in India* (pp. 105−129). Springer, Singapore.

Mauget, S. A., Himanshu, S. K., Goebel, T. S., Ale, S., Lascano, R. J., & Gitz III, D. C. (2021). Soil and soil organic carbon effects on simulated Southern High Plains dryland Cotton production. *Soil and Tillage Research, 212*, 105040. Available from https://doi.org/10.1016/j.still.2021.105040.

MoWR. (2012). *River basin atlas of India*. Ministry of Water Resources, Government of India October 2012. Available from https://indiawris.gov.in/downloads/RiverBasinAtlas_Full.pdf.

Munoth, P., & Goyal, R. (2020). Impacts of land use land cover change on runoff and sediment yield of Upper Tapi River Sub-Basin, India. *International Journal of River Basin Management, 18*(2), 177−189. Available from https://doi.org/10.1080/15715124.2019.1613413.

Palmate, S. S., & Pandey, A. (2021). Effectiveness of best management practices on dependable flows in a river basin using hydrological SWAT model. *Water management and water governance* (pp. 335−348). Cham: Springer. Available from https://doi.org/10.1007/978-3-030-58051-3_22.

Palmate, S. S., Pandey, A., Pandey, R. P., & Mishra, S. K. (2021). Assessing the land degradation and greening response to land s in hydro-climatic variables using a conceptual framework: A case study of central India. *Land Degradation & Development*. Available from https://doi.org/10.1002/ldr.4014.

Pandey, A., Himanshu, S. K., Mishra, S. K., & Singh, V. P. (2016). Physically based soil erosion and sediment yield models revisited. *Catena, 147*, 595−620. Available from https://doi.org/10.1016/j.catena.2016.08.002.

Pandey, A., & Palmate, S. S. (2018). Assessments of spatial land cover dynamic hotspots employing MODIS time-series datasets in the Ken River Basin of Central India. *Arabian Journal of Geosciences, 11*(17), 1−8. Available from https://doi.org/10.1007/s12517-018-3812-z.

Pandey, A., & Palmate, S. S. (2019). Assessing future water−sediment interaction and critical area prioritization at sub-watershed level for sustainable management. *Paddy and Water Environment, 17*(3), 373−382. Available from https://doi.org/10.1007/s10333-019-00732-3.

Ramani, R. S., Patel, P. L., & Timbadiya, P. V. (2021). Key morphological changes and their linkages with stream power and land-use changes in the Upper Tapi River basin, India. *International Journal of Sediment Research*. Available from https://doi.org/10.1016/j.ijsrc.2021.03.003.

Resmi, S. R., Patel, P. L., & Timbadiya, P. V. (2020). Impact of land use-land cover and climatic pattern on sediment yield of two contrasting sub-catchments in upper Tapi Basin, India. *Journal of the Geological Society of India, 96*(3), 253−264. Available from https://doi.org/10.1007/s12594-020-1545-6.

Roy, P. S., Meiyappan, P., Joshi, P. K., Kale, M. P., Srivastav, V. K., Srivasatava, S. K., ... Krishnamurthy, Y. V. N. (2016). Decadal land use and land cover classifications across India, 1985, 1995, 2005. *ORNL DAAC*. Available from https://doi.org/10.3334/ORNLDAAC/1336.

Sargaonkar, A. (2006). Estimation of land use specific runoff and pollutant concentration for Tapi river basin in India. *Environmental Monitoring and Assessment, 117*(1), 491−503. Available from https://doi.org/10.1007/s10661-006-0769-2.

Sharma, P. J., Patel, P. L., & Jothiprakash, V. (2019a). Impact assessment of Hathnur reservoir on hydrological regimes of Tapi River, India. *ISH Journal of Hydraulic Engineering*, 1−13. Available from https://doi.org/10.1080/09715010.2019.1574616.

Sharma, P. J., Patel, P. L., & Jothiprakash, V. (2019b). Impact of rainfall variability and anthropogenic activities on streamflow changes and water stress conditions across Tapi Basin in India. *Science of the Total Environment, 687*, 885−897. Available from https://doi.org/10.1016/j.scitotenv.2019.06.097.

CHAPTER 16

Advances in hydrocarbon bioremediation products: natural solutions

Pankaj Kumar Gupta

Wetland Hydrology Research Laboratory, Faculty of Environment, University of Waterloo, Waterloo, ON, Canada

16.1 Introduction

Shallow aquifers of India are crisscrossed by transportation corridors including roads, rail, and pipelines that make them susceptible to petrochemical spills. According to the annual report 2017−18 of the Ministry of Petroleum and Natural Gas, Government of India, 247.6 million metric tons per year addition capacity of all refineries were developed in 2017. Contamination of soil−water systems due to the release of hydrocarbons, commonly referred to as dense and light nonaqueous phase liquids (LNAPL), is an emerging problem (Gupta & Yadav, 2020; Gupta, Ranjan, et al., 2019; Gupta, Yadav, et al., 2019; Kamaruddin et al., 2011). When released to land, these contaminants migrate downward through the unsaturated zone and consequently light hydrocarbons float and move on top of the water table, while dense hydrocarbons move downward through the water table and penetrate into the saturated zone (Yadav & Hassanizadeh, 2011). Hydrocarbon spills present a significant threat to the environment as they can result in extensive pollution from small spillages (Gupta et al., 2020; Gupta & Yadav, 2019; Kumar et al., 2021; Gupta & Bhargava, 2020; Gupta & Yadav, 2019; Gupta & Yadav, 2017). Soil-water quality management is key to provide safe drinking water supply (Amrit et al., 2019; Dhami et al., 2018; Himanshu et al., 2018, 2021; Pandey et al., 2016).

In saturated zone, biodegradation of dissolved hydrocarbon mass by naturally accruing microbes limits the spreading of its plume. Aerobic degradation of hydrocarbons, especially at groundwater table, causes accelerated biodegradation of such hydrocarbons, while, in deep aquifer zone, an aerobic condition is not sustainable, and anaerobic/methanogenic conditions and oxygen-depleted area might establish, which may reduce the degradation of hydrocarbon. Below the groundwater table, the pore water has a low redox

Advances in Remediation Techniques for Polluted Soils and Groundwater. DOI: https://doi.org/10.1016/B978-0-12-823830-1.00012-2

potential as oxygen is rapidly depleted by microbes. Within this zone, micro-organisms catalyze the redox reactions sequentially using different electron acceptors (nitrate, manganese, iron, sulfate, and organic matter) and grow biomass by the metabolism of the contaminants (Essaid et al., 2015). The cumulative effect of sorption, volatilization, and biodegradation of the dissolved hydrocarbon will eventually match the diminishing rate of dissolution from the pure phase pool, and the dissolved-phase plume will stop growing and will stabilize at a maximum extent in a steady-state condition. Majority of the earlier works focused on fate and transport of pure phase or dissolved hydrocarbon in subsurface either using small-scale laboratory experiments or numerically by considering simplified conditions (Essaid et al., 2015). Variation in subsurface conditions affects soil–water parameters, pure phase hydrocarbon behavior, and dissolved hydrocarbon plume in the underlying/overlying (un)saturated zone. Impact of dynamic groundwater flow regimes on hydrocarbon fate and transport is not investigated in subsurface. Further, most of the earlier studies related to designing the bioremediation approaches are limited to saturated zone and have used mostly data-based optimization approaches (Gupta, 2020). One cannot ignore the role of vadose zone in treating the saturated zone as most of the hydrocarbon leakages take place nearby the surface before polluting the underlying saturated zone.

Natural attenuation of these pollutants has been reported very slow in deep groundwater zone (Basu et al., 2015; Lee et al., 2001; Yadav et al., 2012), especially in cold regions (Ulrich et al., 2010). On the other hand, ex situ treatment may result in high cost and create a lot of disturbances in natural settings (Azubuike et al., 2016). Thus enhanced bioremediation gains a lot of attention to remediate hydrocarbon-polluted sites (Gupta & Yadav, 2017; Yadav et al., 2014). However, microbial degradations were reported slow in cold regions which affects the enhanced bioremediation significantly. Thus thermally enhanced bioremediation was adopted in many hydrocarbons-polluted sites and reported a positive response of native microbes which results a high removal rates of these pollutants (Beyke & Fleming, 2005; Filler & Carlson, 2000; Friis et al., 2006; Němeček et al., 2018; Perfumo et al., 2007). In this chapter, the bioremediation products have been reviewed to highlight their performance. This is an important step in being able to make informed recommendations for spill management and remediation in subsurface under varying environmental conditions.

16.2 Engineered constructed wetlands

A plot-scale self-sustainable constricted wetland, integrated with suitable PV panel, can be an effective tool to treat hydrocarbon. Aeration unit may consist of three diffuser blocks placed at a distance of 50 cm apart in an angle iron frame. Each unit may be supported on legs extending obliquely down-ward, holding the diffuser blocks approximately 0.5 feet from the wetland bottom to prevent stirring and upward transport of bottom materials when the aeration units will be operating. The blocks may be hollow, with small orifices opening into each block that may be joined by a hose to supply air directly to each block. An orifice may be placed in each block to create head loss and assure that each diffuser received approximately the same volume of air, even if they were at different levels on the wetland bottom. Air may be supplied to the diffuser blocks by a portable air compressor. Two submerged and emergent macrophytic species Vetiveria and Salix may be planted around the wetland periphery with a density of 150−200 individual plants/m^2 and an average height of 30−50 cm. These plants were also recommended in wet-lands, as a tertiary unit for wastewater treatments.

16.2.1 Duplex wetland systems

Mustapha et al. (2018) investigated the performance of a duplex constructed wetland (duplex-CW), a hybrid system that combines a vertical flow (VF) CW with an underlying horizontal flow filter (HFF) for a more efficient wastewater treatment. The hybrid system consists of a VFCW in the first stage and an HFF in the second stage. They target diesel range organic (DRO) com-pounds varying from $C_7−C_{40}$ degradation using a series of 4-month/12-week experiments under controlled conditions in a greenhouse. The CWs were planted with *Phragmites australis* and were spiked with different concentra-tions of ammonium−nitrogen (10, 30, and 60 mg L$^{−1}$) and phosphate (3, 6, and 12 mg L$^{−1}$) to analyze their effects on petroleum hydrocarbon degrada-tion. The mean toluene removal efficiencies from the laboratory setup and simulation domain were 58.3($\pm$52.0)%, 89.4($\pm$12.9)%, 88.3($\pm$12.6)%, and 93.3($\pm$11.5)%, 100%, 100% for duplex-CW1, duplex-CW2, and duplex-CW3, respectively. Similar results were also observed in case of other benzene, toluene, ethylbenzene and xylene (BTEX) compounds and DRO compounds. The results confirm that the addition of the nutrient enhanced degradation rate by stimulation of the microbes and modification in CW conditions. Further, the HFFs of duplex CWs were shown the effective removal of access nutrients, which endorse high capability of duplex CWs to sustain system (Fig. 16.1).

16.2.2 Integrated polluted columns and treatment wetlands

Basu et al. (2020) designed integrated wetland and polluted column to study the degradation and transport of toluene, a LNAPL which create major pollu-tion problems in subsurface environments. Two-column setups having an inner diameter of 15 cm and length 120 cm were designed, having a planted

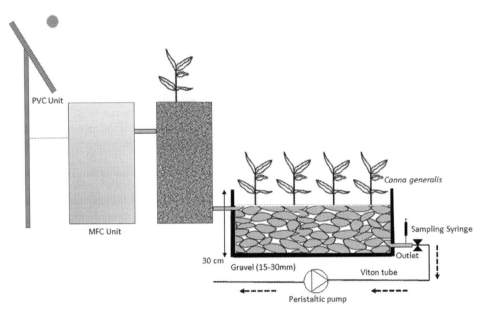

FIGURE 16.1

A self-sustainable constructed wetland integrated to microbial fuel cell (MFC) unit to remove hydrocarbon pollutants.

treatment wetland in the first set and unplanted gravel bed in the second set. A continuous source of mixture (1:1) of root zone water from (un)-planted treatment wetlands and LNAPL contaminated groundwater was allowed to infiltrate into the vertical column. The estimated biodegradation rates show that the toluene degradation was 2.5 times faster in planted wetland setup as compared with the unplanted case.

16.3   Native and specialized microbial communities

Gupta (2020) performed 16S rRNA sequencing to understand microbial dynamics in NAPLs polluted peat soils and to reveal the specialized NAPLs degraders. Microbial communities were predominated by the phyla Acidobacteria and the Proteobacteria. Proteobacteria was comprising 35.44% [STD:8.55%], 60.52% [STD:13.9%], 61.29% [STD:7.26%] in column prior to NAPLs spill in this study, in which γ-*proteobacteria* (30.14%), α-*proteobacteria* (19.26%), followed by δ-*proteobacteria* (1.86%) were dominating classes. Whereas in postexperimental samples (R1 and R2), the average relative abundance of Proteobacteria increases to 83.8% [STD:5.07%], 86.9% [STD:5.9%] in NAPLs contaminated fluctuating and stable water table (WT) columns, respectively. This indicates that the Proteobacteria get benefited most from NAPLs spill in peat soils.

Table 16.1 Native microbial communities and their performance investigated by Gupta (2020).

Phylum	NAPL-polluted dynamic	Control	NAPL-polluted stable
Acidobacteria	0.821799	3.847102	1.547648
Actinobacteria	0.346021	3.551171	1.188155
Bacteroidetes	0	0	0.140141
Chlamydiae	0	0.221948	0.036559
Firmicutes	3.676471	45.74599	4.691689
Planctomycetes	0	0.197287	0
Proteobacteria	94.76644	46.09125	91.75603
Verrucomicrobia	0.389273	0.345253	0.396052

NAPL, *Light nonaqueous phase liquid.*

Acidobacteria was the second most dominating phylum and Acidobacteriaceae family of Acidobacteriia class were survive after exposed to NAPLs (Table 16.1).

16.4 Biodiesels as biostimulators

As (blended-)biofuels are being globally used and expected to increase in near future, their fate, and transport in the soil—water system is an area of major concern. The differences in physical, chemical, and biological characteristics from petrodiesels, and therefore environmental behaviors, can cause a potential pollution risk on receiving environment. A better understanding of fate and transport of (blended)-biodiesels may help to forecast their accurate distributions and to frame effective remediation plans. In this regard, this manuscript presents the review updates and current understanding of fate, transport, and bioremediation of biodiesel in subsurface. The insights discussed in this chapter may help groundwater scientists, environmentalists, and geochemists in predicting the behavior of (blended-)biodiesel contaminants in a different landscape and in assessing the risks of this contamination.

High biodegradability of pure biodiesel attracts geochemists to use it as a source of carbon to enhance the microbial degradation of petrodiesel and blended biodiesel in subsurface (Gupta, 2020). The main reasons behind the use of biodiesel in bioremediation of petrodiesel/crude oil are its high solubility in water which provide more and quick (bioavailability) energy sources for potential microorganisms. This is due to the amphiphilic characteristics of the biodiesel fatty acid esters allowed them to act as surfactant in the water and increase the dissolution of the biodiesel along with blended part of petrodiesel. Furthermore, biodiesel can form aggregates called micelle, which are spherical with hydrophilic head groups facing outward and hydrophobic tails buried in the core. The hydrophobic core of micelle plays a significant role in dissolving petrodiesel and ultimately enhances the dissolved concentration load of petrodiesel in subsurface. These micelles of biodiesel would also fuse into cells of microorganisms and provide sufficient amount of substrates. Addition of biodiesel in subsurface may also increase the surface area of petrodiesel which subsequently enhances microbial biodegradation.

Cyplik et al. (2011) investigated biodegradation of petrodiesel, blended biodiesel (B20), and pure biodiesel using a bacterial consortium under aerobic and nitrate-reducing conditions. Batch experiments were performed using a 250 mL bottle contains 50 mL mineral porous media and fuel at 1.5% (v/v) and by adding 1 mL dense cell suspension of bacterial consortium under aerobic conditions. Later on, nitrate was added as electron acceptor in batches. The results of this study indicate that biodegradation rate increases with increasing biodiesel content (blending) under aerobic conditions while reduced with addition of nitrate. This study also highlighted the significant role of Citrobacter, Comamonadaceae sp. Variovorax sp., on biodegradation enhancement of biodiesels under aerobic conditions. The difference in biodegradations behaviors was negligible for all biodiesels in this study. Pardo et al. (2014) investigated biodegradation of B20 using modified Fenton process by applying ferric ion and chelating agent. The removal of total hydrocarbon was up to 75% where biodiesel degradation was faster than petrodiesel. However, addition of chelating agent increases pH and reduces the biodegradation rate. Although the impact of electron acceptors/nutrients supply on bioremediation performance has been investigated in these studies, no studies have been reported on bioremediation using modification of subsurface variables, such as temperature, soil moisture, etc. Thus there is a need to investigate the combined role of subsurface conditions on performance of biodiesel degraders. In this regard, Sørensen et al. (2011) performed bench-scale experiments to investigate the biodegradation of petrodiesel, soy biodiesel (B100 soy), beef tallow biodiesel (B100 beef tallow). Inoculum containing (10^5 MPN/mL^{-1}) Bacillus amyloliquefaciens, Bacillus sp.; Micrococcus sp.; Candida sp., *Candida dubliniensis, Candida viswamathii*; and *Pichia anomala* was added with all bench setups. Oxygen consumption and CO_2 production were measured for 42 days. High biodegradation of fuel hydrocarbons was observed due to a high performance of 8 fungal and 10 bacterial communities. Likewise, Colla et al. (2014) investigated the combined performance of biostimulation and bioaugmentation for B10 polluted soil. The bacterial inoculum was prepared and added with nutrients (NH_4NO_3 and KH_2PO_4) at moisture level equal to 80% of soil field capacity. The microbial consortium was identified using 16sRNA analysis as *Pseudomonas aeruginosa, Achromobacter xylosoxidans,* and *Ochrobactrum intermedium*. The combined role of nutrient and microbial consortium resulted in a high removal of biodiesel and petrodiesel from soil in this study. Some yeast and fungal groups are also potential for bioaugmentation of biodiesels in soil—water system. However, further studies are required to investigate the performance of potential microbial communities at different types, level of biodiesel-polluted sites under varying conditions.

16.5 Phycoremediation

Algae-supported biodegradation of petrochemical contaminants has been scarcely investigated, and the catabolic pathways of biodegradation of these compounds in algae are still largely unknown (Jacques & McMartin, 2009). There are only few studies that comparatively evaluate the performance of algae and algal—bacterial species for the treatment of petroleum hydrocarbons-polluted sites (Hammed et al., 2016). Initially, Walker et al. (1975) performed experiments with the achlorophyllous alga *Prototheca zopci*, which was found to degrade petroleum hydrocarbons found in Louisiana crude and motor oils. Jacobson and Alexander (1981) grown cultures of Chlamydomonas sp., in the light and dark on acetate, were found to be significant degradation of hydrocarbons. Cerniglia et al. (1979, 1980a,b) reported that both cyanobacteria (blue-green algae) and eukaryotic microalgae were capable of biotransforming major hydrocarbon pollutants (naphthalene) to nontoxic products. Liebe and Fock (1992) found that *Chlamydomonas reinhardtii* is capable to remove some of the iso-octane-extracted polycyclic aromatic hydrocarbons (PAHs) from diesel particulate exhaust. Wolfaardt et al. (1994) show the significant role of a Chlorococcum sp. present in an algal—bacterial consortium to remove the diclofop-methyl. de-Bashan and Bashan (2010) identified *Prototheca zopfii* as the most popular hydrocarbon-degrading microalgae. Ibrahim and Gamila (2004) isolate seven microalgae from the Nile river capable of degrading hydrocarbon pollutants based on their affinity. They showed that *Scenedesmus obliquus* showed higher affinity toward degradation of polycyclic hydrocarbons than *n*-alkanes, whereas *Nitzschia linearis* that demonstrated preference for degradation of *n*-alkanes. The biodegradation of fluoranthene and phenanthrene, typical PAHs, was found accelerated by Nitzschia sp. and *Skeletonema costatum* than natural attenuation.

References

Amrit, K., Mishra, S. K., Pandey, R. P., Himanshu, S. K., & Singh, S. (2019). Standardized precipitation index-based approach to predict environmental flow condition. *Ecohydrology, 12*(7) e2127.

Azubuike, C. C., Chikere, C. B., & Okpokwasili, G. C. (2016). Bioremediation techniques—classification based on site of application: Principles, advantages, limitations and prospects. *World Journal of Microbiology and Biotechnology, 32*(11), 180.

Bashan, L. E., & Bashan, Y. (2010). Immobilized microalgae for removing pollutants: Review of practical aspects. *Bioresource Technology, 101,* 1611–1627.

Basu, S., Yadav, B. K., & Mathur, S. (2015). Enhanced bioremediation of BTEX contaminated groundwater in pot-scale wetlands. *Environmental Science and Pollution Research, 22*(24), 20041–20049.

Basu, S., Yadav, B. K., Mathur, S., & Gupta, P. K. (2020). In situ bioremediation of toluene-polluted vadose zone: Integrated column and wetland study. *Clean–Soil, Air, Water, 48*(5–6), 2000118.

Beyke, G., & Fleming, D. (2005). In situ thermal remediation of DNAPL and LNAPL using electrical resistance heating. *Remediation Journal, 15*(3), 5–22.

Cerniglia, C. E., Gibson, D. T., & van Baalen, C. (1979). Algal oxidation of aromatic hydrocarbons: Formation of 1-naphthol from naphthalene by Agmenellum quadruplicatum, strain PR-6. *Biochemical and Biophysical Research Communications, 88,* 50–58.

Cerniglia, C. E., Gibson, D. T., & van Baalen, C. (1980a). Oxidation of naphthalene by cyanobacteria and microalgae. *Journal of General Microbiology, 116,* 495–500.

Cerniglia, C. E., van Baalen, C., & Gibson, D. T. (1980b). Metabolism of naphthalene by the cyanobacterium Oscillatoria sp., strain JCM. *Journal of General Microbiology, 116,* 485–494.

Colla, T. S., Andreazza, R., Bücker, F., de Souza, M. M., Tramontini, L., Prado, G. R., & Bento, F. M. (2014). Bioremediation assessment of diesel–biodiesel-contaminated soil using an alternative bioaugmentation strategy. *Environmental Science and Pollution Research, 21*(4), 2592–2602.

Cyplik, P., Schmidt, M., Szulc, A., Marecik, R., Lisiecki, P., Heipieper, H. J., & Chrzanowski, è. (2011). Relative quantitative PCR to assess bacterial community dynamics during biodegradation of diesel and biodiesel fuels under various aeration conditions. *Bioresource Technology, 102*(6), 4347–4352.

Dhami, B., Himanshu, S. K., Pandey, A., & Gautam, A. K. (2018). Evaluation of the SWAT model for water balance study of a mountainous snowfed river basin of Nepal. *Environmental Earth Sciences, 77*(1), 1–20.

Essaid, H. I., Bekins, B. A., & Cozzarelli, I. M. (2015). Organic contaminant transport and fate in the subsurface: Evolution of knowledge and understanding. *Water Resources Research, 51*(7), 4861–4902.

Filler, D. M., & Carlson, R. F. (2000). Thermal insulation systems for bioremediation in cold regions. *Journal of Cold Regions Engineering, 14*(3), 119–129.

Friis, A. K., Albrechtsen, H. J., Cox, E., & Bjerg, P. L. (2006). The need for bioaugmentation after thermal treatment of a TCE-contaminated aquifer: Laboratory experiments. *Journal of Contaminant Hydrology, 88*(3–4), 235–248.

Gupta, P. K., & Yadav, B. K. (2017). Bioremediation of non-aqueous phase liquids (NAPLS) polluted soil and water resources. *Environmental Pollutants and their Bioremediation Approaches,* 241–256.

Gupta, P. K., & Yadav, B. K. (2017). Bioremediation of non-aqueous phase liquids (NAPLs) polluted soil and water resources. In R. N. Bhargava (Ed.), Environmental pollutants and their bioremediation approaches. Florida, USA: CRC Press, Taylor and Francis Group, ISBN 9781138628892.

Gupta, P. K., & Yadav, B. K. (2019). Remediation and management of petrochemical polluted sites under climate change conditions. In R. N. Bhargava (Ed.), Environmental contaminations: Ecological implications and management. Springer Nature Singapore Pte Ltd, ISBN 9789811379048.

Gupta, P. K., & Yadav, B. K. (2019). Subsurface processes controlling reuse potential of treated wastewater under climate change conditions. In R. P. Singh, A. K. Kolok, & L. S. Bartelt-Hunt (Eds.), *Water conservation, recycling and reuse: Issues and challenges.* Singapore: Springer, ISBN 9789811331787.

Gupta, P. K., Ranjan, S., & Gupta, S. K. (2019). *Phycoremediation of petroleum hydrocarbon-polluted sites: Application, challenges, and future prospects. Application of microalgae in wastewater treatment* (pp. 145−162). Cham: Springer.

Gupta, P. K., Yadav, B., & Yadav, B. K. (2019). Assessment of LNAPL in subsurface under fluctuating groundwater table using 2D sand tank experiments. *Journal of Environmental Engineering, 145*(9), 04019048.

Gupta, P.K., & Bhargava, R.N. (2020). Fate and transport of subsurface pollutants. Springer, eBook ISBN-978-981-15-6564-9. https://www.springer.com/gp/book/9789811565632.

Gupta, P. K., & Yadav, B. K. (2020). Three-dimensional laboratory experiments on fate and transport of LNAPL under varying groundwater flow conditions. *Journal of Environmental Engineering, 146*(4), 04020010.

Gupta, P. K. (2020). Fate, transport, and bioremediation of biodiesel and blended biodiesel in subsurface environment: A review. *Journal of Environmental Engineering, 146*(1), 03119001.

Gupta, P. K., Kumari, B., Gupta, S. K., & Kumar, D. (2020). Nitrate-leaching and groundwater vulnerability mapping in North Bihar, India. *Sustainable Water Resources Management, 6*, 1−12.

Hammed, S. K., Prajapati, S., & Simsek, H. (2016). Growth regime and environmental remediation of microalgae. *Algae, 31*, 189−204. Available from https://doi.org/10.4490/algae.2016.31.8.28.

Himanshu, S. K., Ale, S., Bordovsky, J. P., Kim, J., Samanta, S., Omani, N., & Barnes, E. M. (2021). Assessing the impacts of irrigation termination periods on cotton productivity under strategic deficit irrigation regimes. *Scientific Reports, 11*(1), 1−16.

Himanshu, S. K., Pandey, A., & Patil, A. (2018). Hydrologic evaluation of the TMPA-3B42V7 precipitation data set over an agricultural watershed using the SWAT model. *Journal of Hydrologic Engineering, 23*(4)05018003.

Ibrahim, M. B. M., & Gamila, H. A. (2004). Algal bioassay for evaluating the role of algae in bioremediation of crude oil: II. Freshwater phytoplankton assemblages. *Bulletin of Environmental Contamination and Toxicology, 73*, 971−978.

Jacobson, S. N., & Alexander, M. (1981). Enhancement of the microbial dehalogenation of a model chlorinated compound. *Applied and Environmental Microbiology, 42*, 1062−1066.

Jacques, N. R., & McMartin, D. W. (2009). Evaluation of algal phytoremediation of light extractable petroleum hydrocarbons in subarctic climates. *Remediation Journal, 20*, 119−132.

Kamaruddin, S. A., Sulaiman, W. N. A., Rahman, N. A., Zakaria, M. P., Mustaffar, M., & Sa'ari, R. (2011). A review of laboratory and numerical simulations of hydrocarbons migration in subsurface environments. *Journal of Environmental Science and Technology, 4*(3), 191−214. Available from https://doi.org/10.3923/jest.2011.191.214.

Kumar, R., Sharma, P., Verma, A., Jha, P. K., Singh, P., Gupta, P. K., Chandra, R., & Prasad, P. V. V. (2021). Effect of physical characteristics and hydrodynamic conditions on transport and deposition of microplastics in riverine ecosystem. *Water, 13*, 2710.

Lee, C. H., Lee, J. Y., Cheon, J. Y., & Lee, K. K. (2001). Attenuation of petroleum hydrocarbons in smear zones: A case study. *Journal of Environmental Engineering, 127*(7), 639−647.

Liebe, B., & Fock, H. P. (1992). Growth and adaption of the green alga Chlamydomonas reinhardtii on diesel exhaust particle extracts. *Journal of General Microbiology, 138*, 973−978.

Mustapha, H. I., Gupta, P. K., Yadav, B. K., van Bruggen, J. J. A., & Lens, P. N. L. (2018). Performance evaluation of duplex constructed wetlands for the treatment of diesel contaminated wastewater. *Chemosphere, 205*, 166−177.

Němeček, J., Steinová, J., Špánek, R., Pluhař, T., Pokorný, P., Najmanová, P., & Černík, M. (2018). Thermally enhanced in situ bioremediation of groundwater contaminated with chlorinated solvents−A field test. *Science of the Total Environment, 622*, 743−755.

Pandey, A., Himanshu, S. K., Mishra, S. K., & Singh, V. P. (2016). Physically based soil erosion and sediment yield models revisited. *Catena, 147*, 595–620.

Pardo, F., Rosas, J. M., Santos, A., & Romero, A. (2014). Remediation of a biodiesel blend-contaminated soil by using a modified Fenton process. *Environmental Science and Pollution Research, 21*(21), 12198–12207.

Perfumo, A., Banat, I. M., Marchant, R., & Vezzulli, L. (2007). Thermally enhanced approaches for bioremediation of hydrocarbon-contaminated soils. *Chemosphere, 66*(1), 179–184.

Sørensen, G., Pedersen, D. V., Nørgaard, A. K., Sørensen, K. B., & Nygaard, S. D. (2011). Microbial growth studies in biodiesel blends. *Bioresource Technology, 102*(8), 5259–5264.

Ulrich, A. C., Tappenden, K., Armstrong, J., & Biggar, K. W. (2010). Effect of cold temperature on the rate of natural attenuation of benzene, toluene, ethylbenzene, and the three isomers of xylene (BTEX). *Canadian Geotechnical Journal, 47*(5), 516–527.

Walker, J. D., Colwell, R. R., Vaituzis, Z., & Meyer, S. A. (1975). Petroleum degrading achlorophyllous alga Prototheca zopfii. *Nature, 254*, 423–424.

Wolfaardt, G. M., Lawrence, J. R., Robarts, R. D., & Caldwell, D. E. (1994). The role of interactions, sessile growth, and nutrient amendments on the degradative efficiency of a microbial consortium. *Canadian Journal of Microbiology, 40*, 331–340.

Yadav, B. K., & Hassanizadeh, S. M. (2011). An overview of biodegradation of LNAPLs in coastal (semi)-arid environment. *Water, Air, & Soil Pollution, 220*(1–4), 225–239.

Yadav, B. K., Ansari, F. A., Basu, S., & Mathur, A. (2014). Remediation of LNAPL contaminated groundwater using plant-assisted biostimulation and bioaugmentation methods. *Water, Air, & Soil Pollution, 225*(1), 1793.

Yadav, B. K., Shrestha, S. R., & Hassanizadeh, S. M. (2012). Biodegradation of toluene under seasonal and diurnal fluctuations of soil-water temperature. *Water, Air, & Soil Pollution, 223*(7), 3579–3588.

Nitrate-N movement revealed by a controlled in situ solute injection experiment in the middle Gangetic plains of India

Pankaj Kumar Gupta[1], Basant Yadav[2], Kristell Le Corre[3] and Alison Parker[3]

[1]Wetland Hydrology Research Laboratory, Faculty of Environment, University of Waterloo, Waterloo, ON, Canada, [2]Department of Water Resources Development and Management (WRD&M), Indian Institute of Technology Roorkee, Roorkee, India, [3]Cranfield Water Science Institute, Cranfield University, Vincent Building, Cranfield, United Kingdom

17.1 Introduction

In many regions around the world, nonpoint source pollutants such as fertilizers, pesticides, livestock wastes, and salts are major sources of pollution for underground water resources (Harter et al., 2005). Nitrate (NO_3^-) is a major nonpoint groundwater pollutant due to its intensive use in agriculture, and its high mobility subsurface (Harter et al., 2005). In the Indian context, high NO_3^- concentrations ranging from 1 to 684 mg L^{-1} (Gupta, 2020) have been observed in the groundwater of densely populated and intensive agricultural areas across 20 of the 29 Indian states (Central Ground Water Board, 2013).

Although the levels of groundwater contamination by NO_3^- have been widely researched across the country (Kumari et al., 2019; Kundu & Mandal, 2009a, 2009b; Reddy et al., 2009; Sankararamakrishnan et al., 2008; Suthar, 2009, 2011), studies on the fate and transport of NO_3^- in these regions remain quasi-inexistent. This is particularly true in the densely populated Ganga river basin that is home to approximately 40% of the Indian population and where groundwater is the main water supply source (Jain, 2002). In the very densely populated north Indian state of Bihar, through which the Ganga river flows, drastic increases in the use of fertilizers have been observed over the past decade, with for instance the application of nitrogen, phosphorus, and potassium (NPK) fertilizer increasing from 21.5 kg ha^{-1} of gross cropped area (GCA) in the early 1980s up to 212.2 kg ha^{-1} in 2013 while the national Indian average was 128.3 kg ha^{-1} of GCA (Fishman et al., 2016).

Advances in Remediation Techniques for Polluted Soils and Groundwater. DOI: https://doi.org/10.1016/B978-0-12-823830-1.00005-5

The region uses several fertilizers such as urea $[CO(NH_2)_2]$, diammonium phosphate $[(NH)_2HPO_4]$, and manure to increase the yield of groundwater-irrigated winter crops such as wheat, paddy, and tobacco. For wheat solely, nitrogen-based fertilizer use on irrigated wheat is almost double that of rainfed wheat with usage rates of 144.9 and 75.9 kg ha^{-1}, respectively. The high tendency of leaching and mobilization of NO_3^- due to its high solubility means that this overuse of fertilizers enhances their transfer subsurface (Wendland et al., 2005). This is particularly true in a basin that mainly consists of sedimentary alluvial deposits which are more vulnerable to nitrogen contamination (Kumari et al., 2019). In addition, dynamic fluctuations of the water table can contribute to further dilution of NO_3^- in the vadose zone, increasing the risk of pollution load on the potable aquifer system of the region (Yadav & Junaid, 2014).

The abovementioned increase in nitrogen-based fertilizers has resulted in high levels of NO_3^- (e.g., up to 70 mg L^{-1}) being measured in the region's water runoffs (Kumari et al., 2019). Such levels are likely to affect the quality of the groundwater in the area, hence potentially increasing health risks for end users. Indeed, Bihar itself uses annually around 9.4 billion cubic meters (BCM) of groundwater out of which 27.4 BCM for irrigation purposes. A recent study by Ahada and Suthar (2018) on health risks associated with the consumption of NO_3^--contaminated groundwater using the USEPA health risk assessment model in 14 districts of Malwa Panjab estimated that 100% of young and 93.4% of adult populations are chronically exposed to toxic levels of NO_3^- that can adversely affect their health causing, for example, methemoglobinemia (or the "blue-baby syndrome") in infants, and stomach cancer in adults. A full understanding of the fate and transport of NO_3^- in agricultural dominating area of the Ganga basin is therefore crucial.

As reported by Musolff et al. (2016), NO_3^- shows a pronounced dynamic movement pattern in the soil—water systems, including significant attenuation by denitrification processes in the deep groundwater zone. Denitrification is the microbial process that converts NO_3^- to nitrogen gases (N_2O or N_2). It can occur in both unsaturated soils and below the water table in the presence of denitrifying bacteria (proteobacteria mainly), where low O_2 concentrations or anaerobic conditions prevail (Singleton et al., 2007). However, water table dynamics, either due to heavy rainfall or excessive pumping, can increase O_2 levels in the deep aquifer zone hence alter NO_3^- attenuation and untimely its movement underground. In addition, several factors, such as soil characteristics, evapotranspiration (ET), topography, infiltration, vegetative cover, runoff, precipitation, freeze/thaw conditions, soil permeability, and porosity, contribute to variations in the NO_3^- contamination (Harter et al., 2005). The movement of NO_3^- in the vadose zone has been widely evaluated (Akbariyeh et al., 2018; Harter et al., 2005;

Joshi & Gupta, 2018). A majority of these studies focus on the understanding of NO_3^- occurrence in groundwater zone using modeling tools such as Geographical Information System (GIS) based DRASTIC modeling (Joshi & Gupta, 2018) and numerical modeling (Akbariyeh et al., 2018; Kumari et al., 2019). A few studies investigated NO_3^- fate and transport using lab-scale experiments (Jemison & Fox, 1994; Thomsen, 2005). However, most of these integrated hydrological models were developed for the analysis of NO_3^- transport in soil zone and unsaturated zone. To the authors' knowledge, in situ solute NO_3^- movement in the groundwater zone, and more specifically in Indian soil–water systems, has never been reported.

Characterizing the spatial extent of the NO_3^- movement zones and their relationship with groundwater pumping and rainfall at catchment level under real conditions (i.e., in situ as opposed to using a modeling approach) is challenging and has rarely been reported in the past (Musolff et al., 2016). Therefore, this study was designed in an attempt to address this knowledge gap by investigating the fate and transport of NO_3^- in groundwater using in situ solute injection experiments. These experiments were performed for a period of 120 days at a site located in North Bihar, 8.5 km away from the Ganga River and looking at the horizontal migration of NO_3^- plumes under various environmental conditions such as high groundwater pumping for irrigation and rainfall. The main objective was to characterize horizontal flow of NO_3^- from one injection well to downgradient locations in a shallow unconfined alluvial aquifer to provide a comprehensive field scale study of NO_3^- contamination of groundwater in the Ganga basin, hence offering a unique opportunity to further understand the fate of NO_3^- in groundwater under real conditions and providing additional information to support the development of management and remediation plan for fresh water and/or polluted sites.

17.2 Study site

The field solute injection experiment was conducted at a site located on a flat agricultural area, about 8.5 km from the North margin of the Ganga river (Fig. 17.1A). On a regional scale, this area has a nearly flat topography with an elevation from 40 to 42 m above mean sea level (amsl), which means that it is more vulnerable to infiltration (Aller et al., 1987). Therefore, fertilizer applied in this area during the irrigation season or the NO_3^- present in surface runoffs will stay on the surface for longer and eventually will leach down to the deeper soil layers. Further, at the water table, the NO_3^- plume starts moving horizontally toward downgradient locations, which may contaminate a large area located far away from the source zone.

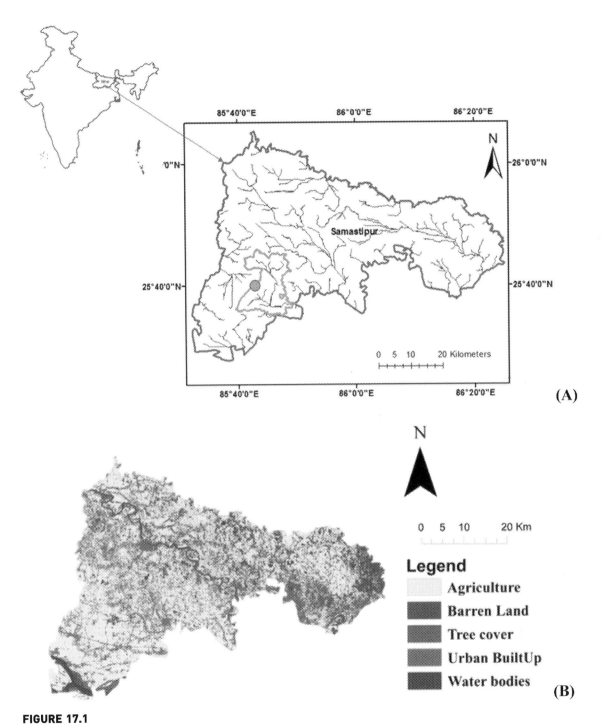

FIGURE 17.1

(A) Location of the experimental site and basin and well (inner outline and dot) in Samastipur District, Bihar, India (25°40'3.46"N; 85°42'51.48"E) (25.66763° 85.7143°) and (B) land use and land cover map generated using the Landsat 8 OLI data (USGS Earth Explorer-http://www.earthexplorer.com).

The Ganga plain foreland basin is a repository of sediments originating from the Himalayas and Peninsular Craton (Acharyya & Shah, 2007). The aquifer in this region is typical of alluvial formation that constitutes prolific aquifers where the tube wells can yield between 120 and 247 $m^3 h^{-1}$ (Central Ground Water Board, 2013; Kumar, Ramanathan, et al., 2016). Tube wells in this region are designed as narrow piezometers made of polyvinyl chloride, containing a slotted screen underneath. The hydraulic conductivity of the aquifer estimated from multiple pumping tests is in the range of 105.2−130 $m\, d^{-1}$. The soil is sandy loam, which is suitable for vegetables and spices cultivation. This middle Gangetic plain region lies in the monsoon tropical zone and has semiarid to subtropical climatic conditions with an average annual rainfall of 1142 mm and average maximum and minimum temperature of 6°C and 45°C, respectively (Verma et al., 2019).

Agriculture is the main source of livelihood in the district and about 83% of the total active population depends on it (Fig. 17.1B). Major crops in this region are rice, maize, wheat, pulses, oilseeds, tobacco, sugarcane, spices, and vegetables. The crops grown in the monsoon season are rainfed; however, the principal winter crops (i.e., wheat) are irrigation dependent. Irrigation is provided by various means such as tube wells, open wells, tank, ponds, and rivers (Sinha et al., 2018). In last decades, this region saw dramatic increase in the number of tube wells and fertilizer applications to increase the crop yield. The average use of fertilizer in this region reported as by Fishman et al. (2016) exceeds 200 $kg\, ha^{-1}$ of GCA. In the particular area of the case study, a survey of 500 farmers was conducted before the start of the field experiment regarding the use of fertilizers in irrigated winter wheat crop which revealed that approximately 350−365 kg of fertilizer per hectare were used.

Regional mapping of the quaternary deposits of the Ganga plain shows that it consists of an active flood plain, river valley terrace surface, and upland interfluves surface (Kumar, Ramanathan, et al., 2016). The alluvial deposit consists of a sequence of clay, silt, and sands, including occasional beds of coarse sand and gravel with kankar interspersed at different depths. The aquifers in this region are prolific and groundwater is tapped from near surface under semiconfined to confined condition (Saha & Shukla, 2013). The average depth of the water table in this region is around 5−10 m below ground level (bgl) (Sinha et al., 2018).

This site was selected for its ease of accessibility, the absence of habitations in the area studied, and the presence of a large number of open piezometric wells located up to 2 km away from the solute injection well serving as observation wells (OWs) (Fig. 17.2). Before the start of the experiment, authorization from the owners of the farming land located within 3 km of the injection wells was requested. The owners of the wells located on the site were also asked not to use them during the monitoring period.

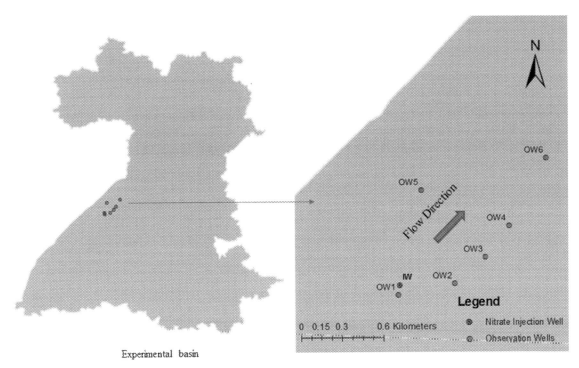

FIGURE 17.2
Experimental site with solute injection and observation well locations.

17.3   Methodology

17.3.1   Hydrological monitoring

The study site was monitored between May 2, 2017 and September 22, 2017. The solute injection experiment was conducted over a period of 120 days, starting from May 25, 2017. Daily meteorological data (i.e., maximum/minimum temperature, humidity, wind speed, sunshine hours, pan evaporation, and rainfall) were obtained from the local meteorological observatory located about 30 km away from the site (25°58′N; 85°41′E) and 53 m amsl. ET rates were estimated using the FAO-56 Penman–Monteith method, described in Kumar, Adamowski, et al. (2016). The hydrogeological investigation of the site was done using sediments collected as drill cuttings obtained from three newly drilled piezometers by the village level administration within 1 km of the site in 2016–17. Sediments were collected at the site using the hand-flapper drilling method (Van Geen et al., 2004) up to a depth of 45 m to develop the lithology of the site.

17.3.2   Solute injection test

Prior to the start of the solute injection test, groundwater samples were collected daily between the May 2 and the May 24, 2017 from all the OWs to measure background NO_3^- concentrations. At the same time, emphasis was placed to establish a network of OWs (Fig. 17.2) and select a solute injection well. As the experiment started, all the irrigation pumps (including private) 2 km around the site were kept off for 30 days (until the June 24, 2017). However, pumping from the private wells started from June 25, 2017 due to the need of preparatory irrigation of pre-monsoon crops. The solute injection experiment was performed during the monsoon season (May–September, 2017) due to the availability of open wells and no use of irrigation wells, which helped in maintaining the nonpumping conditions. Moreover, a high demand of groundwater for crop irrigation (wheat) in the post-monsoon season is very crucial for the crop production and hence the experiment would have not been possible.

On the first day (May 25, 2017), 20 L of solution of ammonium chloride (NH_4Cl) at a concentration of 30 mg L^{-1} was prepared and stored in a carboy as the contaminant source. The source concentration was maintained at 30 mg L^{-1} throughout the experiment to ensure the NO_3^- levels resulting from the injection remained below 45 mg L^{-1}, the permissible limit in India to avoid any health risk. Then, the solute was injected continuously at water table depth (i.e., 5.18 m bgl) using a peristaltic pump at a rate of 50 mL h^{-1} as uniform NO_3^- mass flux starting from May 25, 2017 for the following 16 days. This high rate (50 mL h^{-1}) of injection was maintained to avoid the evaporation and any physical loss of NO_3^- concentration during injection periods. Further, to avoid evaporation loss, the solution container was kept in a cold box and mixed continuously. After the completion of the solute injection, the injection well was kept close. From the first day of solute injection, groundwater samples were collected in 4 mL vials from all six OWs (Fig. 17.2) for NO_3^- concentration analyses. The collected samples were stored at 4°C and analyzed for their NO_3^- content using a single-beam spectrophotometer (Shimadzu, Singapore) at a wavelength of 410 nm.

17.4   Results and discussion

17.4.1   Hydrological characteristics of the experimental site

The study site experienced a wet summer season in 2017 with 38 and 322 mm of rainfall recorded over the period July 1–12 and August 9–16, respectively (Fig. 17.3). The average rainfall over the season was 8.2 mm with a maximum of 157.4 mm on the 47th day of solute injection (i.e., July 11, 2017). During the first month of solute injection test, about 10 rainfall

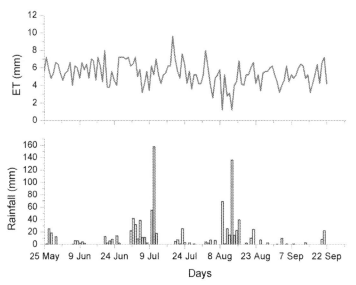

FIGURE 17.3

Daily evapotranspiration/precipitation at the site during solute injection experiment.

events were recorded; however, those rainfall events did not exceed 20 mm. Similarly, no rainfall events exceeding 20 mm were observed at the site during the last month of solute injection. The daily average ET rate was 5.3 mm d^{-1}. The weekly average water table depth observed at the site was 6.6 m bgl compared to a water table depth varying between 5.3 and 6.6 m bgl nearby the Ganga River.

The saturated hydraulic conductivity (K_{sat}) values varied from 105.2 to 130.0 m d^{-1} at the site. The K_{sat} value at the site was high when compared to K_{sat} observed by Bhanja et al. (2019) due to the large sandy (medium size) lithologs as illustrated with the lithofacies of the solute injection site (Fig. 17.4). The top layer (down to an approximately 1.3-m depth) is clay, followed by a 2.7 m of silt layer and a large medium sand layer from a 3.0-m depth.

17.4.2 Nitrate (NO_3^-) transport at solute injection experiment site

The relative concentrations of NO_3^- observed in groundwater with time are presented in Fig. 17.5 for all the OWs. It is important to note that the average background NO_3^- concentration (i.e., over the period May 2 to May 24) was deducted from the observed NO_3^- concentrations to plot the actual relative concentrations at the different OWs. At the OW1 located 374 m away from the solute injection well, a breakthrough in the NO_3^- concentration was

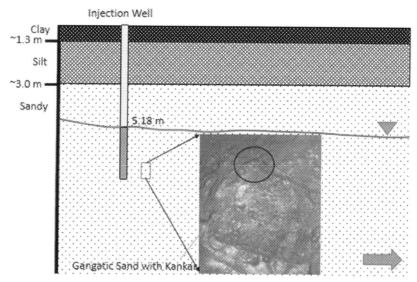

FIGURE 17.4

Lithology observed by sediments collected using the hand-flapper drilling method at site.

observed 9 days after the injection started. The NO_3^- concentration then increased up to relative concentrations of 0.40 on the 40th day after the injection despite three rainfall events at the site over that period. However, the concentration decreased sharply after 40 days suggesting a dilution effect caused by heavy rain (157.4 mm) on the 47th day. Overall, the NO_3^- plumes across the OWs occurred on day 12 at OW1 and OW2; day 22 at OW3, 26 at OW4, 34 at OW5, and 44 at OW6, which were located at a distance of about 593, 806, 648, and 1246 m, respectively, from the solute injection well (Fig. 17.5). It is important to note that the NO_3^- plume reached OW3, OW4, OW5, and OW6 at high concentrations, fortifying advective effects, ranging from 12 to 15 ppm, as opposed to source at OW1 (peak concentrations 12.10 ppm) and OW2 (peak concentrations 11.04 ppm), before it decreased due to two rainfall events following a pattern similar to the one observed at the OW1. The comparatively low concentrations of NO_3^- and OW1 and OW2 suggest that the spatial transport of NO_3^- in groundwater is significantly dependent on the flow direction. Thus, we consider that the groundwater pollution by the impact of agricultural practices is exaggerated along the groundwater flow direction in the study area.

As illustrated in Fig. 17.5, NO_3^- concentrations fluctuated during the later stage of experiment at all the OWs. These highly fluctuating concentrations can be attributed to variations in water flow due to pumping and heavy rainfall. The regions with high average daily precipitation levels experience lower

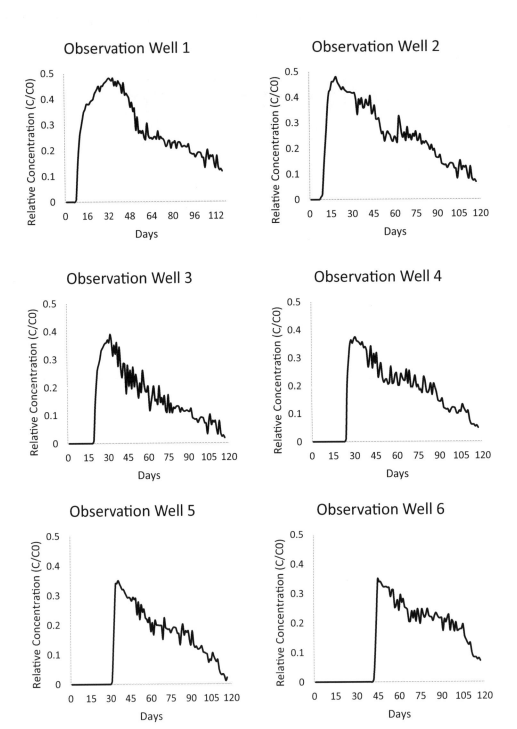

FIGURE 17.5

Relative concentration of NO_3^- for all observation wells (1–6) located around the solute injection well for a period of 120 days.

groundwater NO_3^- concentration as it supports crop growth and subsequent nitrogen uptake as well as dilutes the concentration in the soil and groundwater (in the case of recharge). Furthermore, additional loading from agricultural return flow (pumping and surface ponding irrigation) can cause significant variation in NO_3^- concentrations in the subsurface zone (Fig. 17.5). The variations in NO_3^- concentrations observed at all OWs after 30 days are therefore a likely consequence of uncontrolled private pumping that started around that time (Fig. 17.5). Water table fluctuations due to cumulative effects of rainfall and pumping further affect the fate and transport of NO_3^-. The depth of the water table is also a major factor in determining the time taken for water-soluble contaminants to travel from the land surface down to the aquifer (Aller et al., 1987; Foster et al., 2002). If the water table fluctuations are high, then shallow aquifers are more prone to contamination than deeper ones. Therefore, the aquifers are highly susceptible to NO_3^- contamination in both seasons.

A comparison of the NO_3^- concentrations observed at OW1 and OW2 during nonpumping period and those observed during rainfall events suggest that moderate rainfall events have a dilution effect on NO_3^- concentration with average values decreasing from 11.8 to 13.8 mg L^{-1} at OW1 and OW2, respectively, down to 7.0 and 8.0 mg L^{-1}. A comparison of NO_3^- concentrations measured at OW1 and OW2 during nonpumping periods, pumping periods with heavy rainfall event (45−55th days of experiment), and moderate rainfall event (75−85th days of experiment) presented in Fig. 17.6. The high NO_3^- concentrations observed during nonpumping periods may be due to the early stage experimental conditions where NO_3^- injection was continuous. During the heavy rainfall event, high NO_3^- concentrations ranging from 8.0 to 12.7 mg L^{-1} were observed at OW1, which may be the result of NO_3^- flushing from the crop zone where the NO_3^--based fertilizer is trapped. However, NO_3^- concentrations decreased to values ranging from 8.0 to 11.0 mg L^{-1}, which are slightly lower than the observed concentrations during pumping periods at OW2. High hydraulic conductivity and gradient at this experimental site may contribute to the extent of NO_3^- transport over a large area of about 1.6 km^2.

A deep sandy layer with gravels was observed at the experimental site beneath a thin clay layer that allows a rapid horizontal movement of solute toward down gradient locations. NO_3^- concentrations decreased with increasing distance in the flow direction from the injection well (Fig. 17.6). Table 17.1 lists the average NO_3^- attenuation rates estimated during the periods of nonpumping, heavy rainfall events and moderate rainfall events. Moderate rainfall experienced during the 75−85th days of experiment may have recharged a significant quantity of water in the silt and sandy saturated zone. As the horizontal K_{sat} at this site was very high (i.e., 105.2−132 m d^{-1}), this suggests that the moderate rainfall

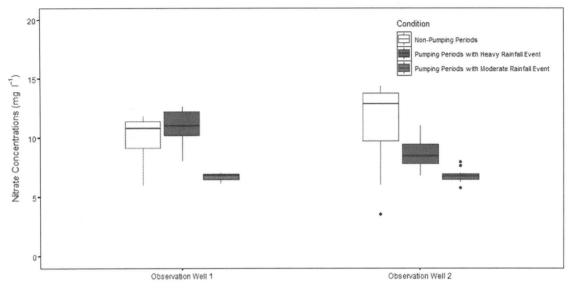

FIGURE 17.6

The NO_3^- concentrations at OW1 and OW2 during nonpumping periods, pumping periods with heavy rainfall event, and moderate rainfall event. *OW*, Observation well.

Table 17.1 Estimated average NO_3^- attenuation rates (mg L^{-1}) during nonpumping periods, heavy and moderate rainfall events.

Experimental durations	OW1	OW2	OW3	OW4	OW5	OW6
Nonpumping (days 10–20)	0.66	0.62	NE	NE	NE	NE
Heavy rainfall (days 45–50)	0.63	0.70	0.77	0.75	0.73	0.67
Moderate rainfall (days 75–75)	0.77	0.77	0.87	0.80	0.82	0.77

NE, *Not experienced;* OW, *observation well.*

events may have accelerated solute movement and subsequent dilution by additional pore water flux. However, the NO_3^- concentrations decreased at a slower rate during the heavy rainfall events (45–50th days), which indicates that the clay layer on the top surface reduced the recharge potential and hence the dilution effect. Thus, a higher average NO_3^- attenuation rate (i.e., 0.77–0.87 mg L^{-1} d^{-1}) was observed during moderate rainfall events than during heavy rainfall events (i.e., 0.63–0.77 mg L^{-1} d^{-1}). This shows that NO_3^- attenuation was dominant in the sandy aquifer layer of the target area. The observed NO_3^- concentrations in this study are in-line with (reactive transport) numerical simulated groundwater

NO_3^- concentrations by Wriedt and Rode (2006) for the Schaugraben catchment, Osterburg (Altmark, Germany). At this site, they observed that denitrification successfully slowed down the advection of the NO_3^- concentration in the deep groundwater zone. Thus, NO_3^- starts attenuating as they move away from the point of injection, resulting in low concentrations at OWs.

Overall, the results of this study indicate that the area nearby the margin of the Ganga river and the Ganga river itself are prone to NO_3^- contamination due to high application rates of nitrogenous fertilizers. It was found that the injection of a small quantity of NO_3^- solute (20 L) could travel in the groundwater zone over long distances, up to 1.6 km^2 from the point of injection. Such threat is more serious as excessive groundwater pumping for irrigation may alter the flow directions, which controls the NO_3^- dynamics (Musolff et al., 2016). Furthermore, groundwater discharge with high NO_3^- concentrations may alter nutrient dynamics of river/lake systems as observed by Jiang et al. (2015). Thus, there is urgent need to manage the irrigation pumping as well as limit the application of N-fertilizer to sustain the groundwater quality in and around this study area.

The current study revealed that the region investigated is very susceptible to NO_3^- contamination, hence potentially affecting the provision of safe drinking water to local communities (Gupta et al., 2020). To provide safe drinking water, methods like raw water source substitution, mixing with low NO_3^- waters, organizational and regional water supply system are suggested, which are local solutions involving no treatment. More advanced techniques for the removal of NO_3^- from pumped water are also available such as adsorption/ion exchange, reverse osmosis, electrodialysis, biochemical denitrification, and catalytic reduction/denitrification (Rezvani et al., 2019). However, unless specific measures to better control the use of N-based fertilizer from a user perspective are taken, the NO_3^- pollution of groundwater sources will remain a major issue. Indeed, Chen (2003) and Ju et al. (2004) suggested that increasing N-application rates to improve crop yields of wheat, maize, greenhouse vegetables, and cotton was not an efficient practice as no significant correlations between yield and N-application rate could be found. Therefore, the options for reducing the level of NO_3^- in water bodies include the optimization of fertilizers uses, organic farming, and crop rotation, regarding which the farmers must be educated. In addition, pumping from deeper aquifer causes significant decline in the groundwater level; hence a faster movement of NO_3^- occurs due to high hydraulic gradient (Musolff et al., 2016). Therefore, a complete study of NO_3^- movement from surface soil through vadose zone and groundwater zone should be further investigated to accurately predict future pollution under varying climatic conditions.

17.5 Conclusion

In this study, a field scale in situ solute injection test was conducted at a sandy aquifer site located in the Samastipur District of Bihar (India) to understand the fate and transport of NO_3^- in groundwater. For this purpose, a mineral nitrogen (NH_4Cl) solution was injected at water table, that is, depth 5.18 m bgl at the rate of 5 mL h^{-1} and its movement was observed in 6 OWs for 120 days (May 25, 2017 to September 22, 2017). Analysis of the results indicates that the horizontal movement of NO_3^- in the saturated zone without pumping is fast (29.5 m d^{-1}). Further, this also suggests that during heavy rainfall events, the movement of NO_3^- may accelerate and concurrently attenuates due to dilution. Its attenuation due to denitrification was comparatively high as NO_3^- moved away from the injection well and thus no breakthrough curves were observed at any OWs. High K_{sat} of the Gangetic sandy saturated zone may enhance the movement of NO_3^- significantly. On the other hand, the Gangetic Kankar may create preferential flow path for the NO_3^- movement. Understanding of nitrate transport within the saturated zone is necessary to predict the plume area and its stability to control future nitrate loading on groundwater and adopt appropriate remediation techniques and hence offering a unique opportunity to further understand the fate of nitrate in groundwater under real conditions and providing additional information to support the development of management and remediation plan for fresh water and/or polluted sites. This study indicates toward the urgent need to manage the irrigation pumping as well as limit the application of N-fertilizer to sustain the groundwater quality in and around this study area.

CRediT authorship contribution statement

Pankaj Kumar Gupta: Conceptualization, design, and conduction of field experiment, data collection, data curation, methodology, formal analysis, writing—original draft. Basant Yadav: Formal analysis, writing-review, and editing. Kristell Le Corre: Formal analysis, writing-review, and editing. Alison Parker: Writing-review and editing.

Conflicts of interest

None.

Acknowledgment

Authors are thankful to local administrations, farmers, and workforce who provided support during this study. Authors would also like to thank Rajendra Central Agricultural University, Pusa for the local meteorological data used in this study.

Data availability

Data, models, or code generated or used during the study are available from the corresponding author by request (list items: NO_3^- concentrations, calibration curves, rainfall, soil types, land use cover map).

References

Acharyya, S. K., & Shah, B. A. (2007). Groundwater arsenic contamination affecting different geologic domains in India—A review: Influence of geological setting, fluvial geomorphology and Quaternary stratigraphy. *Journal of Environmental Science and Health, Part A, 42*(12), 1795–1805.

Ahada, C. P., & Suthar, S. (2018). Groundwater nitrate contamination and associated human health risk assessment in southern districts of Punjab, India. *Environmental Science and Pollution Research, 25*(25), 25336–25347.

Akbariyeh, S., Bartelt-Hunt, S., Snow, D., Li, X., Tang, Z., & Li, Y. (2018). Three-dimensional modeling of nitrate-N transport in vadose zone: Roles of soil heterogeneity and groundwater flux. *Journal of Contaminant Hydrology, 211*, 15–25.

Aller, L., Bennet, T., Lehr, J.H., & Petty, R.J. (1987). DRASTIC. A standardized system for evaluating groundwater pollution potential using hydrogeologic settings. *U.S. EPA report 600/2–87-035*, Oklahoma.

Bhanja, S. N., Mukherjee, A., Rangarajan, R., Scanlon, B. R., Malakar, P., & Verma, S. (2019). Long-term groundwater recharge rates across India by in situ measurements. *Hydrology and Earth System Sciences, 23*(2), 711–722.

Central Ground Water Board (2013). *Ground water information booklet*, Samastipur District, Bihar State, India.

Chen, X. P. (2003). Optimization of the N fertilizer management of a winter wheat/summer maize rotation system in the Northern China Plain. *Ph. D. diss.* Stuttgart: University of Hohenheim.

Fishman, R., Kishore, A., Ward, P.S., Jha, S., & Singh, R.K.P. (2016). Can information help reduce imbalanced application of fertilizers in India? Experimental evidence from Bihar. *International food policy research institute discussion paper 01517 (March 2016)* Washington, DC: International Food Policy Research Institute (IFPRI). <http://ebrary.ifpri.org/cdm/ref/collection/p15738coll2/id/130249.S.C>. Babu et al. Environmental Development 25 (2018) 111–125123.

Foster, S., Hirata, R., Gómez, D., D'Elia, M., & Paris, M. (2002). *Ground water quality protection. A guide for water utilities, municipal authorities and environment agencies*. Washington, DC: The World Bank.

Gupta, P. K. (2020). Pollution load on Indian soil-water systems and associated health hazards: A review. *ASCE Journal of Environmental Engineering, 146*. Available from https://doi.org/10.1061/(ASCE)EE.1943-7870.0001693.

Gupta, P. K., Yadav, B., Kumar, A., & Singh, R. P. (2020). India's major subsurface pollutants under future climatic scenarios: Challenges and remedial solutions. *Contemporary environmental issues and challenges in era of climate change* (pp. 119−140). Singapore: Springer.

Harter, T., Onsoy, Y., Heeren, K., Denton, M., Weissmann, G., Hopmans, J., & Horwath, W. (2005). Deep vadose zone hydrology demonstrates fate of nitrate in eastern San Joaquin Valley. *California Agriculture, 59*(2), 124−132.

Jain, C. K. (2002). A hydro-chemical study of a mountainous watershed: The Ganga, India. *Water Research, 36*(5), 1262−1274.

Jemison, J. M., & Fox, R. H. (1994). Nitrate leaching from nitrogen-fertilized and manured corn measured with zero-tension pan lysimeters. *Journal of Environmental Quality, 23*(2), 337−343.

Jiang, Y., Nishimura, P., van den Heuvel, M. R., MacQuarrie, K. T., Crane, C. S., Xing, Z., & Thompson, B. L. (2015). Modeling land-based nitrogen loads from groundwater-dominated agricultural watersheds to estuaries to inform nutrient reduction planning. *Journal of Hydrology, 529*, 213−230.

Joshi, P., & Gupta, P. K. (2018). Assessing groundwater resource vulnerability by coupling GIS-based DRASTIC and solute transport model in Ajmer district, Rajasthan. *Journal of the Geological Society of India, 92*(1), 101−106.

Ju, X. T., Liu, X. J., Zhang, F. S., & Roelcke, M. (2004). Nitrogen fertilization, soil nitrate accumulation, and policy recommendations in several agricultural regions of China. *Ambio, 33*, 300e305.

Kumar, D., Adamowski, J., Suresh, R., & Ozga-Zielinski, B. (2016). Estimating evapotranspiration using an extreme learning machine model: Case study in north Bihar, India. *Journal of Irrigation and Drainage Engineering, 142*(9)04016032.

Kumar, M., Ramanathan, A. L., Rahman, M. M., & Naidu, R. (2016). Concentrations of inorganic arsenic in groundwater, agricultural soils and subsurface sediments from the middle Gangetic plain of Bihar, India. *Science of the Total Environment, 573*, 1103−1114.

Kumari, B., Gupta, P. K., & Kumar, D. (2019). In-situ observation and nitrate-N load assessment in Madhubani district, Bihar, India. *Journal of Geological Society of India (Springer), 93*(1), 113−118. Available from https://doi.org/10.1007/s12594-019-1130-z.

Kundu, M. C., & Mandal, B. (2009a). Agricultural activities influence nitrate and fluoride contamination in drinking groundwater of an intensively cultivated district in India. *Water, Air, and Soil Pollution, 198*(1−4), 243−252.

Kundu, M. C., & Mandal, B. (2009b). Nitrate enrichment in groundwater from long-term intensive agriculture: Its mechanistic pathways and prediction through modeling. *Environmental Science and Technology, 43*(15), 5837−5843.

Musolff, A., Schmidt, C., Rode, M., Lischeid, G., Weise, S. M., & Fleckenstein, J. H. (2016). Groundwater head controls nitrate export from an agricultural lowland catchment. *Advances in Water Resources, 96*, 95−107.

Reddy, A. G. S., Kumar, K. N., Rao, D. S., & Rao, S. S. (2009). Assessment of nitrate contamination due to groundwater pollution in north eastern part of Anantapur District, AP, India. *Environmental Monitoring and Assessment, 148*(1−4), 463−476.

Rezvani, F., Sarrafzadeh, M. H., Ebrahimi, S., & Oh, H. M. (2019). Nitrate removal from drinking water with a focus on biological methods: A review. *Environmental Science and Pollution Research, 26*(2), 1124−1141.

Saha, D., & Shukla, R. R. (2013). Genesis of arsenic-rich groundwater and the search for alternative safe aquifers in the Gangetic Plain, India. *Water Environment Research, 85*(12), 2254−2264.

Sankararamakrishnan, N., Sharma, A. K., & Iyengar, L. (2008). Contamination of nitrate and fluoride in ground water along the Ganges Alluvial Plain of Kanpur district, Uttar Pradesh, India. *Environmental Monitoring and Assessment, 146*(1−3), 375−382.

Singleton, M. J., Esser, B. K., Moran, J. E., Hudson, G. B., McNab, W. W., & Harter, T. (2007). Saturated zone denitrification: Potential for natural attenuation of nitrate contamination in shallow groundwater under dairy operations. *Environmental Science and Technology*, *41*(3), 759−765.

Sinha, R., Gupta, S., & Nepal, S. (2018). Groundwater dynamics in North Bihar plains. *Current Science*, *114*(12), 2482−2493.

Suthar, S. (2011). Contaminated drinking water and rural health perspectives in Rajasthan, India: An overview of recent case studies. *Environmental Monitoring and Assessment*, *173*(1−4), 837−849.

Suthar, S., Bishnoi, P., Singh, S., Mutiyar, P. K., Nema, A. K., & Patil, N. S. (2009). Nitrate contamination in groundwater of some rural areas of Rajasthan, India. *Journal of Hazardous Materials*, *171*(1−3), 189−199.

Thomsen, I. K. (2005). Nitrate leaching under spring barley is influenced by the presence of a ryegrass catch crop: Results from a lysimeter experiment. *Agriculture, Ecosystems and Environment*, *111*(1−4), 21−29.

Van Geen, A., Protus, T., Cheng, Z., Horneman, A., Seddique, A. A., Hoque, M. A., & Ahmed, K. M. (2004). Testing groundwater for arsenic in Bangladesh before installing a well. *Environmental Science and Technology*, *38*(24), 6783−6789.

Verma, R. R., Srivastava, T. K., & Singh, P. (2019). Climate change impacts on rainfall and temperature in sugarcane growing Upper Gangetic Plains of India. *Theoretical and Applied Climatology*, *135*(1−2), 279−292.

Wendland, F., Bogena, H., Goemann, H., Hake, J. F., Kreins, P., & Kunkel, R. (2005). Impact of nitrogen reduction measures on the nitrogen loads of the river Ems and Rhine (Germany). *Physics and Chemistry of the Earth*, *30*(8−10), 527−541, Parts A/B/C.

Wriedt, G., & Rode, M. (2006). Modelling nitrate transport and turnover in a lowland catchment system. *Journal of Hydrology*, *328*(1−2), 157−176.

Yadav, B. K., & Junaid, S. M. (2014). Groundwater vulnerability assessment to contamination using soil moisture flow and solute transport modeling. *Journal of Irrigation and Drainage Engineering*, *141*(7), 04014077.

CHAPTER 18

Integrated water resources management in Sikta irrigation system, Nepal

Rituraj Shukla[1], Ishwari Tiwari[2], Deepak Khare[2,3] and Ramesh P. Rudra[1]

[1]School of Engineering (Water Resources Engineering), University of Guelph, Guelph, ON, Canada, [2]Irrigation Department, Nepal, [3]Indian Institute of Technology, Roorkee (IITR), India

18.1 Introduction

Global population is rising rapidly. The rate of population growth is higher in developing countries than in developed countries. In 1992 world's population was 5.7 billion and in 2012 it reached up to 7 billion (United Nations, 2012). UN has projected that the world's population will reach around 10 billion in 2050. This rapid growth of population is due to high birth rate in the third world countries. This increasing population and higher living standard have increased per capita demand of water. Growing population needs more food and forces to increase crop yield which needs adequate and timely application of water. Environmental provision set by national and international laws will require more water for aquatic life, wildlife refuges, recreation, and scenic value and riparian habitats. Besides above mentioned facts, social, economic, and political landscape changes have further put more thrust on water resources.

The stress on management of water resource, the precious amenity of world, will increase with the elapse of time. However, ample water is available on the planet, there is large variation in availability in space and time. Seawater which is unfit for most human uses occupies 97% of the Earth's water. The remaining 3% is inaccessible, either locked in polar icecaps or in deep underground aquifers. Hence, only 0.4% of all of the water on the Earth in a form that is usable and accessible by human beings. Therefore water resources management became extremely essential.

Many management concepts have been practiced for water resources management. BHIWA (Basin-wide holistic Integrated Water Assessment), European Water Frame work Directive along with DSS have been used for water

Advances in Remediation Techniques for Polluted Soils and Groundwater. DOI: https://doi.org/10.1016/B978-0-12-823830-1.00002-X

management of basin level (Bazzani et al., 2004; Khan et al., 2005; Harmancioglu et al., 2008). However, a popular and widely accepted tool for groundwater management is numerical groundwater simulation which embodies the principles of mass balance and momentum theory. Its flexible modular nature allows selective use of packages for specific simulation tasks (Xu & Chen, 2005) 2005, Abdulla & Tamer, 2006; Hollander et al., 2009). Strenuous efforts have been focused on the groundwater management in the single case study of the aquifer of a basin (Abdulla & Tamer, 2006; Mao et al., 2005; Rejani et al., 2008; San Juan et al., 2010). Raul et al. (2011) studied Hirakunda Canal command area of eastern India together with the effects of irrigation water to the aquifer and its response due to various stress conditions. This study had also provided possible water management practices to overcome the upcoming extremes in the aquifer system. Groundwater pollution is another serious issue where pollutants were ejected into the system and intrusion of seawater polluted the aquifer due to overexploitation of groundwater (Rejani et al., 2008).

The prediction accuracy of model depends on the perfectness in interpreting the natural system in model. Surrogating the natural groundwater system with initial condition, boundary conditions, and stresses on the system is the basis of efficient decision otherwise misleading would happen. Therefore both calibration and validation are absolutely necessary for the prediction of accurate hydraulic parameters. The model has been calibrated in two sequential stages: a steady-state calibration followed by a transient calibration (Raul et al., 2011; Sakiyana & Yazicigil, 2004). Groundwater models were validated in different methods. Raul et al. (2011) had validated the model using data of stress periods different from calibrated data using same observation wells. But Xu et al. (2011) had validated his model using same stress period that used for calibration of model.

Among available groundwater model, Visual MODFLOW 4.2 has been selected as a tool for groundwater modeling of the study area because it is well documented; has easy accessibility and versatility; is widely used and user-friendly; and provides fairly accurate prediction results (Kashaigili et al., 2003). Visual MODFLOW can simulate both the confined and unconfined aquifer using MODFLOW 2000 (Harbaugh et al., 2000). In this study, a transient groundwater model was developed using Visual MODFLOW 4.2 for simulating different boundary and stress conditions of the unconfined aquifer.

18.2   Study area

Geographically the area comprises lands between $27°58'07''$ and $28°13'12''$ N latitudes and $81°30'07''$ and $81°43'55''$ E longitudes. A location map of the project area is shown in Figs. 18.1–18.3. Out of total cultivable area of 58,990 ha in the district, the study area covers about 37,352 ha enclosed by the East-West Highway (Mahendra Rajmarg) to the north, the national boundary to the south, the Duruwakholato to the east, and Man Khola to the west (Fig. 18.1). The study area is situated in a climate zone of

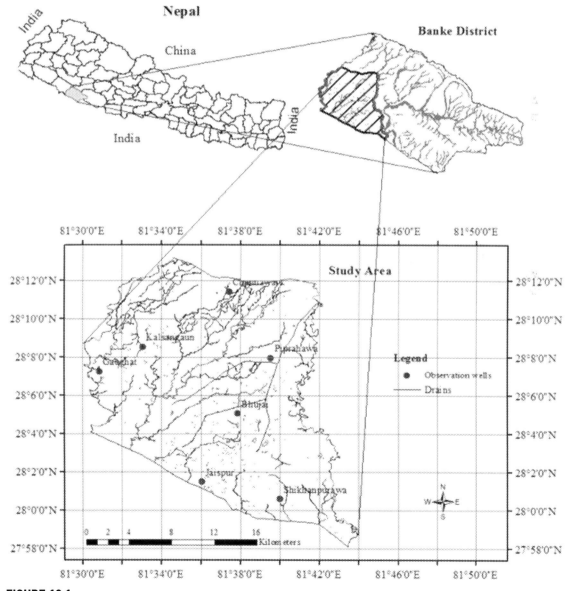

FIGURE 18.1

Location of study area.

subtropical monsoon. The mean annual rainfall is1400 mm of which more than 80% occurs during the month of June to September. The temperature in the project area ranges from minimum of 5°C in the month of January to the maximum of 44°C in the month of June. The range of humidity falls between 60% in May and 85% in January. The Terai is underlain by a thick sequence of saturated detrital sediments of alluvial and colluvial origin. The study area is an alluvial plain formed by material transported by several rivers which rise in the highlands to the north of the command area and flow southward to the Indian border. The elevation of the area ranges from 134 to 174 m. The slope of the land is gentle toward south.

The overall cropping intensity in the proposed command area has been estimated at 168%, which is slightly higher than the Banke district that is an average of 161%. Although cropping intensity is slightly higher in the command area than the district average, it can be considered still low. The low rate of overall cropping intensity (168%) in the area has to be seen in the context of very low irrigation coverage in the area.

18.3 Methodology/philosophy

18.3.1 Conceptual model of the study area

The conceptual model of the study area was developed based on the surface elevation and litho logs of the area. The surface elevation of the study area was obtained from the SRTM (Shuttle Radar Topography Mission) DEM (Digital Elevation Model) downloaded from GLCF website. The SRTM DEM with resolution of 90×90 m which represents fairly accurate ground surface than other DEM (Frey and Frank, 2012). The Visual MODFLOW has the facility to import the DEM directly to represent the ground surface of the area of concern. The DEM showed that the study area has downward slope toward south.

An important tool to characterize the aquifer is hydrogeological profiles. Hydrogeological profile was conceptualized from the data obtained from litho logs. There are alternative distinct confined/semiconfined aquifers separated by clay layers of thickness ranging from 2 to 41 m. First aquifer having thickness 3−51 m exists at height from mean sea level is 191−80 m.

Only unconfined aquifer was taken for the study purpose. The unconfined aquifer has been exploited by extracting water using shallow tube wells of 610 of different village development committees. These tube wells have less discharging capacity and has utilized for irrigation purpose. For drinking water, small hand pumps were installed in almost all individual houses. Due to overexploitation of groundwater from the first aquifer, the water table of southern part of the

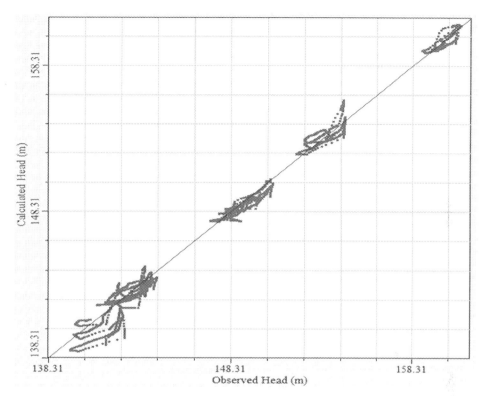

FIGURE 18.2

A plot of observed versus simulated heads in calibration.

area is gradually decreasing. Some deep tube wells have been installed for irrigation and drinking purposes and some for industrial institution.

For model simulation, some observed values of known points were necessary. Those observed values were compared with the simulated value to ascertain theoretical representation of actual natural system within prescribed computer code so that the predicted values are reliable. In groundwater model, groundwater head is a threshold for comparing measured and calculated values. Historical data for groundwater level were available in seven observation points. Those points were scattered within the study area (Figs. 18.2–18.3).

18.3.2 Abstraction of groundwater

The well package of MODFLOW is designed to simulate the inflow and outflow through recharge and pumping wells (Rejani et al., 2008). The residual abstraction amounts in the first aquifer was assigned to 610 wells. Average pumping

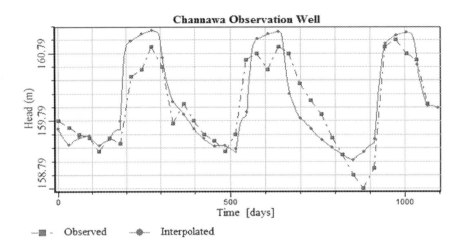

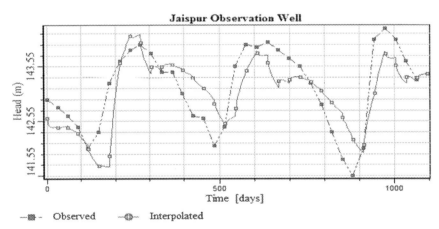

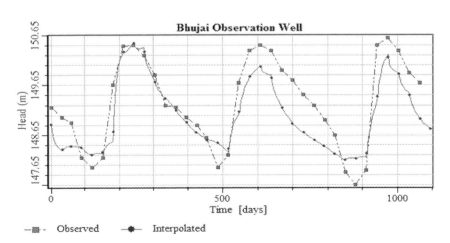

FIGURE 18.3

Time series plot of observed and simulated groundwater level at individual observation points in calibration.

from shallow tube wells located in the first aquifer ranges from 5 to 10 lps and that from deep tube wells which extract water from second and third aquifers ranges from 15 to 40 lps. No data are available to assess the actual withdrawal of groundwater from first and other semiconfined aquifers. A thumb rule is prevailing in Nepal is that a shallow tube well would irrigate 2.5 ha of land and that of a deep tube well would irrigate 40 ha. Actual withdrawal of groundwater had been assumed to be equal to the amount lesser than the average of crop water required for existing cropping pattern or maximum discharge of the tube well during that very month. The average crop water requirement had been calculated by averaging the crop water requirement of crops prevailing in respective months. There are nine nonirrigating wells in the study area. They supply water for industrial, municipal, and rural usages. The amount of groundwater extraction from these deep tube wells other than irrigation tube wells was taken as the amount of groundwater extraction in 10 h of their running in 1 day with their average discharging capacity. The average discharging capacity of irrigation deep wells was taken as 25 lps and of water supply well was taken as 15 lps for the purpose of this study.

The total population of the study area is 0.227 million (Central Bureau of Statistics Government of Nepal (2012). According to the GIS shape file provided by the Department of Survey of Nepal, there are 311 villages within the study area. On an average, the number of population in each village except municipality is 731. There is no other source of drinking water other than groundwater. Almost all people extract groundwater for their daily use. In rural area, one person needs 45 L of water per day. Therefore, from one village average amount of water extracted for drinking and other purpose is 32.85 m^3 per day and one well in each village was assumed to abstract the stated amount of water from unconfined aquifer for drinking purpose.

18.3.3 Discretization of study area

The study area was discretized by dividing it into finite difference cells. Parameters representing physical characteristics and flow conditions were assigned to each cell. Visual MODFLOW calculates the hydraulic head and aquifer parameter at the center of the cells. The area was divided into 56 rows and 46 columns (Figs. 18.4−18.5). The grid spacing is 500×500 m in whole area. Thus the total number of cells over the area is 2576 in each layer. Based on the data availability of the groundwater level , month was chosen as the time step within which all hydrological stresses can be assumed to be constant.

18.3.4 Assigning boundary conditions

There are two rivers: Duduwa river to the east and Man river to the west which serve as two Cauchy (rivers) boundaries of the basin. The north and

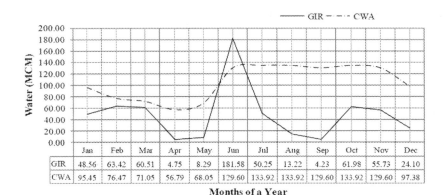

FIGURE 18.4

Comparison of GIR and CWA for Proposed Cropping Pattern. *CWA*, canal water availability; *GIR*, gross irrigation requirement.

south boundaries were assumed as the Neumann boundary of thin wall which allows frequent entry and exit of water flux into the system. The effect of flow between the rivers and aquifer was simulated by dividing the rivers into reaches containing single cells. The river bed conductance, C_{river}, was computed by using the following equation:

$$C_{river} = \frac{K_r L W_r}{B}$$

where K_r is the hydraulic conductivity of the river bed $(m\,s^{-1})$, W_r is the width of the river (m), L is the length of the reach/grid size (m), and B is the thickness of the riverbed (m). River bed conductivity had been obtained from similar study conducted by Raul et al. (2011). These values have been given in Table 18.1.

18.3.5   Initial conditions

Five-year continuous data were available for groundwater table on these observation wells. If the initial condition for groundwater levels was provided, model would easily converge the calculated groundwater level. That is why initial condition is necessary to start the calculation. The water level in observation wells of January 2007 was taken as the initial condition.

18.3.6   Hydrogeological parameters

The top layer of the area is composed of alluvial soil. The unconfined aquifer is of fine-to-medium sand. Initially, aquifer properties were taken from literature matching the property of the soil obtained from litho logs. The top and second layers of the study area have divided into three soil zones to assign soil properties.

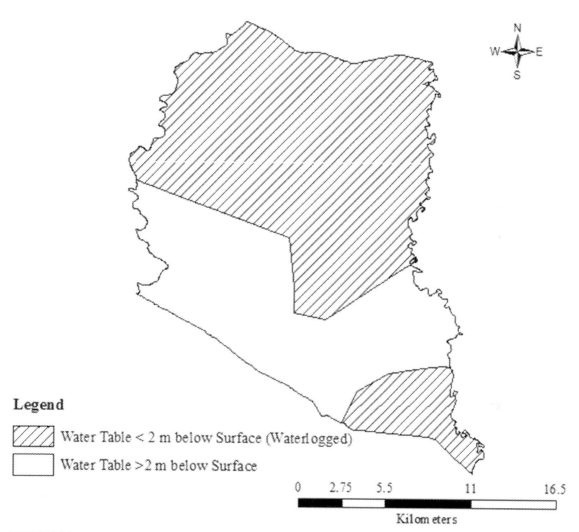

FIGURE 18.5
Waterlogged area in monsoon in existing condition.

18.3.7   Groundwater recharge

Groundwater replenishment is the major event for groundwater development. Recharge is that phenomenon which increases the groundwater storage. Groundwater recharge is related to precipitation, land coverage, river discharge, and evaporation (Mao et al., 2005, Myrab, 1997; Rao et al., 1996). The principal sources of groundwater recharge of the aquifers of the study area are precipitation, rivers, and interflow from the Siwalik hill side. The

Table 18.1 Conductivities of riverbed.

S. N.	Month	K_f (m s^{-1})
1	January	2.5×10^{-7}
2	February	2.5×10^{-7}
3	March	2.5×10^{-7}
4	April	2.5×10^{-7}
5	May	2.5×10^{-7}
6	June	1.3×10^{-7}
7	July	1.3×10^{-7}
8	August	1.3×10^{-7}
9	September	1.3×10^{-7}
10	October	1.3×10^{-6}
11	November	7.5×10^{-7}
12	December'	4.4×10^{-7}

recharge package in MODFLOW is designed to simulate aerial distributed recharge to the groundwater system (Rejani et al., 2008). Percentage precipitation that contributes in groundwater recharge was taken from the formula developed by R.S. Chaturvedi in similar study area of Ganga–Yamuna Doab in India. Chaturvedi's formula has been widely used in the area of tropical region. Therefore the modified recharge formula of Chaturvedi (1973) was used for preliminary calculation of groundwater replenishment (R) in the study area. This empirical formula is as follows:

$$R = 1.35(P - 14)^{0.5}$$

where R is the net recharge in inch due to precipitation and P is the average annual rainfall in inch.

The precipitation data of the area had been taken from three meteorological stations located within the study area. These stations were Sikta (0409), Naubasta (0414), and Khajura (0419).

18.3.8 Evapotranspiration

The evapotranspiration (ET) package of the numerical groundwater flow model simulates the effects of plant transpiration via capillary rise from the saturated zone, that is, directly from the water table. The ET package approach is based on the assumptions that (1) when the water table is at or above the ground surface (top of layer 1), ET loss from the water table occurs at the maximum rate as specified by the user; (2) when the elevation of the water table is below the extinction depth or is beneath the layer 1, ET from the water table is negligible;

and (3) between these limits, ET from the water table varies linearly with water table elevation (Raul et al., 2011).

The extinction depth depends upon the root zone depth and capillary rise. Root zone depth is the depth of the soil reservoir that the plant can reach to get water. The average effective root zone depth for summer crops was taken as 0.7 m and for winter crops 0.5 m for the study purpose. The capillary rise in average condition of soil is about 1.4 m (Raul et al., 2011). The sum effective root depth and capillary rise is the extinction depth.

This ET package represents only a minor part of ET in the phreatic area where the groundwater is close to the surface. Therefore extinction depth could be taken zero in monsoon season because rainfall accumulated over crop field satisfies ET. There is always downward flow of water to groundwater table from surface where rainfall accumulation occurs in dyke of field in monsoon. The maximum extinction depth is the summation of average effective depth and capillary rise. Hence, extinction depth varies from 0, during peak monsoon, to 1.9 and 2.1 m below the ground surface.

ET of monsoon season would satisfy from the ponded water and ET package does not deal with this value when the water surface is above the ground surface. Therefore it is reasonable to adopt this value equal to zero during monsoon. The average reference ET of remaining months was calculated based on daily measured reference ET at Khajura Station (station index 409) within the study area, crop coefficients of recommended by the FAO in humid region. Monthly maximum ET and root zone depths have been tabulated in Table 18.2.

Table 18.2 Maximum evapotranspiration and extinction depth.

S. No.	Month	Maximum ET (mm day^{-1})	extinction depth(m)
1	January	1.87	2.1
2	February	3.13	1.9
3	March	3.13	1.9
4	April	3.44	1.9
5	May	6.95	1.9
6	June	7.26	1.9
7	July	0	0
8	August	0	0
9	September	0	0
10	October	3.75	2.1
11	November	1.93	2.1
12	December	1.15	2.1

ET, evapotranspiration.

18.3.9 Computation of irrigation water requirement

Within rainfed area of the study, crops experience stress during its growing period due to irregularities of rainfall. After knowing that the extra water is to be supplied to the crops during scarcity of the rain water, water requirement for the existing and future cropping pattern is necessary. The gross irrigation requirement of a crop can be calculated by using the following equation (Raul et al., 2011).

$$GIR = \frac{(ETc - RFef)}{\eta a} + SPR - SMC - GWC$$

where GIR is the gross irrigation requirement in field (mm), ETc is the crop ET (mm) which is equal to ($ETo \times Kc$), ETo is the reference ET (mm), Kc is the crop coefficient, SPR is the special-purpose requirements (mm), SMC is the soil moisture contribution, and GWC is the capillary contribution from groundwater, RFef is the effective rainfall (mm), and ηa is the field application efficiency (fraction). SMC and GWC may be considered negligible under certain localized condition. In this study, these two components are not taken into calculation.

In the above equation, ETo is necessary. Various methods of reference crop ET computation are available (Allen et al., 1998; Doorenbos & Pruitt, 1977). The FAO Penman-Monteith method (Allen et al., 1998) has been recommended as the standard method. However, the measured reference ETo was available at Khajura Station (index no. 409) which was used for the calculation of crop water requirement. FAO-recommended crop coefficient values (Kc) of humid region were adopted for each crop. Effective rainfall was calculated using the US Department of Agriculture Soil Conservation Service method (Smith et al., 1998). Field application efficiency was taken as 32% for rice and 58% for nonrice crops (Doorenbos & Pruitt, 1977).

18.3.10 Special-purpose water requirements

For all crops, land water is required for land preparation, that is, presowing irrigation. Land preparation includes easy ploughing, land smoothening, and disking the undulated land. Nursery raising and transplanting of rice also required irrigation. These requirements are categorized as special-purpose water requirements (SPR). For all crops, land preparation requires water of 70 mm (Tyagi, 1980), whereas rice requires 200 to 250 mm of water for nursery raising and transplanting. In this study, average value of 225 mm of water is required for SPR of rice.

18.4 Groundwater modeling

18.4.1 Calibration

The model has been calibrated in two sequential stages: a steady-state calibration followed by a transient calibration (Gaaloul, 2014; Raul et al., 2011;

Sakiyana & Yazicigil, 2004). The steady-state condition is a condition that existed in the aquifer before any development had occurred. Matching the initial heads observed for the aquifer with the hydraulic heads simulated by MODFLOW is called steady-state calibration. Steady-state calibration was made against the observed water levels of January 2007. The model is successfully simulated with a root mean square error (RMSE) of 0.346 m and standard error of estimation (SEE) of 0.126 m (Fig. 18.2).

Continuous monthly observation of well data from January 2007 to December 2011 were available. Three years monthly groundwater head data ranged from January 2007 to December 2009 were taken for transient state calibration purpose. This period was divided into 36 stress periods. Boundary conditions and stresses of the aquifer system were the input of the model in monthly basis. The calculated groundwater heads were compared with the observed head at observation points. For proper consistency, Winpest subroutine of Visual MODFLOW was run to adjust the sensitive parameter so that the objective function would become minimum providing good match of calculated heads with observed one at particular observation points.

A plot of observed versus simulated groundwater level of six observation well points was shown in Figs. 18.2−18.5 which had shown satisfactory match with the SEE 0.01 m, RMSE 0.57 m, and correlation coefficient (r) 0.996. The time series plots of observed and simulated groundwater level of each individual observation points have shown in Fig. 18.3.

18.4.2   Validation

Groundwater models were validated in different methods. Raul 2011 had validated the model using data of stress periods different from calibrated data using same observation wells. But Xu et al. (2011) had validated his model using same stress period that used for calibration of model. However, they used different observation wells which were not used for calibration. In this study, different stress periods were used same observation wells both in calibration and validation processes.

The groundwater observation data, other boundary conditions and stresses of December 2009 to December 2011 were used for validation purpose. The other aquifer properties remained the same that were obtained from the calibrated model. The validation result showed the goodness of fit parameters as SEE 0.01 m and RMSE 0.556 m. Hence, it indicated that the parameters were calibrated properly.

Time series plots of measured and simulated head at all observation wells showed a fair match of observed and calculated head at that particular point

Table 18.3 Comparison of model evaluation parameters.

S. N.	Location	SEE (m)		RMSE (m)	
		Calibration	Validation	Calibration	Validation
1	Bhujai	0.02	0.023	0.47	0.515
2	Channawa	0.016	0.019	0.307	0.369
3	Gaughat	0.05	0.037	1.117	0.722
4	Jaispur	0.024	0.027	0.47	0.546
5	Kalsangaun	0.016	0.018	0.324	0.539
6	Piprahawa	0.027	0.036	0.537	0.693
7	Sikhanpurwa	0.017	0.016	0.456	0.49

RMSE, *root mean square error;* SEE, *standard error of estimation.*

at different stress periods. The value of performance criteria at all individual observation well has been shown in Table 18.3. These values showed that the model has been calibrated satisfactorily and can be used to predict the future scenario of different boundary and stress conditions of the system.

18.5   Result and discussion

18.5.1   Gross irrigation requirement

Seasonal gross irrigation water requirement of the crops was estimated. The monsoon net irrigation water requirement of crops from sowing to harvesting varies from 32 to 430 mm, whereas in the case of nonmonsoon crops, net irrigation water requirement ranges from 160 to 320 mm.

Seasonal weighted average depth of irrigation water requirement for whole crop period was computed from the gross irrigation requirement of all the crops grown in the command area for existing cropping pattern. The weighted average crop delta varies from 650 mm in winter to 979 mm in monsoon season. Based on this weighted average delta, water for irrigation was estimated as 165 MCM in winter and 341 MCM in monsoon season. The annual gross irrigation water requirement for existing cropping pattern was found to be 506 MCM. This value is equal to the annual GIR based on average weighted methods.

When the rainfed command area receives irrigation facility, the existing cropping pattern will change and cropping intensity will reach more than 200%. Demand of irrigation water will increase. If the cropping pattern would increase up to 195%, the monsoon and nonmonsoon net water requirement would reach up to 196.6 and 380 MCM, respectively.

18.5.2   Irrigation water supply and demand

Sikta irrigation project is under construction. The main canal of the system is supposed to carry 50 cumec of water. Therefore diverted water from head work to main canal is 50 cumec or as available in the river after aquatic requirement.

For proposed cropping pattern, same scenario was seen as in the existing cropping pattern. Scarcity (50 MCM) is only in the month of June (Fig. 18.4). This month is dry and generally rice nursery raising and paddy transplanting methods are practiced during this month. Hence, the requirement of water for crop plant is high and crucial. Groundwater extraction would be high in this month because other surface irrigation sources are not available.

18.5.3   Calibration and validation output

Model calibration showed a reasonably good match between the observed water table and simulated output. A scatter plot (1:1 plot) of the measured groundwater level against the calibrated head, as illustrated in Fig. 18.2, has shown reasonably a better fit between these two data sets. SEE and RMSE of calibrated model for different well points varied from 0.016 to 0.05 m. and 0.307 to 1.117 m, respectively, whereas those for validation model varied from 0.016 to 0.037 m and 0.369 to 0.722 m, respectively. SEE and RMSE are tabulated in Table 18.3 per observation points. These values are within acceptable limits.

After successful calibration and validation of model, the hydraulic conductivities of first layer ranged from 3.57×10^{-3} to 1.06×10^{-6} m s^{-1}. This value for second layer varied from 1.83×10^{-5} to 5.35×10^{-5} m s^{-1}. The third layer had the conductivity of 3.08×10^{-7} m s^{-1}. The specific yield of the first, second, and thirdlayers are 0.06, 0.3, and 0.1, respectively, whereas specific storage of these layers is 0.001, 0.0001, and 0.001 m^{-1}, respectively. The annual recharge for different recharge zones was found to be varied between 194 to 380 mm, providing an average annual recharge of 217 mm.

18.5.4   Groundwater availability

In existing situation, study of water table fluctuation showed that the water table remains within 2 m below the ground surface in around 24,000 ha of study area during peak monsoon months (Fig. 18.5). Just after the monsoon period, the water table starts depleting and reached up to 5.2 m below the ground surface. About 6% of the area has water table within 2 m of depth in winter, which, therefore, is in waterlogged condition in nonmonsoon season as well.

Present groundwater withdrawal from unconfined aquifer was maximum in southern part of the study area. An observation well, Gaughat, in the south west part of the area (Fig. 18.6) showed that groundwater depth from the surface is declining with year. This declination is more pronounce in non-monsoon (May) season than in monsoon season (September). If this

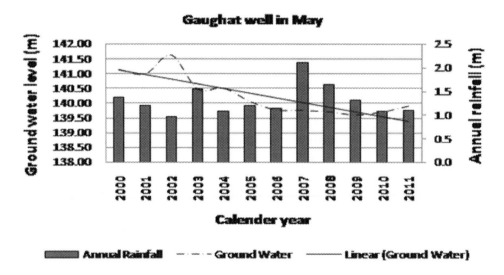

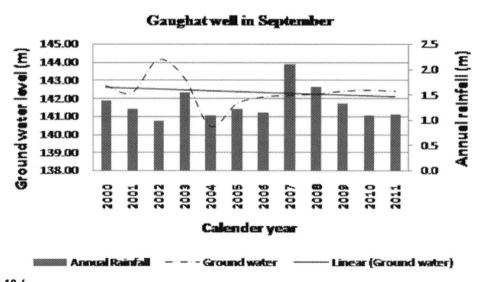

FIGURE 18.6

Groundwater-level fluctuation in Gaughat observation well.

scenario exists or increases, the groundwater table will decrease alarmingly in this area, whereas the northern part of the area has stable groundwater table (Fig. 18.7).

The thickness of aquifer is more in the northern part where groundwater table has increasing trend. However, groundwater level in the northern part of the study area, where aquifer thickness is less, is decreasing due to high groundwater exploitation.

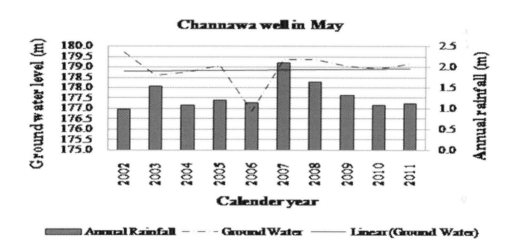

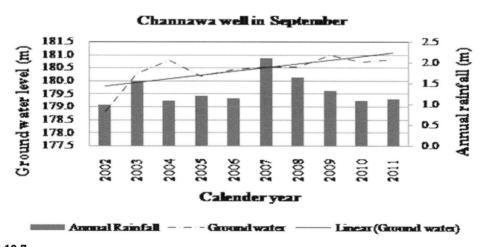

FIGURE 18.7

Groundwater-level fluctuation in Channawa observation well.

The model had provided total annual recharge of groundwater from precipitation as 14.7 MCM in prevailing situation. Recharge per annum from river was 0.2 MCM. Annual extraction of groundwater was 7.4 MCM. Return flow to the river was 1.6 MCM.

18.5.5 Sensitivity of model parameters

Winpest package of Visual MODFLOW 4.2 analyses the sensitivity of model parameter. The record of sensitivity of Pest had shown that hydraulic conductivities and recharge were most sensitive parameters of the aquifers at almost all sites of the command area. The model was comparatively less sensitive to specific storage and specific yield, but river bed conductivity did not show any visible effect except at nearby area to the river (Fig. 18.8).

18.5.6 Scenario analysis

The 64% of area is in water logged condition in peak monsoon season in prevailing condition of aquifer. This level would further rise after irrigation water application. Groundwater recharge due to irrigation water varies from 60%−67% of irrigation water (Scot et al., 2005). Therefore it would be reasonable to adopt the recharge due to application of irrigation water in the study area after the commencement of ongoing Sikta irrigation system as 65% of applied irrigation water.

In the calibrated model of the study area, the above recharge boundary condition was applied to get probable groundwater scenario for 3 years. The model outputs showed that the groundwater level remains within 2 m below the ground surface in entire area in peak monsoon period.

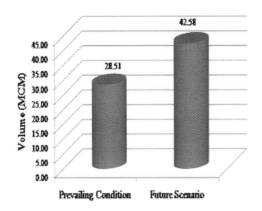

FIGURE 18.8
Total available groundwater.

In winter season, groundwater table would decrease up to the depth of 2.5 m from ground surface (Fig. 18.9). Predicting scenario showed that the water table remains within 2 m in 64% of the area up to 1 year. After 3 years of irrigation water application, whole area would be in waterlogged situation in monsoon. However, the water table would rise to reach within root zone depth for 85% of area in winter within 3 years. Only 5800 ha of land remain with water table below 2 m. But the water table rising scenario warns the whole area would be in waterlogged condition within 4 to 5 years of irrigation water application (Fig. 18.9).

Annual availability of groundwater after irrigation water application provided by model would be 42.58 MCM, whereas this value was 28.51 MCM in existing situation of stresses (Fig. 18.8). June is the driest month of the year and canal supply is insufficient. The supplement of deficit amount for irrigation regarding proposed cropping pattern in June was 52.0 MCM (Fig. 18.4) and should be balanced from reliable groundwater. But the available water

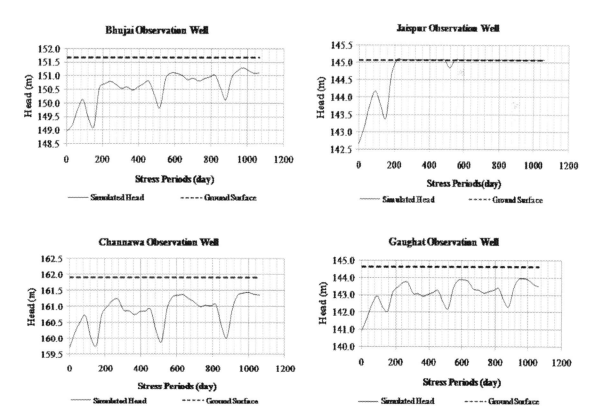

FIGURE 18.9

Time series plot of observed versus simulated head after irrigation.

during this month in unconfined aquifer after 1 year of irrigation application would be almost 41.56 MCM. This entire quantity could not be extracted and economically and environmentally available water is even less which is insufficient to meet the irrigation water demand in the driest month of the year.

18.6   Conclusion and recommendations

The salient features of the study area are summarized are as follows:

1. The study of groundwater fluctuation revealed the existence of groundwater table within 2 m below the ground surface in more than 64% of area in monsoon season, whereas the same declined up to 5.2 m below the surface level in nonmonsoon season in existing stress and boundary condition. Therefore there is enough scope of groundwater development.

2. The evaluation of calibration and validation was performed by means of plots of measured and simulated groundwater level as well as statistical goodness of fit parameters such as SEE and RMSE. SEE and RMSE for calibrated model varied from 0.016 to 0.05 m and 0.307 to 1.117 m, respectively. Hydraulic conductivity, specific yield, and specific storage of the layers were 3.57×10^{-3} to 3.08×10^{-7} m s^{-1}, 0.06 to 0.3, 0.009 to 0.001 m^{-1}, respectively. Annual recharge had spatial and temporal variation in different zone. It varied from 194 to 380 mm per annum (average 217 mm).

3. Groundwater recharge was the most sensitive parameter to model output groundwater flow flowed by conductivities of first and second layer (aquifer) in almost all sites of the study area. The model outputs were comparatively less sensitive to specific storage and specific yield. The river bed conductivity has no effect to the sites except the sites nearby rivers.

4. After the application of canal water to the study area, most of the (85%) area would be in waterlogged condition within 3 years of irrigation water application. Only 15% of land in south west part of the area has water table up to 2.5 m below the ground surface in monsoon. However, the increasing scenario of groundwater table in all observation wells predicts that the entire area would be in waterlogged condition after 4 to 5 years of irrigation water application.

5. In the month of June, canal water is insufficient to meet the crop water requirement. The deficit would be 52 MCM. This requirement could not meet from unconfined aquifer. So, extraction of groundwater from semiconfined and deep aquifer at its driest condition without exploiting the environment of aquifer is necessary.

The following strategy should be adopted for preventing the waterlogging, increase in water storage in aquifer, and integrated water resources management within the aquifer level.

1. Maximum groundwater extraction should be undertaken in nonmonsoon period so that ample space would be available in aquifer for storing recharge water in monsoon season.
2. A conjunctive use water management for irrigation should be practiced to reduce the water table level below the root zone depth to get rid of water logging problem after the irrigation water application begin.
3. Combined simulation and optimization modeling study has to be undertaken to find out the maximum possible pumpage.
4. Water resource planning and management of canal command area should be done to reduce the future risk after application of irrigation water.

Acknowledgments

The authors would like to express their sincere thanks to the Hydrology and Meteorology Department of Nepal, the Groundwater Resources Development Board of Nepal, and GWRDB Field Office, Nepalgunj for providing necessary data and information. Special thanks to Indian Institute of Technology (IIT) Roorkee for providing the license software of Visual Modflow and work space to accomplish this work.

References

Abdulla, F., & Tamer, A.-A. 'D. (2006). Modeling of groundwater flow for Mujib aquifer, Jordan. *Journal Earth System Science, 113*, 289–297.

Allen, R. G., Pereira, L. S., Raes, D., & Smith, M. (1998). *Guidelines for computing crop water requirements. Irrigation and Drainage Paper 56*. Rome, Italy: FAO.

Bazzani, G. M., Di Pasquale, S., Gallerani, V., & Viaggi, D. (2004). Irrigated agriculture in Italy and water regulation under the European Union Water Framework Directive. *Water Resources Research, 40*. Available from https://doi.org/10.1029/2003WR002201.

Central Bureau of Statistics, Government of Nepal. 2012. http://cbs.gov.np/wp-content/uploads/2012/11/VDC_Municipality.pdf.

Chaturvedi, R. S. (1973). *A note on the investigation of ground water resources in western districts of Uttar Pradesh: Annual report*. U. P. Irrigation Research Institute, pp. 86–122.

Doorenbos, J., & Pruitt, W. O. (1977). *Guidelines for predicting crop water requirements: Irrigation and Drainage Paper 24*, FAO: Rome.

Frey, H., & Frank, P. (2012). On the suitability of the SRTM DEM and ASTER GDEM for the compilation of topographic parameters in glacier inventories. *International Journal of Applied Earth Observation and Geo Information, 18*, 480–490.

Gaaloul, N. (2014). GIS-based numerical modeling of aquifer recharge and saltwater intrusion in arid southeastern Tunisia. *Journal of Hydrologic Engineering, 19*(4), 777–789.

Harbaugh, A. W., Banta, E. R., Hilly, M. C., & McDonald, M. G. (2000). *MODFLOW-2000, The US Geological Survey modular ground-water model — User guide to modularization concepts and the ground-water flow process. US Geological Survey open file report 92*.

Harmancioglu, NB, Fedra, K, & Barbaros, F (2008). Analysis for sustainability in management of water scarce basins: the case of the Gediz River Basin in Turkey. *Desalination, IWA Publishing, 226*, 175–182.

Hollander, H. M., Mull, R., & Panda, S. N. (2009). A concept for managed aquifer recharge using ASR-wells for sustainable of ground water resources in an alluvial coastal aquifer in eastern India. *Physics and Chemistry of the Earth, 34*, 270–278.

Kashaigili, J. J., Masauri, D. A., & Abdo, G. (2003). Ground water management by using mathematical modelling: Case study of Makutupora groundwater basin in Dodoma Tanzaniya. *Bostawana Journal of Technology, 2*, 19–24.

Khan, S., Jianxin, M., Rahimi Jamnani, M. A., Hafeez, M., & Gao, Z. (2005) *Modeling water futures using food security and environmental sustainability approaches conference paper.* http://www.mssanz.org.au/modsim05/papers/khan_2.pdf.

Mao, X., Jia, J., Changming, L., & Hou, Z. (2005). A simulation and prediction of agricultural irrigation on groundwater in well irrigation area of the piedmont of Mt. Taihang, North China. *Hydrological Processes, 19*, 2071–2084.

Myrab, S. (1997). Temporal and spatial scale of response area and groundwater variation in till. *Hydrological Processes, 11*, 1861–1880.

Rahul, S. K., Panda Sudhindra, N., Hollander, H., & Billib, M. (2011a). Integrated water resource management in a major canal command area in eastern India. *Hydrological Processes, 25*, 2551–2562, Wiley.

Rahul, S. K., Panda Sudhindra, N., Hollander, H., & Billib, M. (2011b). Integrated water resource management in a major canal command area in eastern India. *Hydrological Processes, 25*, 2551–2565.

Rao, R., Mohan, M. S., Adhikari, R. N., Chittaranjan, S., & Chandrappa, M. (1996). Influence of conservation measures on groundwater regime in semi arid tract of South India. *Agricultural Water Management, 3*, 301–312.

Raul, S. K., Panda, S. N., Holländer, H., & Billib, M. (2011). Integrated water resource management in a major canal command in eastern India. *Hydrological Processes, 25*(16), 2551–2562.

Rejani, R., Jha Madan, K., Panda, S. N., & Mull, R. (2008). Simulation modeling for efficient groundwater management in Balasore Coastal Basin, India. *Water Resources Management, 22*, 23–50.

Sakiyana, J., & Yazicigil, H. (2004). Sustainable development and management of an aquifer system in western Turkey. *Hydrogeology Journal, 12*, 66–80.

San Juan, C. A., Belcher, W. R., Laczniak, R. J., & Putnam, H. M. (2010). Hydrologic components for model development. *Death Valley regional ground-water flow system, Nevada and California: Hydrogeologic framework and transient ground-water flow model. Professional Paper, 1711*, 95–132.

Scot, K. I., Delwyn, S. O., & Chen, C. -H. (2005). *Effects of irrigation and rainfall reduction on ground water recharge in the Lihue Basin Kauai, Hawaii. Scientific investigation report, 2005–5146.*

Smith, M., Allen, R., & Pereira, L. (1998). Revised FAO methodology for crop-water requirements.

Tyagi, N. K. (1980). *Crop planning and water resources management in salt affected soils—A systems approach* (Ph.D. dissertation). Jawaharlal Nehru Technological University, Hyderabad, India.

United Nations. (2012). Available from http://www.unwater.org/downloads/UNW_status_report_Rio2012.pdf.

Xu, C. Y., & Chen, D. (2005). Comparison of seven models for estimation of evapotranspiration and groundwater recharge using lysimeter measurement data in Germany. *Hydrological Processes: An International Journal, 19*(18), 3717–3734.

Xu, G. H., Zhongi, Q., & Pereira, L. S. (2011). Using MODFLOW and GIS to assess changes in ground water dynamics in response to water saving measures in irrigation districts of the Upper Yellow River basin. *Water Resources Management, 25*, 2035–2059, Springer.

CHAPTER 19

Hydrocarbon pollution assessment and analysis using GC–MS

Pankaj Kumar Gupta
Wetland Hydrology Research Laboratory, Faculty of Environment, University of Waterloo, Waterloo, ON, Canada

19.1 Introduction

Gas chromatography mass spectrometer (GC–MS) is synergistic combination of two powerful micro analytical techniques. It integrates the GC and MS detector (MSD) for qualitative and quantitative identification of organic compounds. The GC separates the representative chemical from compound, and MSD provides the information that aids in structural identification using the mass spectrum of each components. The key feature of the GC–MSD is the systems that heat the injector and oven, carrier gases, detector, transfer line and allow programmed for temperature control. An important facet of the gas chromatograph is the use of a carrier gas, such as hydrogen or helium, to transfer the sample from the injector, through the column, and into the detector. The column, or column packing, contains a coating of a stationary phase. Separation of components is determined by the distribution of each component between the carrier gas (mobile phase) and the stationary phase. A component that spends little time in the stationary phase will elute quickly. The MS uses selective ion monitoring (SIM) and SCAN technique to identify the ion current during dwell time in column.

GC–MSD system is controlled by a computer having software tools to control the physical parameters of the system such as the temperature zones, gas flows, injection, and pumping system. The computer also having the software tools to handle the data generated during a run, that is, compares mass spectra to a library to identify compounds and performs quantitation of peaks generated by the GC–MSD.

19.2 Previous works

Soil-water quality analysis is one of the intresting topics of research among environmentalist, geochemist and so on (Amrit et al., 2019; Dhami et al.,

Advances in Remediation Techniques for Polluted Soils and Groundwater. DOI: https://doi.org/10.1016/B978-0-12-823830-1.00010-9

2018; Gupta et al., 2020; Gupta & Sharma, 2018; Himanshu et al., 2018, 2021; Kumar et al., 2021; Kumari et al., 2019; Pandey et al., 2016). GC—[flame ionization detector (FID)] MSD is recommended analytical instrument to quantitative and qualitative analysis of hydrocarbons [nonaqueous phase liquids (NAPLs)] from soil—water samples by EPA (Method No. 8260, 1996), USGS and many other agencies. Therefore previously (Table 19.1) most of investigation used it to identify such representative chemical from soil—water samples. The MSDs are the more advance techniques for the qualitative (SCAN mode) and quantitative (SIM mode) analysis of environmental samples. In contaminant transport studies, many researchers conducted and analyzed the contaminated soil, water, air samples using the GC—(FID) MSD techniques. Gupta and Yadav (2019b, 2019a, 2020); Basu et al. (2020) used GC FID for measuring toluene concentration from different batch system in bioremediation studies. Furthermore, Gupta & Yadav (2017); Gupta & Bhargava (2020) used GC—FID to quantify the toluene concentration during biodegradation under seasonal and diurnal fluctuations of soil—water temperature in batch system. Basu et al. (2015) analyzed water samples collected from the rhizospheric zones of the mesocosms using GC—FID for toluene compound. Similarly, Ronen et al. (2005) analyzed toluene by GC—MSD according to EPA SW-846, Method 8260 during VOCs investigation in the Saturated—Unsaturated interface region of a contaminated phreatic aquifer. A headspace-analysis approach to assess the sorption by soils using GC—MSD conducted by Balseiro-Romero and Monterroso (2013). Kim et al. (2005) investigated the gas transport and analyzed samples using GC—FID during soil column experiments. Similarly, (Picone et al., (2013) analyzed the soil vapor by GC during NAPLs vapor transport in column system. Toluene in the liquid from the column was analyzed with a gas chromatograph equipped with an FID by Møller et al. (1996). Cotel et al. (2011) investigated the vertical diffusive mass transfer of trichloroethylene (TCE) in a low water-saturated porous medium, diffusion experiments were conducted in soil columns and TCE analyzed by GC—ECD. Furthermore, the GC—MSD techniques were also integrated with some more advance techniques like 2/3D GC system for the multiphase NAPLs (contaminant) transports studies. (Møller et al. (1996) integrated the GC—MS investigation to biofilm hybridization and scanning confocal laser microscopy.

In this study also the GC—MSD is used for soil—water and soil—air samples analysis collected from the multidimensional domains.

19.3 GC—MS system: specification

GC—MSD system consists of the integration of GC system and MSD equipped with carrier gases, automatic sampler along with work station

Table 19.1 Summary of previous studies performed to investigate hydrocarbon pollution using gas chromatography mass spectrometer (GC–MS).

Sr. no.	References (LNAPL compound)	Study domain	Specification	Detector	Carrier gas	Gas flow	GC column	Oven temp.	Analytical methodology finding/comments
1	Morgan et al. (1993)	Groundwater wells samples	Vega GC600 gas chromatograph	FID	Helium, nitrogen (99.99% pure)	10 mL min[1]	30 m × 0.54 mm (i.d.)	190°C	• 5 mL of water were added and sparged at 20°C with nitrogen and helium gas • Dry purge (water removal) 2 min at 20°C; desorb 5 min at 180°C; tube clean 30 min at 180°C
2	Alvarez et al. (1991)	Batch system	Packard 5890 gas chromatograph	FID	Helium, nitrogen, hydrogen (99.99% pure)	—	30 m × 0.5 mm (i.d.) silicone	—	• 0.5 mL water was taken from vials by auto samples • Detection limits 0.01 mg L[1]
3	Dorea et al. (2006)	Groundwater wells samples	ShimadzuQP5050A MS GC-17A	MSD	Helium (99.99% pure)	1.5 mL min[1]	30 m × 0.25 (i.d.) silicone	280°C	Performed in the SIM mode
4	Almeida and Boas (2003)	Groundwater samples	Fisons MFC 800 series II gas chromatograph	FID	Helium, oxygen, hydrogen	—	30 m × 0.53 mm (i.d.)	210°C	The detector and injector temperatures were set at 300°C and 200°C, respectively. The split less time was 90 s
5	(Kubinec et al., (2004)	Water samples	GC-HP 5890 SERIES II Hewlett-Packard	FID	Helium (99.99% pure)	—	30 m × 0.53 mm (i.d.)	250°C	The detectors temperature was maintained at 250°C. Injection flow rate 200 mL min[1] at various temperatures of injection port in the range of 120°C–320°C, and oven temperature 250°C for each analysis
6	Boonsaner et al. (2011)	Soil–water	Perkin Elmer, model Autosystem XL	FID	Helium (99.99% pure) air, hydrogen	5 mL min[1]	30 m × 0.32 mm (i.d.)	240°C	Injector temperature 280°C, oven temperature 50°C–240°C with ramp rate 5°C min[1] and 240°C–280°C with ramp rate 10°C min[1], detector temperature 280°C

Continued

Table 19.1 Summary of previous studies performed to investigate hydrocarbon pollution using gas chromatography mass spectrometer (GC–MS). *Continued*

Sr. no.	References (*LNAPL compound*)	Study domain	Specification	Detector	Carrier gas	Gas flow	GC column	Oven temp.	Analytical methodology finding/comments
7	BianchinNunes et al., (2012)	Water samples	Shimadzu (GCMS-QP2010 Plus)	Rtx-5MS column	Helium (99.99% pure)	1 mL min^{-1}	30 m × 0.25 mm (i.d.)	80°C–300°C	Injection and ion source temperatures were both maintained at 260°C
8	Sieg et al. (2008)	Groundwater samples	Thermoquest CE-GC 2000	MS (Voyager)	Helium (purity 5.0)	–	60 m × 0.32 mm (i.d.)	190°C	• The injector temperature to 230°C • SIM and SCAN mode • Selective ion mass 91 for toluene • Retention time 13.0 min
9	Lovanh et al. (2002)	Batch system culture experiment	Hewlett Packard 5890	FID	Helium (99.99% pure)	–	30 m × 0.32 mm (i.d.)	–	Detection limits were approximately 1 mg L^{-1} for BTEX
10	Coulon (2009)	Coal Tar (glass holder)	Perkin Elmer, Wellesley, MA AutoSystem XL	MS	Helium (99.99% pure)	–	30 m × 0.40 mm (i.d.)	50°C	SCAN mode (range *m/z* 33–350)
11	Coulon and Delille (2006)	Sub-Antarctic soils	Thermo Trace GC coupled with Thermo Trace DSQ^R MS	MS	Helium (99.99% pure)	–	30 m × 0.25 mm (i.d.)	300°C	SCAN mode (range *m/z* 60–650)
12	Baedecker et al. (2011)	Groundwater sample	Agilent 6890A GC	Agilent 5973 MS	Helium (99.99% pure)	–	30 m × 0.25 mm (i.d.)	Variable up to 250°C	The MS was operated with an ionizing energy of 70 eV at 250°C in full scan mode (*m/z* 50–250)
13	Kristensen et al. (2010)	Soil–water	Agilent GC WCOT CP-624CB	FID	N$_2$ gas	5.2 mL min^{-1}	30 m × 0.53 mm (i.d.)	80°C	Injector, and detector temperature 150°C and 200°C, respectively
14	Balseiro-Romero and Monterroso (2013)	Soil–water	Agilent-Varian 450-GC	MS (Varian 220-MS)	Helium (99.99% pure)	1 mL min^{-1}	30 m × 0.25 mm (i.d.)	Variable up to 200°C	The injector was operated at 250°C and in split/splitless mode, with a 1/10 split ratio
15	Picone et al. (2013)	Soil–water	Fisons AS 8000	FID	–	–	50 m × 0.32 (i.d.)	190°C	The temperatures of the injector and detector were 200°C, and 250°C
16	Holden and Noah (2005)	Soil	Hewlett Packard 5890A	FID	–	–	2.0 m × 1/8 in.	70°C	All standards and samples were in thermal equilibrium prior to GC analysis

#	Reference	Sample	Instrument	Detector	Carrier gas	Flow rate	Column dimensions	Temperature	Notes
17	Nerantzis and Mark (2010)	Soil vapor	ProGC, Unicam	FID	Helium air, N₂, hydrogen	—	25 m × 0.32 mm (i.d.)	215°C	Detector temperature = 245°C, injector temperature = 245°C
18	Meyer-Monath et al. (2014)	Meconium samples	Thermo Fisher Scientific, Austin, TX, USA	MS DSQ II detector	Helium (99.99% pure)	1.2 mL min⁻¹	30 m × 0.25 mm (i.d.)	Variable up to 240°C	• The transfer line and ionization source temperature were set at 250 C • MS acquisition was in SIM mode • Solvent delay of 4.20 min
19	(Niri et al. (2009))	Car exhaust	Varian 3800 Mississauga. ON, Canada	FID	Helium air, N₂, hydrogen	1.8 mL min⁻¹	30 m × 0.25 mm (i.d.)	110°C	• Injector temperature was 250 C • Retention times for toluene 2.1 min
20	Djozan and Assadi (1997)	Water samples	Shimadzu (Japan) GC-15A	FID	Helium air, N₂, hydrogen	30 mL min⁻¹	25 m × 0.33 mm (i.d.)	Variable up to 150°C	The analytes were injected in the splitless mode at 280°C the splitter was opened after 1 min
21	Zhou et al. (2013)	Soil	Agilent Technologies 7890A GC-5975 GC MS	PID/MS	Helium (99.99% pure)	2.5 mL min⁻¹	30 m × 0.25 mm (i.d.)	50 C	• The injection was operated in split mode with a split ratio of 20:1, and the injector temperature was 260°C • The GC–MS transfer line temperature was 280°C, source 230 C, ionization potential 70 eV, and scan range 35–210 m/z
22	Han et al. (2007)	Seabed sediment	Perkin Elmer, USA	PID	N₂ gas (99.99% pure)	1 mL min⁻¹	60 m × 0.32 (i.d.)	Variable up to 180°C	Injector temperature 180°C
23	Mangani et al. (2000)	Water	Carlo Erba HRGC series 5300, (Milano, Italy)	FID (Graphite)	—	—	11.2 cm × 3 mm (i.d.)	180°C	The stationary phase is GCB modified
24	Assadi et al. (2010)	Water	Shimadzu GC 2010	FID	Helium (99.99% pure)	30 cm s⁻¹	30 m × 0.22 mm (i.d.)	Variable up to 100°C	The injection port was held at 200°C and used in splitless mode with splitless time 0.5 min

Continued

Table 19.1 Summary of previous studies performed to investigate hydrocarbon pollution using gas chromatography mass spectrometer (GC—MS). *Continued*

Sr. no.	References (LNAPL compound)	Study domain	Specification	Detector	Carrier gas	Gas flow	GC column	Oven temp.	Analytical methodology finding/comments
25	Andreoli (1999)	Blood and urine	Hewlett Packard HP 6890	MS (Palo Alto, CA, USA)	Hydrogen	–	30 m × 0.25 mm (i.d.)	Variable up to 120°C	• MS acquisition was performed in SIM • Temperatures of injector and MS detector were 280°C and 230°C, respectively
26	Buddhadasa et al. (2002)	Soil	HP 5890 GC	MSD HP 5970	Helium (99.99% pure)	–	30 m × 0.25 mm (i.d.)	Variable up to 180°C	The GC–MSD was operated using an injector temperature of 220°C an interface temperature of 250°C an inlet split at 20:1
28	Tumbiolo et al. (2004)	Air	Varian 3800 GC	MS (Varian Saturn 2200)	Helium (99.9C% pure)	1 mL min⁻¹	30 m × 0.25 mm (i.d.)	Variable up to 250°C	5 μ scans mode
29	Telgheder et al. (2009)	Water sample	HP 5890 (Agilent, Waldbronn, Germany)	MS	Nitrogen 5.0	0.8 mL min⁻¹	30 m × 0.25 mm (i.d.)	80°C	The GC–DMS transfer line is made of a copper tube, which can be electrically heated to the desired temperature 80°C
30	Nardi (2006)	Water sample	Perkin Elmer 8500 series	FID	Hydrogen	–	4 m × 0.21 mm (i.d.)	200°C	Detection was by FID, kept at 200°C
31	Prikryl et al. (2006)	Water sample	HP 5890 series (Hewlett-Packard)	FID	Helium (99.99% pure)	–	0.75 mm (i.d.)	40°C	The detector temperature was maintained at 250°C
32	(Jiang et al. (2015)	Soil	Agilent Technologies, 6890N	FID	H₂, air	(H₂) 40 mL min⁻¹	–	Variable up to 220°C	The detector temperature was 300°C, the injection port temperature was 250°C, and the 1 μL samples were loaded with an auto sampler with a split mode (5:1)

#	Reference	Sample	Instrument	Detector	Gases	Carrier gas	Column	Temperature	Notes
33		Soil–water	Nucon model 5765	FID	N$_2$, H$_2$, air 20 mL min^{-1}	N$_2$ (25 mL min^{-1})	6 ft × 1/8 in.	120°C	The temperature of both the injection port and detector port was kept at 150°C
34	Basu et al. (2015)	Water samples	Varian GC model CP-3800	FID	N$_2$, H$_2$, air (H$_2$, min^{-1})	N$_2$ (25 mL min^{-1})	30 m × 0.25 mm (i.d.)	100°C	The temperatures of the GC inlet and detector are kept isothermal at 161°C and 100°C, respectively, during the analysis
35	(Moller et al. (1996)	Water	GC-9A; Shimadzu, Kyoto, Japan	FID	—	—	80/120 Carbopack	80°C	Online injection was conducted automatically and carried out isothermally at 180°C

FID, *Flame ionization detector*; GCB, *graphitized carbon black*; LNAPL, *light nonaqueous phase liquid*; MS, *mass spectrometer*; SIM, *selected/single ion monitoring*.

Table 19.2 Detail specification of gas chromatography and mass spectrometer detector (GC–MSD) system used to analyze the toluene samples during entire experiments.

Sr. no.	Component	Description	Specification
1	GC	Column type: Chrompack capillary column	
		Column dimension	30 m × 0.25 mm
		Column coating	Silicone, 0.25 μm
		Column working temperature ranges	−60°C to 350°C
		Oven dimension	30 × 28 cm
		Oven temperature ranges	−60°C to 350°C
2	MSD	Mode: SCAN and SIM	
		MassHunter IP	10.1.1.100
		High vacuum pressure	$4.87e^{-05}$
		Turbo pump speed	0%–100%
		Operating system	1.2/26 T
3	ALV system	ALV system sampling point = 16	
		Valve	16
		Wash valve	02
		Solvent valve	01
4	Carrier gases		
	1. Helium	Purity	99.999%
	2. Nitrogen	Capacity	14 kg
	3. Hydrogen	Initial pressure	2000 psi
	4. Air	Regulator pressure	100 psi
5	Gas clean filter	Maximum pressure for all filter	15 bar/219 psi
6	Turbo pump	Power	50.60 Hz
7	Work station	System type	64 bit operating system
			NIST MS Search 2.0 (MS Library)
			MSD ChemStation Data Analysis

SIM, *Selected/single ion monitoring.*

having mass hunter software for control and data analysis. Table 19.2 describes the component wise specification of GC–MSD and following are the section wise details of GC–MSD system.

19.3.1 GC column

The toluene analyzed in GC equipped with an MSD and FID. The chrompack capillary column having dimension 30 m × 0.020 mm and inner diameter 0.25 μm used for analysis. The column was having −60°C to 350°C working temperature range. For this analysis the GC oven temperature was maintained at 45°C and constant pressure 5.5 psi. The inlets temperature was maintained 250°C, at pressure 9.4667 psi, and total flow 244.2 mL min^{-1}.

Similarly, the septum purge flow rates were 3 mL min^{-1} with split flow and ratio 240 mL min^{-1}, 240 mL min^{-1}, respectively.

19.3.2   MS detector

The GC system was equipped with an MSD, which was used in the sample analysis in entire experiment. The MSD was connected with front via thermal aux nickel chamber. The MSD temperatures were maintained at 230°C and 150°C for MS source and MS Quad, respectively. The MSD integrates two vacuum pumps, filament chamber and detector for SCAN and SIM, for qualitative and quantitative analysis. SIM mass spectrometry is a technique used to improve the sensitivity of a mass spectral analysis.

19.3.3   Carrier gases

In the GC–MSD system, 99.999% pure grade nitrogen and/or 99.999% pure helium gas are used as carrier gas. Similarly, air and hydrogen (99.999% pure grade) are also integrated with the GC for tune process and FID analysis. The flow rate of the carrier gases was 20 mL min^{-1} via filtration system.

19.3.4   ALS system

ALS system is automatic sampler integrated with GC system in front plunger. In ALS system the GC syringe (mentioned earlier) is fitted at proper position for injection. The ALS system is controlled for injection volume, plunger speed, dwell time, and viscosity delay for different analysis.

19.3.5   Gas filter

The four gas filter units are installed between carrier gas cylinder and GC/MS setup via copper pipe pathways. These are oxygen filter, hydrocarbon filter, carrier gas filter, and moisture filter under saturated condition. All fitters are having the maximum capacities of pressure 15 bar/291 psi.

19.3.6   Vacuum pump

One Agilent Vacuum (Turbo) pump is integrated with MS system to maintain vacuum in MS chamber.

19.4   Method of toluene analysis

The combinations of different parameters are integrated to control the (pre/post)-analysis using method loading panels. The separate method is developed for the Toluene analysis using recommended parameters listed in Table 19.3.

Table 19.3 Summery of date-wise autotune report.

Abund	H$_2$O%	Nitrogen%	Oxygen%	CO$_2$%	N$_2$/H$_2$O
218,880	1182.22	772.63	28.07	38.23	65.35
109,496	54.44	5.72	1.45	4.66	10.5
131,520	39.52	2.64	0.81	3.18	6.67
468,736	283.23	4.63	1.17	4.11	1.63
305,664	27.65	1.77	0.47	1.04	6.41
379,840	30.25	1.61	0.36	1.24	5.33
460,416	23.19	1.89	0.36	1.14	8.14
502,592	13.21	1.61	0.32	0.77	12.21
453,376	15.55	1.8	0.28	0.79	11.59
435,520	6.16	1.69	0.21	0.42	27.41
378,240	176.16	2.25	0.42	1.76	1.28
633,088	228.95	2.77	0.61	1.94	1.21
364,928	35.18	2.31	0.38	0.98	6.56
321,920	26.58	2.01	0.27	0.71	7.55

19.4.1   Method creation

Methods are created using the Method editor, the Method Information, and the Edit GC Parameters dialog boxes. These dialog boxes can be accessed using icons or menus on the instrument control panel. The method creation consists of the organization of different input and/or control parameters in one program. The GC parameters are oven temperature, inlet/outlets temperature, column pressure and temperature, ALS control and injection, etc. Similarly the MS parameters are MS Source temperature, MS Quad temp, Turbo pump speed, vacuum statues, etc. Once these inputs are filled along with an indication of maximum limits/ranges, the method file is saved for further analysis. The summery of the component-wise input parameters for method creation for the analysis of Toluene by GC—MSD are listed in Table 19.4.

19.4.2   MSD method

SIM mode is a data acquisition technique where only selected ion fragments are monitored to obtain maximum sensitivity. In SIM/Scan mode, to complete one cycle, the MSD acquires a single group of SIM data followed by a single group of Scan data. Therefore In Toluene analysis SIM/Scan Mode is applied as MSD method.

Table 19.4 Summery of the method component-wise input parameters for the analysis of Toluene by gas chromatography and mass spectrometer detector (GC–MSD).

Sr. no.	Components	Parameter's	Input
1	Oven	Temperature	45°C
		Temp. rate	off
		Equilibration time	0.1 min
		Maximum oven temp	275°C
		Hold time	5 min
		Run time	5 min
2	Column	Pressure	5.5 psi constant pressure
		Inlet	Aux PCM A
		Outlet	Front detector
		Heated by	Oven
3	Thermal Aux 1	Type	Nickel crystal
		Temperature	375°C
4	Thermal Aux 1	Types	MSD transfer line
		Temperature	250°C
5	Inlets	Types	Split-splitless inlet
		Heater	250°C
		Pressure	9.4667 psi
		Total flow	244.2 mL min^{-1}
		Septum purge flow	3 mL min^{-1}
		Mode	Split
		Split ratio	200:1
		Split flow	240 mL min^{-1}
6	ALS	Injection type	Standard
		Injection	10 μL syringe
		Injection volume	1 μL
		Solvent A washes	Preinjection 1 μL
		Solvent B washes	Postinjection 1 μL
		Sample washes	2
		Sample pimps	2
		Dwell time	0 min
		Plunger speed	Fast
		Viscosity delay	0 s
7	MS parameter	MS source temp	250°C
		MS quad temp	150°C
		Turbo speed	100%
		Hi vacuum	1.2×10^4
		Turbo pump speed must be increasing upto 100%	
8	Carrier gases	Helium (99.999% pure) at 100 psi regulator pressure	

19.4.3 Tune MSD

Tuning is the process that adjusts the MS for good performance over the entire mass range. Using a known compound as a calibrator, the tune parameters are set to achieve sensitivity, resolution, and mass assignments for the known calibration ions. Tuning is performed using either the autotune or manual tune features. *Autotune* adjusts all MS parameters to predetermined target values. It finds the mass peaks, obtains adequate response and resolution, optimizes the response of three masses, optimizes response and resolution of other masses, and obtains accurate mass assignments. There are needs of autotune, when the MS system is first set up, when the acquisition software is reinstalled or upgraded, and when maintenance is performed on the MSD. The resulting parameters from an autotune or quicktune are saved in a tune log file.

19.4.4 Tune report assessment

MassHunter Data Acquisition includes a report template that allows to graphically view the trend of each parameter tuned over the number of tunes saved in the log file. The tune reports consisting the air/water ratio: $H_2O\%$, $O_2\%$, $N_2\%$, $N_2/H_2O\%$ along with abundance mass chromatogram. The recommendation for $H_2O\%$, $O_2\%$, $N_2\%$, $N_2/H_2O\%$ is less than 20% for good preformation of MSD. This trend data help to schedule maintenance and troubleshoot the MS.

19.4.5 Method for running a sample sequence

A sequence is a list of samples to be analyzed automatically without operator intervention. Each sample is assigned a method to be used for analysis. The sample injection dialog appears to identify the injection style and locates the data files. The sequence methods consider injection volume, level of analysis, samples location vial numbers, GC method for analysis, and types of sample.

19.5 Calibration

The GC–MSD was calibrated with known standard toluene (from stock solution) concentration for the same method loaded in GC for toluene. The stock solution was prepared by diluting the stock solution in 120 mL batch bottles. The prepared samples having different concentration were taken in 2 mL vials (sterile vials) and analyzed in triplicate at room temperature ($250°C \pm 20°C$). The response areas of the peak for different concentration was plotted as calibration curve shown in Fig. 19.1. The plotted calibration curve was linear fitted and gives the 0.994372 value of coefficient of determination (Fig. 19.2). The associated response areas and bias% are tabulated in Table 19.5.

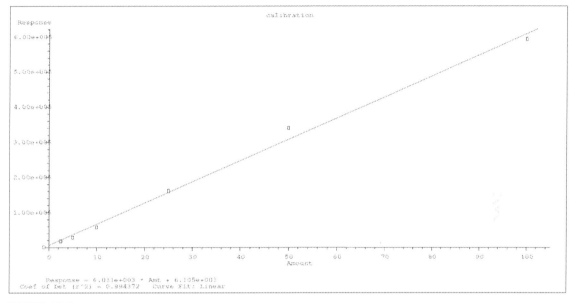

FIGURE 19.1

Calibration curve obtained from the analysis of standard known concentration of toluene for GC–MSD. *GC–MSD*, Gas chromatography and mass spectrometer detector.

Sr. No	Description	Information
1	Library	NIST Search 2.0
2	Targeted Mass	92,91
3	Contributor	Japan AIST/NIMC Database-Spectrum MS-NW-67
4	Spectrum	MW:92 CAS#: 108-88-3 NIST#: 227551 ID#: 51664 DB: mainlib
5	10 Largest peaks	91 999 92 776 65 121 39 107 63 74
		51 64 63 54 50 41 89 39 62 32

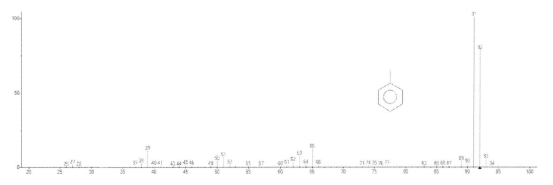

FIGURE 19.2

The mass spectrum of the Toluene obtain form NIST search library used in GC–MSD. *GC–MSD*, Gas chromatography and mass spectrometer detector.

Table 19.5 The concentration-associated response areas and bias% obtained from gas chromatography and mass spectrometer detector (GC—MSD) analysis of known concentration of standard toluene for calibration.

Sr. no.	Sample code	Concentration (ppm)	Response (ratio)	Bias%
1	000S1	2.5	17,491.0000	− 17.42
2	005S1	5	28,421.0000	− 21.61
3	010S1	10	57,409.0000	− 13.55
4	025S1	25	160,923.0000	2.58
5	050S1	50	339,840.0000	10.47
6	100S1	100	593,427.0000	− 2.58

19.6 Mass spectrum of toluene (Fig. 19.2)

References

Amrit, K., Mishra, S. K., Pandey, R. P., Himanshu, S. K., & Singh, S. (2019). Standardized precipitation index-based approach to predict environmental flow condition. *Ecohydrology*, *12*(7) e2127.

Andreoli, R (1999). Solid-phase microextraction and gas chromatography-mass spectrometry for determination of monoaromatic hydrocarbons in blood and urine: Application to people exposed to air pollutants. *Chromatographia*, *50*(3—4), 167—172.

Assadi, Y., Fardin, A., & Mohammad, R. M. H. (2010). Determination of BTEX compounds by dispersive liquid—liquid microextraction with GC—FID. *Chromatographia*, *71*(11—12), 1137—1141.

Baedecker, M. J., Eganhouse, R. P., Barbara, A. B., & Geoffrey, N. D. (2011). Loss of volatile hydrocarbons from an LNAPL oil source. *Journal of Contaminant Hydrology*, *126*(3—4), 140—152. Available from http://doi.org/10.1016/j.jconhyd.2011.06.006.

Balseiro-Romero, M., & Monterroso, C. (2013). A headspace-analysis approach to assess the sorption of fuel volatile compounds by soils. *Soil Science Society of America Journal*, *77*(3), 800—808. Available from http://www.scopus.com/inward/record.url?eid = 2-s2.0-84880161912&partnerID = 40&md5 = 2ccd784411a471772dae0c7bc29ad33d.

Basu, S., Kumar Yadav, B., & Mathur, S. (2015). Enhanced bioremediation of BTEX contaminated groundwater in pot-scale wetlands. *Environmental Science and Pollution Research*, *22*, 20041—20049. Available from http://link.springer.com/10.1007/s11356-015-5240-x.

Basu, S., Yadav, B.K., Mathur, S., & Gupta, P.K. (2020). In situ bioremediation of LNAPL polluted vadose zone: Integrated column and wetland study. *CLEAN - Soil, Air, Water*, 2000118.

BianchinNunes, J., Nardini, G., Merib, J., Dias, A. N., Martendal, E., & Carasek, E. (2012). Simultaneous determination of polycyclic aromatic hydrocarbons and benzene, toluene, ethylbenzene and xylene in water samples using a new sampling strategy combining different extraction modes and temperatures in a single extraction solid-phase microextraction-gas chromatography—mass spectrometry procedure. *Journal of Chromatography A*, *1233*, 22—29.

Boonsaner, M., Borrirukwisitsak, S., & Boonsaner, A. (2011). Phytoremediation of BTEX contaminated soil by Canna × generalis. *Ecotoxicology and Environmental Safety*, *74*(6), 1700—1707.

Buddhadasa, S. C., Barone, S., Gibson, E., Bigger, S. W., & Orbell, J. D. (2002). Method dependency in the measurement of BTEX levels in contaminated soil. *Journal of Soils and Sediments, 2*(3), 137–142.

Cotel, S., Gerhard Schäfer, V. B., & Baussand, P. (2011). Effect of density-driven advection on trichloroethylene vapor diffusion in a porous medium. *Vadose Zone Journal, 10*(2), 565.

Coulon, F. (2009). Understanding the fate and transport of petroleum hydrocarbons from coal tar within gasholders. *Environment International, 35*(2), 248–252.

Coulon, F., & Delille, D. (2006). Influence of substratum on the degradation processes in diesel polluted sub-Antarctic soils (Crozet Archipelago). *Polar Biology, 29*(9), 806–812.

Dhami, B., Himanshu, S. K., Pandey, A., & Gautam, A. K. (2018). Evaluation of the SWAT model for water balance study of a mountainous snowfed river basin of Nepal. *Environmental Earth Sciences, 77*(1), 1–20.

Djozan, D., & Assadi, Y. (1997). A new porous-layer activated-charcoal-coated fused silica fiber: Application for determination of BTEX compounds in water samples using headspace solid-phase microextraction and capillary gas chromatography. *Chromatographia, 45*, 183–189.

Gupta, P. K., & Yadav, B. K. (2017). Bioremediation of non-aqueous phase liquids (NAPL$_s$) polluted soil and water resources. In R. N. Bhargava (Ed.), Environmental pollutants and their bioremediation approaches. Florida, USA: CRC Press, Taylor and Francis Group, ISBN 9781138628892.

Gupta, P. K., Kumari, B., Gupta, S. K., & Kumar, D. (2020). Nitrate-leaching and groundwater vulnerability mapping in North Bihar, India. *Sustainable Water Resources Management, 6*, 1–12.

Gupta, P. K., & Sharma, D. (2018). Assessments of hydrological and hydro-chemical vulnerability of groundwater in semi-arid regions of Rajasthan, India. *Sustainable Water Resources Management, 1*(15), 847–861.

Gupta, P. K., & Yadav, B. K. (2019a). In R. P. Singh, A. K. Kolok, & L. S. Bartelt-Hunt (Eds.), *Subsurface processes controlling reuse potential of treated wastewater under climate change conditions". Water conservation, recycling and reuse: Issues and challenges*. Singapore: Springer, 9789811331787 (ISBN).

Gupta, P. K., & Yadav, B. K. (2019b). Remediation and management of petrochemical polluted sites under climate change conditions. In R. N Bhargava (Ed.), Environmental contaminations: Ecological implications and management. Springer Nature Singapore Pte Ltd., ISBN 9789811379048.

Gupta, P. K., & Yadav, B. K. (2020). Three-dimensional laboratory experiments on fate and transport of light NAPL under varying groundwater flow conditions. *ASCE Journal of Environmental Engineering, 146*(4)04020010.

Gupta, P.K., & Bhargava, R.N., (2020). Fate and transport of subsurface pollutants, Springer, eBook ISBN-978-981-15-6564-9. Available from https://www.springer.com/gp/book/9789811565632.html.

Han, D., Ma, W., & Chen, D. (2007). Determination of biodegradation process of benzene, toluene, ethylbenzene and xylenes in seabed sediment by purge and trap gas chromatography. *Chromatographia, 66*(11–12), 899–904.

Himanshu, S. K., Ale, S., Bordovsky, J. P., Kim, J., Samanta, S., Omani, N., & Barnes, E. M. (2021). Assessing the impacts of irrigation termination periods on cotton productivity under strategic deficit irrigation regimes. *Scientific Reports, 11*(1), 1–16.

Himanshu, S. K., Pandey, A., & Patil, A. (2018). Hydrologic evaluation of the TMPA-3B42V7 precipitation data set over an agricultural watershed using the SWAT model. *Journal of Hydrologic Engineering, 23*(4)05018003.

Holden, P. A., & Noah, F. (2005). Microbial processes in the vadose zone. *Vadose Zone Journal, 4*(1), 1.

Jiang, B., Zhou, Z., Dong, Y., Tao, W., Wang, B., Jiang, J., & Guan, X. (2015). Biodegradation of benzene, toluene, ethylbenzene, and o-, m-, and p-xylenes by the newly isolated bacterium Comamonas sp. JB. *Applied Biochemistry and Biotechnology, 176*(6), 1700–1708.

Kim, H., Seungjae Lee, J.-W. M., & Rao, P. S. C. (2005). Gas transport of volatile organic compounds in unsaturated soils. *Soil Science Society of America Journal, 69*(4), 990.

Kristensen, A. H., Henriksen, K., Mortensen, L., Scow, K. M., & Moldrup, P. (2010). Soil physical constraints on intrinsic biodegradation of petroleum vapors in a layered subsurface. *Vadose Zone Journal: VZJ, 9*(1), 137.

Kubinec, R., Berezkin, V. G., Górová, R., Addová, G., Mračnová, H., & Soják, L. (2004). Needle concentrator for gas chromatographic determination of BTEX in aqueous samples. *Journal of Chromatography B, 800*(1-2), 295–301.

Kumar, R., Sharma, P., Verma, A., Jha, P. K., Singh, P., Gupta, P. K., Chandra, R., & Prasad, P. V. V. (2021). Effect of physical characteristics and hydrodynamic conditions on transport and deposition of microplastics in riverine ecosystem. *Water, 13*, 2710.

Kumari, B., Gupta, P. K., & Kumar, D. (2019). In-situ observation and nitrate-N load assessment in Madhubani District, Bihar, India. *Journal of Geological Society of India, 93*(1), 113–118.

Lovanh, N., Craig, S. H., & Pedro, J. J. A. (2002). Effect of ethanol on BTEX biodegradation kinetics: Aerobic continuous culture experiments. *Water Research, 36*(15), 3739–3746.

Mangani, F., Lattanzi, L., Attaran Rezai, M., & Cecchetti, G. (2000). Graphite-lined capillary column for the separation of BTEX mixture and other volatile organic compounds (VOCs). *Chromatographia, 52*(3–4), 217–220.

Meyer-Monath, M., Beaumont, J., Morel, I., Rouget, F., Tack, K., & Lestremau, F. (2014). Analysis of BTEX and chlorinated solvents in meconium by headspace-solid-phase microextraction gas chromatography coupled with mass spectrometry. *Analytical and Bioanalytical Chemistry, 406*(18), 4481–4490.

Møller, S., Pedersen, A. R., Poulsen, L. K., Arvin, E., & Molin, S. (1996). Activity and three-dimensional distribution of toluene-degrading Pseudomonas putida in a multispecies biofilm assessed by quantitative in situ hybridization and scanning confocal laser microscopy. *Applied and Environmental Microbiology, 62*(12), 4632–4640.

Nardi, L. (2006). Post-extraction losses of volatile aromatic hydrocarbons during capillary extraction–HRGC analysis: A quantitative assessment. *Chromatographia, 65*(1–2), 51–57.

Nerantzis, P. C., & Mark, R. D. (2010). Transport of BTEX vapours through granular soils with different moisture contents in the vadose zone. *Geotechnical and Geological Engineering, 28*(1), 1–13.

Niri, V. H., Mathers, J. B., Musteata, M. F., Lem, S., & Pawliszyn, J. (2009). Monitoring BTEX and aldehydes in car exhaust from a gasoline engine during the use of different chemical cleaners by solid phase microextraction-gas chromatography. *Water, Air, and Soil Pollution, 204*(1), 205–213.

Pandey, A., Himanshu, S. K., Mishra, S. K., & Singh, V. P. (2016). Physically based soil erosion and sediment yield models revisited. *Catena, 147*, 595–620.

Picone, S., Grotenhuis, T., van Gaans, P., Valstar, J., Langenhoff, A., & Rijnaarts, H. (2013). Toluene biodegradation rates in unsaturated soil systems versus liquid batches and their relevance to field conditions. *Applied Microbiology and Biotechnology, 97*(17), 7887–7898.

Přikryl, P., Kubinec, R., Jurdakova, H., Ševčík, J., Ostrovský, I., Sojak, L., & Berezkin, V. (2006). Comparison of needle concentrator with SPME for GC determination of benzene, toluene, ethylbenzene, and xylenes in aqueous samples. *Chromatographia, 64*(1), 65–70.

Ronen, D., Ellen, R. G., & Laor, Y. (2005). Volatile organic compounds in the saturated–unsaturated interface region of a contaminated phreatic aquifer. *Vadose Zone Journal, 4*(2), 337.

Sieg, K., Fries, E., & Wilhelm, P. (2008). Analysis of benzene, toluene, ethylbenzene, xylenes and N-aldehydes in melted snow water via solid-phase dynamic extraction combined with gas chromatography/mass spectrometry. *Journal of Chromatography. A, 1178*(1−2), 178−186.

Telgheder, U., Malinowski, M., & Jochmann, M. A. (2009). Determination of volatile organic compounds by solid-phase microextraction-gas chromatography-differential mobility spectrometry. *International Journal for Ion Mobility Spectrometry, 12*(4), 123−130.

Tumbiolo, S., Gal, J. F., Charles Maria, P., & Zerbinati, O. (2004). Determination of benzene, toluene, ethylbenzene and xylenes in air by solid phase micro-extraction/gas chromatography/ mass spectrometry. *Analytical and Bioanalytical Chemistry, 380*(5−6), 824−830.

Zhou, Y.-Y., Yu, J.-F., Yan, Z.-G., Zhang, C.-Y., Xie, Y.-B., Ma, L.-Q., Gu, Q.-B., & Li, F.-S. (2013). Application of portable gas chromatography-photo ionization detector combined with headspace sampling for field analysis of benzene, toluene, ethylbenzene, and xylene in soils. *Environmental Monitoring and Assessment, 185*(4), 3037−3048.

Index

Note: Page numbers followed by "*f*" and "*t*" refer to figures and tables, respectively.

A

Abiotic process, 238, 245–246
Absolute risk aversion coefficient (ARAC), 55
Acetaminophen, 238
Acetate, 114–116
Acetoclastic methanogens, 113–114
Achromobacter, 72–73
   A. xylosoxidans, 314–315
Acidic magmatites, 260
Acidobacteria, 313
Acidobacteriaceae, 313
Acinetobacter, 69, 72–73
AdaBoost algorithm, 91–92
Adams–Bohart kinetics model, 36–38
   comparison of breakthrough curves, 39*f*
   constants, 38*t*
Adsorbent
   preparation of, 23–26
      fixed-bed column studies, 26–29
      procedure for preparation of adsorbent from sludge, 23–24
Adsorption capacity (N_o), 40–41
Advection–dispersion equation (ADE), 142
Aerobic biodegradation, 107
Aerobic degradation, 309–310
Agilent model 5977A, 369
Algae, 106
Alkanindiges, 72–73
All India Soil and Land Use Survey (AISLUS), 291–292
Alluvial deposit, 323
α-*proteobacteria*, 312
ALS system, 369

Alternative tillage practices, comparison of, 58–59, 59*f*
Alteromonas, 72–73
Alum sludge, 23
2-amino-musk ketone, 239
Ammonia volatilization, 133–135
Ammonium chloride (NH_4Cl), 325
Analytical contaminant transport modeling, 148–149
Analytical–numerical approach, 157–158
Anionic surfactants, 240
Anthropogenic contaminants, 233. *See also* Emerging contaminants (ECs)
Anthropogenic processes, 219–220
Anthropogenic sources, 259
Aquatic biota caused by microplastics, 195–196
Aquatic-based sources, 193–194
Aquifer
   confinement, 240–241
   system, 282–283
ArcGIS environment, 296–298
Arthrobacter, 72–73
Artificial intelligence (AI), 87
Artificial neural network (ANN), 157–158
   ANN-GWO model, 158
Astragalus, 262–263
Atmospheric boundary, impacts on, 173–174
Autotune, 372
Azithromycin, 238

B

Bacillus sp., 68–69, 71–73, 314–315

   B. amyloliquefaciens, 75, 314–315
   B. licheniformis, 75
   B. mojavensis, 75
   B. pumilus, 75
   B. subtilis, 75
Bacteria, 67, 69, 106
Bacterial biosurfactants (BBSs), 68
   anionic BBSs, 77
   bioremediation of contaminated soils, 70–78
   classification, 68–69
      high molecular weight biosurfactants, 69
      low molecular weight biosurfactants, 69
      structure of biosurfactant molecule, 68*f*
Bacterial communities, 110
Bagging method, 91–92
Basin units, assigning weights to, 299–300, 300*t*
Baye's rule, 91–92
Bed depth effect, 32–33
   breakthrough curve, 32*f*
   column data analysis for parameters, 33*t*
Bed-depth service time model (BDST model), 38–41
   parameters, 40*t*
   plot, 40*f*
Below ground level (bgl), 323
Benzene, toluene, ethylbenzene, and xylene (BTEX), 111–113
Benzothiazoles, 239
Benzotriazoles, 239
Beta-cypermethrin, 71
Bicarbonate, 260
Billion cubic meters (BCM), 320

379

Bioaccumulation of pharmaceutical compounds, 238
Bioaugmentation, 314—315
Bioavailable P (BAP), 49
Biodegradation, 107, 309—310
Biodiesels as biostimulators, 313—315
BIOEFGM, 157—158
Biofilter, 245—246
Biological methods, 245
Biological oxygen demand (BOD), 22
Biomonitoring studies, 246
Bioremediation
   of contaminated soils
      bacterial biosurfactants in, 70—78
      biosurfactants in bioremediation of heavy metals, 75—78
      biosurfactants in bioremediation of hydrocarbons, 72—75
      biosurfactants in bioremediation of pesticides, 71—72
   of PAHs, 107—110
Biosurfactants
   in bioremediation of heavy metals, 75—78
   in bioremediation of hydrocarbons, 72—75
   in bioremediation of pesticides, 71—72
   mechanism, 72
Biotic process, 238, 245—246
Bisphenol-A (BPA), 239
Black microplastics, 208—210
Blue-baby syndrome.
   See Methemoglobinemia
Boosting algorithm, 91—92
Braunuer—Emmett—Telller (BET), 24—26
   multipoint BET adsorption parameters for waste alum sludge, 26t
   surface area analysis, 24—26
Breakthrough curves (BTCs), 27, 158, 160—162
Brine—rock system, 6—7
Burkholderia, 72—73

C
Caffeine, 239—240
Calcium carbonate, 260

Calcium chloride, 197—198
Calcium ions (Ca²⁺ ions), 275—276
Calibration, 348—349, 351, 372—373
   concentration-associated response areas, 374t
   curve, 373f
Candida sp, 314—315
   C. dubliniensis, 314—315
   C. viswamathii, 314—315
Canna generalis, 111
Carbamazepine, 238
Carbamodithioic acid, 239
Carbon capture and storage (CCS), 1
Carbon-dioxide dynamics (CO₂ dynamics), 1
   CO₂ geo-sequestration projects, 2
   CO₂—brine—rock interaction in subsurface, 8—9
      CO₂—brine—rock interaction, 9t
   factors affecting CO₂ migration in subsurface, 6—8
      absolute and relative permeability, 7
      change in fluid properties, 8
      change in stress pattern, 8
      dissolution and precipitation of minerals, 7
      existence of third-phase saturation, 6—7
      saturation history, 6
      wettability, 6
   gaseous CO₂ in subsurface, 3—6
      different formation of geological sequestration process, 4f
   modeling of CO₂ in subsurface, 12—14
      coupling of flow and transport equation, 13f
      model domain for performing simulation, 14f
      parameters in simulation, 13t
   numerical modeling for investigating CO₂ dynamics, 11—12
   potential risk associated with CO₂ leakage, 10
   worldwide CO₂ storage projects, 2—3
Carbonic acid, 260
Carcinogenesis, 195—196
Carrier gases, 369

Cartesian coordinate system, 7
Cationic BBSs, 77
Certainty equivalent (CE), 55
Chaturvedi's formula, 345—346
Chemical methods, 245
Chemical nonequilibrium, 146—147
Chemical oxygen demand (COD), 21—22
   analysis, 29
Chlamydomonas sp., 315
   C. reinhardtii, 315
Chlorococcum sp, 315
Chlorpyrifos, 71
Chromatographic techniques, 237
Citrobacter, 314—315
Clark kinetics model, 41—43
   comparison of breakthrough curves, 42f
   parameters, 41t
Clark's solution, 41
Climate change, 182, 282—283
   methodological framework for evaluating climate change impacts on subsurface, 182—183
Climatic variability, 171—172
   impacts on
      atmospheric boundary, 173—174
      ground—surface water interactions, 177—178
      subsurface water quality, 178—182
      water storage and flow pattern, 175—177
   methodological framework for evaluating climate change impacts on subsurface, 182—183
Clustering analyses, 89—90
Coastal hydrogeology, 269
Coefficient of variance, 206—207
Color of microplastics, 208—210
Column adsorption studies
   fish pond wastewater, 21—22
   kinetics studies on adsorption of chemical oxygen demand on sludge adsorbent, 33—43
   materials and methods, 23—29
   preparation of adsorbent, 23—26
   methods of removing chemical oxygen demand, 22

results, 29–43
  bed depth effect of, 32–33
  dynamic adsorption studies, 30
  influence of operational variables, 31–33
  influent flow rate, effect of, 31–32
  waste alum sludge, 23
Column data analysis, 27–28
Comamonadaceae sp., 314–315
Composite erosion intensity map, 299, 301–302
  erosion intensity classification, 302t
  of UTRB, 301f
Conservation tillage, 49
Constructed wetlands (CWs), 111, 131–132
Contaminant transport modeling, 141–142
  categorization of mathematical modeling studies related to Indian groundwater and soil systems, 148–158
  for saturated porous media, 143–148
  in subsurface environment using mobile–immobile model, 158–162
Contaminated soils, bioremediation of, 70–78
Continuous-time random walk model (CTRW model), 148
Convention on Biological Diversity (CBD), 195–196
Conventional ADE model, 142
Conventional advection–dispersion model, 143–144
Correlated uniform standard deviates (CUSDs), 53
Correlation matrix, 53
Corrosion inhibitors, 239
Cost estimation, 54–55
Coumarin, 238
Crank–Nicolson FDM, 148–149
Crenarchaeota, 113
Critical micelle concentration (CMC), 68, 72, 76–77
Cropland, 301–302
Cyanobacteria, 315

D
Darcy's law, 5, 12
December, January, February seasons (DJF seasons), 223
Decision trees, 91–92
Deep-well current meters, 220–221
Deferribacter, 114–116
Delivery ratio, assigning, 298–299, 299t
Denitrification, 133, 320–321
Dense non-aqueous phase liquids (DNAPLs), 125, 150
Density variation, 273–274
Descriptive statistics, 203
Diammonium phosphate ($[NH]_2HPO_4$), 319–320
Diclofenac, 238
Diesel range organic compounds (DRO compounds), 311
Dietzia, 72–73
Digital elevation model data (DEM data), 293–294
  location map of UTRB, 294f
Dimethyl ester, 239
Dinitrogen (N_2), 133
Dispersive SPME, 237
Dissolved P (DP), 49
δ-proteobacteria, 312
Downhole televiewer, 220–221
Drainage system, 282–283
Drometrizole, 239
Duduwa river, 343–344
Duplex constructed wetland (duplex-CW), 311
Duplex wetland systems, 311
  self-sustainable constructed wetland integrated to microbial fuel cell, 312f
Dupuit–Forchheimer assumption, 279
Dynamic adsorption studies, 30

E
Edit GC Parameters dialog boxes, 370
El Nino Southern Oscillation (ENSO), 173–174
Electrical resistivity tomography (ERT), 277–278
Electromagnetic method (EM method), 277–278
Elemental selenium (Se^0), 259

Emerging contaminants (ECs), 233, 247
  challenges and scope, 246
  detection and analysis, 237
  fate of emerging contaminants in groundwater, 240–242
  health effects of, 243t
  methods to remove, 245f
  potential risks associated with, 242–243
  remediation of, 244–246, 244t
  sources in groundwater, 234–236, 235f
    point and nonpoint sources of, 236t
  types, 238–240, 241f
    industrial chemicals, 239
    lifestyle products, 239–240
    personal care products, 239
    pesticides, 238
    pharmaceutical products, 238
    surfactants, 240
Empirical distribution function, 54
Endocrine disrupting chemicals or compounds (EDCs), 233
Engineered constructed wetlands, 310–312
  duplex wetland systems, 311
  integrated polluted columns and treatment wetlands, 311–312
Enhanced oil recovery (EOR), 2
Enterobacter, 72–73
Environmental contaminations, 72–73
Environmental pollution, 105
Equilibrium linear sorption isotherm, 143–144
Erosion intensities, 301–302
  basin units, 299
Error analysis, 29
Ethylhexyl methoxycinnamate, 239
Eukaryotic microalgae, 315
Euryarchaeota, 113–114
Evaluate model, 54
Evapotranspiration (ET), 171–172, 320–321, 346–347
  maximum evapotranspiration and extinction depth, 347t
Extraction process, 237
Extreme learning machine (ELM), 157

F

Farm management practices for water quality improvement
data, 50−52
mean annual runoff, sediment, and nutrient discharge, 52t
resource areas, 51f
tillage treatments and fertilizer application, 51t
methods, 53−56
cost and profit estimation, 54−55
Monte Carlo simulation, 53−54
stochastic efficiency approach, 55−56
results, 56−61
comparison of alternative tillage practices, 58−59
evaluation of water quality improvement, 59−61
profit distributions, 57−58
validation results, 56
Fate and transport of PAHs
laboratory experimental studies on, 130t
mechanisms, 127−128
in subsurface, 125−126, 126f
Fenofibrate, 238
Fenton process, 314−315
Fermentation, 135
Fertilizers, 319
Fibers, 204
microplastic, 194
Filaments, 204
Films, 204
Finite-difference method (FDM), 142, 279
Finite-element method (FEM), 142, 279
Fire retardants, 233
Firmicutes, 113
Fixed-bed column studies, 26−29
COD analysis, 29
column data analysis, 27−28
error analysis, 29
Flame ionization detector (FID), 361−362
Flow pattern, impacts on, 175−177
Fluid−rock interaction, 8
Foam, 204
Food & Agriculture Organization (FAO), 21−22

FAO Penman-Monteith method, 348
FAO-56 Penman−Monteith method, 324
Fourier-transform infrared analysis (FT-IR analysis), 24, 25f
Fragments, 204
Freshwater discharge, 273−274
Freshwater environment, microplastic in, 193−195
Freundlich adsorption isotherm, 41
Freundlich equation, 41
Fungi, 67, 106

G

Galaxolide, 239
γ-proteobacteria, 312
Ganga plain foreland basin, 323
Ganga river basin, 319−320
Gangetic plain, 226−228
Gas chromatography mass spectrometer (GC−MS), 361−369
ALS system, 369
calibration, 372−373
carrier gases, 369
gas filter, 369
GC column, 368−369
mass spectrum of toluene, 374
MS detector, 369
previous studies to investigate hydrocarbon pollution using, 363t
specification, 368t
toluene analysis method, 369−372
vacuum pump, 369
Gas chromatography−mass spectrometer detector (GC−MSD), 361
Gas filter, 369
GC column, 368−369
Genetic algorithm (GA), 157
Geo-hydrological characteristics, 173
Geo-sequestration, 1
Geobacillus, 114−116
Geographic Information System (GIS), 291
Geological carbon sequestration, 11
Geophysical methods, 277−278
Ghyben−Herzberg formula, 279

Global atmospheric circulation models (GCMs), 183
Global Land Data Assimilation System (GLDAS), 221
GLDAS-2, 221−222
GLDAS-2.0, 221−222
GLDAS-2.1, 221−222
GLDAS-2.2, 221−223
reanalysis, 221
data used and methodology, 221−222
rainfall variations, 223
seasonal variations in rainfall and groundwater, 223−226
trend analysis of precipitation and groundwater, 226−228
Global warming, 1
Gravity Recovery and Climate Experiment satellite (GRACE satellite), 221
Gray wolf optimizer (GWO), 157−158
Gray−Richardson−Klose−Schumann model (GRKS model), 53
Greenhouse gas (GHG), 1
Grid discretization, 279
Gross cropped area (GCA), 319−320
Gross irrigation, 350
Gross irrigation requirement (GIR), 348
Ground−surface water interactions, impacts on, 177−178
Groundwater, 219−220. See also Subsurface water
calibration, 348−349
contamination, 319−320
emerging contaminants in, 234−236, 235f
fate of, 240−242
point and nonpoint sources of, 236t
extraction, 341−343
fluctuation, 220−221
head, 341
Ml techniques used in, 91−93
supervised ML techniques, 91−92
unsupervised ML techniques, 92−93
modeling, 348−350
pollution, 337−338
quality, 219−220
recharge, 345−346

replenishment, 345–346
seasonal variations in, 223–226
    monthly groundwater variations
        of period, 225f
    seasonal rainfall variations of
        period, 226f
systems, 87
trend analysis of, 226–228, 229f
validation, 349–350
variations, 223

H

Harmonized World Soil Database
    website, 296
Hatnur dam reservoir, 292–293,
    298–299
Head measurement, 277
Heavy metals (HMs), 67, 105
    biosurfactants in bioremediation
        of heavy metals, 75–78
        mechanism, 76–78
        mechanism of HM removal, 78f
        remediation of heavy metal
            contaminated soil, 77t
Henry problem, 273–274, 279
Heterogeneity in seawater intrusion
    process, 275
HHCB-lactone, 239
Hierarchical clustering, 92–93
High molecular weight (HMW), 69
    BBSs, 70f
    biosurfactants, 69
Hirakud project, 292
Horizontal flow filter (HFF), 311
Horizontal subsurface flow method
    (HSSF method), 111–113
Human health, microplastic and,
    196–197
Hybrid system, 311
Hydraulic retention time (HRT),
    111–113
Hydrocarbons, 106–107, 127
    bioremediation products
        biodiesels as biostimulators,
            313–315
        engineered constructed
            wetlands, 310–312
        native and specialized microbial
            communities, 312–313
        phycoremediation, 315
    biosurfactants in bioremediation
        of, 72–75

degradation of hydrocarbons,
    74t
mechanism, 74–75
spills, 309
Hydrochemical processes in seawater
    intrusion process, 275–276
Hydrogen sulfide (H_2S), 113
Hydrogenotrophic methanogens,
    113–114
Hydrogeological parameters, 344
Hydrogeological profiles, 340
Hydrological monitoring, 324
Hydrolysis, 135

I

Ibuprofen, 238
Imbibition, 6
In-Salah project, 2–3
Indian groundwater and soil systems
    analytical contaminant transport
        modeling, 148–149
    categorization of mathematical
        modeling studies related to,
        148–158
    integrated analytical–numerical or
        simulation–optimization
        approach, 157–158
    mathematical modeling-based
        studies, 151t
    numerical flow and contaminant
        transport modeling, 150–156
Indian monsoon, 223
Indian soil–water system
    field scale implications, 264
    NORM, 261–262
    remedial measures, 262–264
    selenium, 260–261
Industrial chemicals, 239
Industrial effluent discharge, 234
Influent flow rate
    effect of, 31–32
    breakthrough curve, 31f
Inorganic pollutants, 67
Inorganic selenites, 259
Insect repellents, 233
Integrated polluted columns and
    treatment wetlands, 311–312
Iron-reducing bacteria, 114–116,
    116t
Irrigation water requirement,
    computation of, 348
Isopropyl myristate, 239

J

June, July, August seasons (JJA
    seasons), 223

K

K-means clustering, 92–93
Karst aquifers, 240–241
Ketoprofen, 238
Kinetic constant (K_a), 40–41
Kinetics studies
    Adams–Bohart kinetics model,
        36–38
    on adsorption of chemical oxygen
        demand on sludge adsorbent,
        33–43
    BDST model, 38–41
    Clark kinetics model, 41–43
    Thomas kinetics model, 33–34
    Yoon–Nelson kinetics model,
        35–36
Klebsiella, 71
Kocuria, 72–73
Kyoto Protocol, 3

L

Laboratory experiments, 129
Lactate, 114–116
Land use land cover data (LULC
    data), 293–295, 295f, 295t
Least square of error, 29
Levenberg–Marquardt optimization,
    157
Lifestyle products, 239–240
Light nonaqueous phase liquids
    (LNAPL), 125, 309
    entrapped, 127–128
    pools, 128–129
Linear alkylbenzene sulfonate (LAS),
    240
Linear correlation matrix, 53
Linear ion trap–MS, 237
Linear regression, 91–92
Liquid chromatography (LC), 237
Liquid-phase microextraction, 237
Livestock wastes, 319
Local transverse dispersion, 274
Logistic regression, 91–92
Longitudinal dispersion, 274–275
Low molecular weight (LMW), 69
    BBSs, 70f
    biosurfactants, 69

M

Machine learning techniques (ML techniques), 87−90
  comparison between statistical methods and, 88t
  efficacy of, 93−96, 94t
  generic steps in, 90−91
  in groundwater modeling, 91−93
  reinforcement learning algorithm, 90
  supervised learning algorithm, 89
  unsupervised learning algorithm, 89−90
Magnesium ions (Mg^{2+} ions), 275−276
Maithon project, 292
Man river, 343−344
Manganese-reducing bacteria, 114−116, 116t
Mann−Kendall test, 222
Mapping erosion intensity units, 299
March, April, May seasons (MAM seasons), 223
Marinobacter, 72−73
Mass of microplastics, 204−205
Mass spectrometer detector (MSD), 361, 369−371
Mass spectrometry technique (MS technique), 237
Mass transfer zone (Z_m), 32−33
MassHunter Data Acquisition, 372
Mathematical models, 142
Mechanical dispersion, 273−274
Mehanopyrales, 113−114
Membrane bioreactors (MBRs), 245−246
Metal ions, 77
Metamorphites, 260
Methanobacteriales, 113−114
Methanococcales, 113−114
Methanogenesis, 113−114, 135
Methanogens, 113−114, 115t
Methanomicrobiales, 113−114
Methanosarcinales, 113−114
Methemoglobinemia, 320
Method editor, 370
Method Information, 370
Methyl ester, 239
Methyl paraben, 239
Methylotrophic methanogens, 113−114
Microbes

and consortium to degrade PAHs, 113−116
  iron-, nitrate-, and manganese-reducing bacteria, 114−116
  methanogens, 113−114
  sulfate-reducing bacteria, 113
Microbial communities, 263−264
  native and specialized, 312−313, 313t
Microbial degradation kinetics models, 116−117
Microbial growth kinetics, 116−117
Micrococcus sp, 314−315
Microorganisms, 113−114
Microplastics, 192−193, 233
  color of microplastics at different locations, 208−210
  and human health, 196−197
  mass of, 204−205
  method developed for extraction of, 203−204
  number of, 206−207
  objectives of study, 198
  occurrence, fate, types, and source in freshwater environment, 193−195
  pollution, 194
  potential concern for pollution and damage to aquatic biota caused by, 195−196
  sample collection practices and estimation of, 197−198
  sediments treatment and extraction of, 200−203
  size of, 210−211
  source of, 211−212
  types, 204
    in different streams, 207−208
Micropollutants, 233
Minerals, dissolution and precipitation of, 7
Mobile−immobile model (MIM), 142, 144−146, 145f
  contaminant transport modeling in subsurface environment using, 158−162
MODFLOW, 150
Molecular diffusion, 273−274
Monod's kinetics, 117
Monooxygenases, 107
Monte Carlo simulation, 53−54
MT3DMS, 150

Multiprocesses nonequilibrium model (MPNE model), 142, 146−147, 146f
Multivariate empirical distribution (MVE distribution), 53−54
Musgrave equation, 291−292
Mycobacterium, 72−73

N

N-butylbenzenesulfonamide, 239
Naïve Bayes method, 91−92
Nanomaterials, 233
National Environmental Standards and Regulation Enforcement Agency (NESREA), 22, 29−30
Native microbial communities, 312−313
Natural solutions
  biodiesels as biostimulators, 313−315
  engineered constructed wetlands, 310−312
  native and specialized microbial communities, 312−313
  phycoremediation, 315
Naturally occurring microorganisms, 105−106
Naturally occurring radioactive material (NORM), 261−262
Nearest neighbor algorithms, 91−92
Nicotine, 239−240
Nitrate (NO$_3^-$), 319
  concentrations at OW1 and OW2, 330f
  estimated average NO$_3^-$ attenuation rates, 330t
  relative concentration of NO$_3^-$ for observation wells, 328f
  transport, 326−331
Nitrate-N movement
  methodology, 324−325
    hydrological monitoring, 324, 326f
    solute injection test, 325
  results, 325−331
    hydrological characteristics of experimental site, 325−326
    nitrate (NO$_3^-$) transport, 326−331
  study site, 321−323, 324f
Nitrate-reducing bacteria, 114−116, 116t

Nitrification, 131–132
Nitrobacter, 131–132
Nitrogen (N), 49
  nitrogen-based fertilizer, 319–320
Nitrogenous fertilizers, 331
Nitrosomonas, 131–132
Nitrospira, 113
Nitzschia sp, 315
  N. linearis, 315
Nizamsagar project, 292
Nonaqueous phase liquids (NAPLs), 117, 180–182, 361–362
Nonionic metals, 77
Nonionic surfactants, 240
Nonparametric Mann–Kendall test, 222
Nonpoint source pollutants, 319
Nonpoint sources of ECs in groundwater, 234, 236t
Nonylphenol ethoxylates (NPEOs), 240
Northeastern Sivalik foothill zone (NSFZ), 260
Numerical flow and contaminant transport modeling, 150–156
Numerical modeling of seawater intrusion, 279, 280t
Nutrient prices, 52

O

Observation points (OBS), 160–162
Observation wells (OWs), 323
Ochrobactrum, 71
  O. intermedium, 314–315
Octocrylene, 239
Oil spills, 72–73
One-way ANOVA test, 203
Oonopsis, 262–263
Optimization models, 279
Organic pollutants, 67
Organochlorine pesticides, 196–197
Organophosphate, 71
Otway Basin Pilot Project, 3
Oxybenzone, 239

P

P-value test, 206–207
Pacific Decadal Oscillation (PDO), 173–174
Paired sample *t*-test, 206–207
Panchet project, 292

Pandorea, 72–73
Particle swarm optimization (PSO), 157
Particulate N (PN), 49
Particulate P (PP), 49
Patna, 191, 199–200
  sampling sites at bank of River Ganga near Patna, Bihar, 199f
PE terephthalate (PET), 192
Péclet number (P_e), 273–274
Pellets, 204
Personal care products (PCPs), 233, 239
Pesticides, 105, 233, 238, 319
  biosurfactants in bioremediation of, 71–72, 71t
Petroleum
  derivatives, 67
  hydrocarbon degraders, 72–73
Pharmaceutical compounds, 233
Pharmaceutical products, 238
Phenazone, 238
2-phenoxy-ethanol, 239
Phosphate fertilizers, 260
Phosphorus (P), 49
Phragmites australis, 311
Phthalates, 239
Phycoremediation, 315
Phyla, 113
Physical hazards, 195–196
Physical methods for ECs, 245
Physical nonequilibrium, 146–147
Physical remediation techniques, 263
Physical–chemical techniques, 263–264
Physics-guided ML techniques, 93–96
Pichia anomala, 314–315
Pink microplastics, 208–210
Plant-assisted bioremediation technique, 111
Plant–microbes interactions, 111–113
Plastic, 191–192
  productiveness of, 203
  waste, 191
Point sources of ECs in groundwater, 234, 236t
Pollutants, 21–22
Pollution, 234
  to aquatic biota caused by microplastics, 195–196

Polybrominated diphenyl ethers, 195–196
Polychlorinated biphenyls (PCBs), 105, 195–197
Polycyclic aromatic hydrocarbons (PAHs), 72–73, 105, 125, 196–197
  bioremediation of, 107–110
  degradation, 116–117
  fate and transport mechanisms, 127–128
  laboratory and numerical investigation on biodegradation of, 108t
  microbes and consortium to degrade, 113–116
  microorganisms used to remediate PAHs-polluted sites, 106
  polishing PAH-polluted site using subsurface-constructed wetlands, 131–135
  pollutants, 106–107
  studies investigated PAHs behaviors in laboratory domain, 128–130
Polycyclic musks, 239
Polyethylene (PE), 192
Polymeric BBSs, 69
Polymers, 191–192
Polypropylene (PP), 192
Polystyrene (PS), 192
Polyvinylchloride, 192
Pore-filling technique, 281–282
Pore-scale heterogeneity, 141–142
Precipitation, trend analysis of, 226–228, 228f
Principal Component Analysis (PCA), 92–93
Prioritization of erosion prone areas data, 293–298
  DEM data, 293–294
  LULC data, 294–295
  slope data, 296–298
  soil data, 296
  methodology, 298–301
    assigning delivery ratio, 298–299
    assigning weights to basin units, 299–300, 300t
    computation of sediment yield index, 300–301
    mapping erosion intensity units, 299

Prioritization of erosion prone areas
(*Continued*)
results, 301–305
composite erosion intensity
map, 301–302
conservation treatments,
304–305
subbasin wise treatment
prioritization, 302–304
study area, 293
Products
lifestyle, 239–240
personal care, 239
pharmaceutical, 238
Profit distributions, 57–58
cumulative distribution functions,
57f
operating costs, 62t
stoplight chart, 58f
Profit estimation, 54–55
Propranolol, 239
Propyl paraben, 239
Proteobacteria, 113, 312
Prototheca
P. zopci, 315
P. zopfii, 315
Pseudo-solubility, 72
Pseudomonas, 68–69, 71–73, 77–78,
263–264
MGF-48, 263–264
P. aeruginosa, 31–32, 68–69,
314–315
P. alcaligenes, 68–69
P. cepacia, 68–69
P. chlororaphis, 68–69
P. clemencea, 68–69
P. collierea, 68–69
P. fluorescens, 68–69
P. luteola, 68–69
P. putida, 31–32, 68–69
P. rhizophila, 68–69
P. stutzeri, 68–69
P. teessidea, 68–69

R
R-square, 206–207
Radial point collocation method
(RPCM), 150
Radon (^{222}Rn), 261–262
Rainfall
seasonal variations in,
223–226

monthly rainfall variations on
period, 224f
seasonal rainfall variations of
period, 226f
variations, 223
groundwater variations, 223
Ramganga project, 292
Regression, 29
modeling problems, 89
Reinforcement learning algorithm,
90, 90f
Relative risk aversion coefficient
(RRAC), 55
Remediation of emerging
contaminants, 244–246
Remote-sensing technologies,
220–221
Respiration, 135
Rhodococcus, 72–73
Risk premium (RP), 55–56
Risk-adjusted farm profits, 59–61
Risk-averse producers, 55
River Ganga of Patna, Bihar
materials and method
sampling of soil sediments, 200
sediments treatment and
extraction of microplastics,
200–203
statistical analysis, 203
study area, 199–200
microplastic and human health,
196–197
objectives of study, 198
occurrence, fate, types, and source
of microplastic in freshwater
environment, 193–195
plastic waste, 191
potential concern for pollution
and damage to aquatic biota,
195–196
results, 203–212
sample collection practices and
estimation of microplastics,
197–198
Root mean square error (RMSE),
348–349
Root zone depth, 347

S
Salicylic acid, 238
Saline water, 275–276
Salts, 319

Sample collection practices and
estimation of microplastics,
197–198
Sampling of soil sediments, 200
Sandy loam (soil), 323
Saturated porous media
contaminant transport models for,
143–148
conventional
advection–dispersion model,
143–144
CTRW model, 148
mobile–immobile model,
144–146, 145f
multiprocesses nonequilibrium
model, 146–147, 146f
SCAN technique, 361
Scenedesmus obliquus, 315
Sea level
fluctuations, 275
rise, 282–283
Seasonal variations in rainfall and
groundwater, 223–226
Seawater intrusion (SI), 269, 275
climate change, and sea level rise,
282–283
management of, 280–282
measurement and monitoring of,
276–278
geophysical methods, 277–278
head measurement, 277
tracer techniques for seawater
intrusion studies, 278
modeling and prediction,
278–279
analytical solutions for seawater
intrusion problems, 279
numerical modeling of seawater
intrusion, 279
process, 270–276, 282–283
heterogeneity in, 275
hydrochemical processes
involved in, 275–276
sea level fluctuations and, 275
upconing, 276
Sediment
deposition, 298–299
detachment, 298–299
flow, 298
transportation, 298–299
Sediment yield index method (SYI
method), 291–292
computation of, 300–301

Sediment Yield Prediction equations, 291–292
Sediments treatment and extraction of microplastics, 200–203
Selective ion monitoring (SIM), 361
  mass spectrometry, 369
  mode, 370
Selenates (Se^{6+}), 259
Selenides (Se^{2-}), 259
Selenites (Se^{4+}), 259
Selenium (Se), 259–261
  contamination, 259
  leaching of, 259
Sensitivity of model parameters, 354
September, October, November seasons (SON seasons), 223
Sequence methods, 372
Shallow alluvial aquifers, 240–241
Shallow unconfined aquifer, 127–128
Shewanella putrefaciens CN32, 263–264
Sikta irrigation system
  groundwater modeling, 348–350
  methodology/philosophy, 340–348
    abstraction of groundwater, 341–343
    assigning boundary conditions, 343–344, 346t
    computation of irrigation water requirement, 348
    conceptual model of study area, 340–341
    discretization of study area, 343
    evapotranspiration, 346–347
    groundwater recharge, 345–346
    hydrogeological parameters, 344
    initial conditions, 344
    special-purpose water requirements, 348
  result, 350–356
    calibration and validation output, 351
    gross irrigation requirement, 350
    groundwater availability, 351–354, 353f
    irrigation water supply and demand, 351
    scenario analysis, 354–356, 355f

sensitivity of model parameters, 354
Siltation, 292–293
Simulation–optimization approach-based studies (S/O approach-based studies), 148, 157–158
Single-beam spectrophotometer, 325
Skeletonema costatum, 315
Sleipner project, 2–3
Slope data, 296–298
  area-wise distribution of slope classes, 298t
  percent slope map of UTRB, 297f
Sludge
  characterization of sludge-derived adsorbent, 24–26
    BET surface area analysis, 24–26
    FT-IR analysis, 24
  chemical oxygen demand on sludge adsorbent, 33–43
  procedure for preparation of adsorbent from, 23–24
Sodium chloride (NaCl), 197–198, 200–201
Sodium iodide (NaI), 197–198
Sodium ions (Na$^+$ ions), 275–276
Sodium polytungstate, 197–198
Soil
  data, 296
    area-wise distribution of soil textural classes, 297t
    soil texture map of UTRB, 296f
  erosivity factor, 291–292
  microplastic in soil sediments, 198
    of freshwater ecosystem, 199–200
    sampling of, 200
  pollution, 67
  soil–water systems, 259, 309
  surface, 49
  types, 143
Soil contamination, 67
Solid-phase extraction, 237
Solid-phase microextraction (SPME), 237
Solute injection test, 325
Sonar, 220–221
Southern High Plains (SHP), 49–50
Spatial heterogeneity, 274
Spatial modeling of ground surface, 221
Special-purpose requirements (SPR), 348

Specialized microbial communities, 312–313
16sRNA analysis, 314–315
Standard deviation (SD), 204–207
Standard error of estimation (SEE), 348–349
Stanleya, 262–263
Staphylococcus, 72–73
Statistical analysis, 87, 203
Steady-state condition and calibration, 348–349
Stenotrophomonas, 71
Stochastic efficiency approach, 55–56
Stochastic efficiency with respect to function approach (SERF approach), 50
Stoplight charts, 57–58
Storm surges, 275
Streptobacillus, 72–73
Streptococcus, 72–73
Stygofauna, 282
Subbasin wise treatment prioritization, 302–304
  subbasin-wise priority rating, 304t
  SYI value-based priority rating, 303t
  treatment prioritization for UTRB, 303f
Subsurface flow (SSF), 111
Subsurface water
  impacts on subsurface water quality, 178–182
  resources, 171, 173
  storage, 175
Subsurface-constructed wetlands (SSCW), 131–132
  polishing PAH-polluted site using, 131–135
Sufactin, 72–73
Sulfamethazine, 238
Sulfamethoxazole propyphenazone, 238
Sulfate (SO$_4^{2-}$), 113
Sulfate-reducing bacteria (SRB), 113
  in soil–water system, 114t
Sulfide, 113
Supervised learning algorithm, 89, 89f
Supervised ML techniques, 91–92
Support vector machines (SVMs), 91–92
Surface flow (SF), 111

Surfactants, 240
Surfactin, 75
Synthetic organic compounds, 239–240

T

t-test, 54
Tapi River, 293
Terrestrial water storage (TWS), 221
Theil–Sen median trend analysis, 222
Thermodesulfobacterium, 113
Thermoterrabacterium ferrireducens, 263–264
Thomas kinetics model, 33–34
  comparison of breakthrough curves, 35f
  constants of Thomas adsorption model, 34t
Thomas model kinetic constant (k_{TH}), 34
Thumb rule, 341–343
Time series analysis, 222
Time-of-flight–MS, 237
Toluene analysis method, 369–372
  date-wise autotune report, 370t
  method creation, 370, 371t
  method for running sample sequence, 372
  MSD method, 370–371
  tune MSD, 372
  tune report assessment, 372
Tonalide, 239
Total Kjeldahl N (TKN), 52
Total N (TN), 52
Total P (TP), 52
Total petroleum hydrocarbon (TPH), 105
Total variation diminishing method, 279
Trace organic contaminants (TrOCs), 195–196

Tracer techniques for seawater intrusion studies, 278
Transformation products (TPs), 238
Transient flow process, 269
Transparent microplastics, 208–210
Trend analysis of precipitation and groundwater, 226–228
Trichloroethylene (TCE), 142–143, 361–362
Triclosan, 238
Trimipramine, 238
Tune MSD, 372
Tune parameters, 372
Tune report assessment, 372
Tungabhadra project, 292
Tuning process, 372

U

Unconfined aquifer, 340–341
Universal Soil Loss Equation, 291–292
Unsupervised learning algorithm, 89–90, 89f
Unsupervised ML techniques, 92–93
Upconing, 274–276
Upper Tapi River basin (UTRB), 292–293
Uranium contamination (U contamination), 259
Urea ($CO[NH_2]_2$), 319–320
US Department of Agriculture Soil Conservation Service method, 348
UV filters, 233

V

Vacuum pump, 369
Validation, 349–350, 350t
  output, 351
Variovorax sp., 314–315
Vertical flow (VF), 311

Vertical subsurface flow method (VSSF method), 111–113
Visual MODFLOW 4.2 tool, 338

W

Waste alum sludge, 23
Wastewater, 21–22
  treatment, 311
Wastewater treatment plants (WWTPs), 193–194
Water, 219–220. *See also* Subsurface water
  evaluation of water quality improvement, 59–61
  price assumptions, 62t
  risk premiums relative to moldboard plowing, 60f
  impacts on water storage, 175–177
  security, 219–220
Weyburn project, 2–3
Winter wheat production, 50, 58
World Health Organization (WHO), 261–262

X

Xylorhiza, 262–263

Y

Yeast, 67
Yellow microplastics, 208–210
Yoon–Nelson kinetics model, 35–36
  comparison of breakthrough curves, 37f
  constants, 36t

Z

Zero-order kinetics, 117
Zinc chloride ($ZnCl_2$), 197–198

Printed in the United States
by Baker & Taylor Publisher Services